C3

Speaking of Yangzhou

A Chinese City, 1550–1850

Harvard East Asian Monographs 236

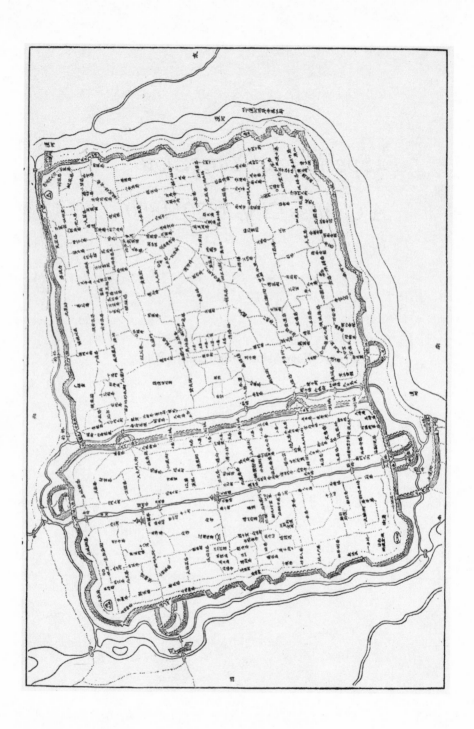

Speaking of Yangzhou

A Chinese City, 1550–1850

Antonia Finnane

Published by the Harvard University Asia Center
Distributed by Harvard University Press
Cambridge (Massachusetts) and London, 2004

Printed in the United States of America

The Harvard University Asia Center publishes a monograph series and, in coordina-
tion with the Fairbank Center for East Asian Research, the Korea Institute, the Reischauer Institute of Japanese Studies, and other faculties and institutes, administers
research projects designed to further scholarly understanding of China, Japan, Vietnam, Korea, and other Asian countries. The Center also sponsors projects addressing
multidisciplinary and regional issues in Asia.

Library of Congress Cataloging-in-Publication Data
Finnane, Antonia.
  Speaking of Yangzhou : a Chinese city, 1550-1850 / Antonia Finnane.
     p. cm. -- (Harvard East Asian monographs ; 236)
  Includes bibliographical references and index.
  ISBN 0-674-01392-1 (alk. paper)
  1. Yangzhou Shi (China)--History. I. Title. II. Series.
  DS797.56.Y355F56 2004
  951'.136--dc22

                              2004007384

Index by the author

⊗   Printed on acid-free paper

Last figure below indicates year of this printing
14   13   12   11   10   09   08   07   06   05   04

Frontispiece: Street map of Yangzhou, 1883. Zhou Cun, *Taipingjun zai Yangzhou*, frontispiece, adapted from *GXJDXZ*.

In memory of my father

Peter Finnane

1926–2001

ొ

Acknowledgments

I thank the following people for their inspiration, support, and assistance at various stages of the research for this book: Wang Gungwu, Helen Dunstan, Lü Zuoxie, Qi Longwei, Wei Minghua, Xu Minhua, Patricia Grimshaw, Charles Sowerwine, Xiong Fan, Kaiyoko Hashimoto, Bick-har Yeung, and the readers for the press. William Kirby kindly referred the manuscript to the Publications Program of the Harvard University Asia Center. Research for this book was aided by research grants from the Australian Research Council and the Faculty of Arts at the University of Melbourne. The maps, prepared from Chinese originals, were created by Chandra Jayasuriya, and the illustrations were produced with the aid of Lee McRae. Funding for the maps and illustrations was generously provided by the University of Melbourne.

I owe a special debt of gratitude to John Fout, who read the entire manuscript in draft and gave me the benefit of his considerable experience as writer and editor.

My husband, John Fitzgerald, and my parents, Peter and Patricia Finnane, have taken a deep interest in the book's progress. I take this opportunity to acknowledge in print my indebtedness to them all. My father died in 2001. I lovingly dedicate the book to his memory. The lives of my children, Daniel, Siobhan, Therese, Genevieve, and Bernard, are intertwined with the book's history. To quote Joseph Levenson, they added years to the work and joy to the years.

A.F.

ॐ

Contents

Tables, Maps, and Figures   xiii

Preface   xvii

Part I   Foundations

1   Introduction   3

2   A Name and a Place   18
    City and Empire 24/   City and Prefecture 27/
    A Central Place? 34

Part II   From Ming to Qing

3   City of Merchants   43
    The Lianghuai Salt Monopoly and Ming Yangzhou 47/
    The West Merchants 49/   The Huizhou Merchants 57/
    A Huizhou Family in Late Ming Yangzhou 62

4   Yangzhou's Ten Days   69
    The Manchus and the Fall of the Ming 70/   "Yangzhou
    shiri ji": The First Six Days 72   Alternative Histories 1:
    Zheng Yuanxun 78/   Alternative Histories 2: Zong
    Hao 84/   "Yangzhou shiri ji": The Last Four Days 88

5   The Loyalist City   90
    *Reconstruction and Rehabilitation   96/   Rewriting
    Yangzhou   101/   Transitions   110*

Part III   City and Hinterland

6   Managing the Salt   117
    *Monarchs and Merchants   118/   Monopoly Merchants
    and Their Money   121/   Merchants and Salters   127/
    Salt Officials   131/   Smugglers   135/   City, Salt,
    Infrastructure   145*

7   Controlling the Waters   148
    *The Hydraulic Infrastructure   152/   Governing the
    Waters   155/   The Salt Administration and Hydraulic
    Management   164/   City and Hinterland   169*

8   Shaping the City   172
    *Spatial Differentiation in the New City   176/   Yangzhou's
    Gardens and Extramural Expansion   188/   The Gardens
    and Urban Culture   199/   The Great Age of Tourism   204/
    The City in Retreat   209*

Part IV   Hui City, Yang City

9   City of Women   213
    *Women on View   215/   Consumers and Producers   222/
    Yangzhou and Its "Others"   228*

10 The Huizhou Ascendancy   236
    *The Huizhou Colony   238/   Native-Place Associations 240/
    Philanthropic Activities   243/   City and Prefecture   250/
    The Blurring of Social Boundaries?   253/   Who Was Not at the
    Ma Brothers' Party?   256/   Redefining Salt Merchants   261*

11 Native Place in an Immigrant City   265
    *Merchants and Scholars   267/   The Literary Inquisition   270/
    Genealogies of Scholarship   274/   The Social Origins of the
    Yangzhou School   276/   Native Place on the Agenda   283/
    Home and the World   292*

Postscript

12 Rather Like a Dream   297
    Salt, Waterways, and Regional Decline   300/   The
    Ruined City   308/   The View from Shanghai   312

Appendixes

A   Distribution of Market Towns in Yangzhou
    Prefecture, ca. 1733   319

B   Ranked Officials of the Lianghuai Salt
    Administration in the Eighteenth Century   321

C   Natural Disasters in Gaoyou Department, 1645–85   324

D   Selected Items of Expenditure on Water Control
    by Salt Merchants in Jiangbei, 1727–1806   325

E   Private Gardens Visited by the Qianlong Emperor   327

F   Individualist Painters Associated with Yangzhou
    in the Eighteenth Century   328

G   Scholars Associated with the Yangzhou School   329

Reference Matter

Notes   331

Works Cited   391

Character List   429

Index   443

CR

Tables, Maps, and Figures

Tables

6.1	Salt patrols under the Yangzhou field administration, 1731–98	139
10.1	Huizhou and Yangzhou surnames	239
10.2	*Jinshi* graduates from Yangzhou prefecture under the Qing to 1806	252
10.3	Ancestral places of men portrayed in *The Ninth Day Literary Gathering*	258
A	Distribution of market towns in Yangzhou prefecture, ca. 1733	319
B	Ranked officials of the Lianghuai salt administration in the eighteenth century	321
C	Natural disasters in Gaoyou department, 1645–85	324
D	Selected items of expenditure on water control by salt merchants in Jiangbei, 1727–1806	325
E	Private gardens visited by the Qianlong emperor	327
F	Individualist painters associated with Yangzhou in the eighteenth century	328
G	Scholars associated with the Yangzhou School	329

Maps

1	Administrative cities, salt yards, and important waterways in Jiangbei	28
2	Salt monopoly sectors under the Qing	46
3	Sources of migrants to Yangzhou	51
4	Canals and major market towns in Jiangbei	151
5	Southwest quarter of the New City	180
6	Southeast quarter of the New City	182
7	Northwest quarter of the New City	184

Figures

1	Lane in Yangzhou, 1980	xviii
2	Old garden buildings near Slender West Lake, 1980	xix
3	City gate and wharf, ca. 1946	14
4	The tomb of Puhading	45
5	Layout of the Garden of Retirement	66
6	Zheng Yuanxun, *Landscape after Shen Zhou*	67
7	Yu Zhiding, *Wang Shizhen Releases the Quail*	103
8	Boiling up the salt, 1693 and 1806	129
9	The salt transport bypassing Yangzhou	146
10	The Grand Canal at Baoying Lake, 1793	150
11	Song city and Qing city compared	174
12	A peep show in Yangzhou, late nineteenth century	187
13	Tianning Temple and the imperial travel palace	188
14	Sites along Slender West Lake	196
15	The White Stupa (Baita)	197
16	The Gaoyong Mansion	198
17	Wang Yun, *Garden of Retirement*	202
18	Luo Ping, *Drinking in the Bamboo Garden*	203
19	Huang Shen, *Lady Carrying a Qin*	220
20	Early eighteenth-century fashions	224

21	The façade of the Hunan *huiguan*	242
22	The Ma brothers' garden party, 1743	254
23	Portrait of Ruan Yuan	267
24	The Fountain of Letters Pavilion	269
25	The Wenchang Tower	310

Preface

I first visited Yangzhou in early 1980, not long after it had been opened to foreign visitors for the first time in many years. It was a quiet, gray, rather dusty town, but charming. In the lanes of the eastern part of the city, the small tiled houses looked as though they had been standing for centuries, which was probably the case (see Fig. 1). The grander buildings, although dilapidated, hinted at better days in some distant past. Hawkers sold good things to eat on the street.

Chinese New Year was approaching, and visitors were strolling by the old mansions and pavilions that line the shores of the little lake outside the city (see Fig. 2). They were treading in the footsteps of earlier generations of tourists. In the second half of the eighteenth century, as I was to discover, Yangzhou boasted a lively tourist industry, and it continued to attract visitors long after the wellsprings of its wealth had been exhausted. As a form of economic activity, however, tourism was not strongly promoted in the People's Republic of China before the 1980s. The most recent tourist guide to the city had been published in 1954.

A copy of that guide was given to me by someone I met then, a man who struck up a conversation with my husband and me as we wandered the streets. He invited us to his home, a dark room with a stamped earthen floor, plied us with cups of hot, sweetened water, and told us, sardonically, about his experience of

Fig. 1   Lane in the "New City" of Yangzhou, 1980
(photograph by the author).

being "struggled" during the Cultural Revolution. I took photographs on that trip, but to my regret I did not keep a diary and cannot now record his name. He was in his forties at a guess, poorly dressed, but knew some English. He must have had a bad class background. As we were leaving, he presented me with a small, tattered book, that bore on its pages his own occasional notes and marks of emphasis. The title was *Yangzhou mingsheng* — "Famous sites of Yangzhou." The paper was coarse, and the black and white photos of famous sites rather blurred. A color

Fig. 2 Old garden buildings on the banks of Slender West Lake, 1980 (photograph by the author).

print on the cover showed the Five Pavilion Bridge, also known as the Lotus Flower Bridge, constructed in 1757 in honor of a forthcoming visit by the emperor. At that time Yangzhou was one of the most prosperous cities in the empire, and people came from near and far to see it in all its glory. I have used this cover as the illustration for the jacket of this book, with thanks to our kind host of many years ago.

ა

Speaking of Yangzhou

A Chinese City, 1550–1850

CʒB

PART I

Foundations

T his book is about one of the great cities of late imperial China, a place well known to historians because of the enormous wealth accumulated within its walls during the eighteenth century. To foreigners it may be less familiar than Suzhou or Hangzhou, the famed centers of Jiangnan culture that feature on so many tourist routes, but to people in China its name carries rich historical and cultural associations. Zhu Ziqing (1898–1948), a well-known essayist of the Republican era, found in his adult years that allusion to his hometown elicited a common response among fellow scholars in Beijing. "Mention the name 'Yangzhou' in front of someone," he wrote, "and the person will nod or wag his head saying, 'A fine place! A fine place!'"[1]

Famous places lend themselves readily to anthropomorphism, and writing a book about a city is in some respects not unlike writing a biography. A place can acquire an imagined personality, even a gender, and become an actor its own history. Such was the case with Yangzhou, and how it was imagined is among the concerns of the present study. We can physically survey the city of the past by using maps, textual records, and historical remnants, and we can describe its economic and administrative functions on the basis of surviving records; to locate the city in the worldview of the society that nourished it, however, we must turn to a cultural study of the place.

The name of this book derives from the title of an essay by Zhu Ziqing, "Speaking of Yanzghou" ("Shuo Yangzhou"). Zhu wrote a number of essays on his hometown, most of which reflect on the tensions between the popular images of Yangzhou and his own lived experience of the city. The name of the book correspondingly refers both to the idea of the city and to the complexities of the history that produced the idea.

The book is divided into four parts, of which the first lays the historiographical and historical foundations for the remainder. Chapter 1 introduces the city within the discursive structures supplied by cultural iconography and historical scholarship, identifying the tropes governing knowledge of the city within Chinese society and the problems of historical analysis presented by the idiosyncratic urban society that emerged in the eighteenth century. Chapter 2 locates the city in time and place, establishing the layers of historical reference that informed later understandings of Yangzhou and mapping out the terrain within which it took shape.

☸

ONE

Introduction

In the courtyard of the Yangzhou Museum stands a monument to Marco Polo, merchant of Venice, who claimed in his famous book to have served as governor of Yangzhou for three years during the reign of Kublai Khan, in the late thirteenth century.[1] Polo chose his Chinese city well. Even then Yangzhou's place in the Chinese imagination was comparable to that later held by Venice in the European mind. It owes its place to precisely the mix of wealth, power, and cultural productivity that has so firmly imprinted that other city in another tradition. The timing of the cultural achievements of the two places was different. Venice's "Golden Age" was between the thirteenth and sixteenth centuries. Yangzhou had two such ages, the first in the Sui-Tang period, between the sixth and ninth centuries, and the second, on which this book focuses, between the sixteenth and the nineteenth. The poetry of Tang dynasty Yangzhou and paintings from the Qing period continue to circulate through the Chinese intellectual world, sustaining the name of the city as a point of reference in Chinese cultural history. Like Venice, Yangzhou evokes images of artists, men of letters, great merchant families, waterways, an urban environment of considerable charm, a past imbued with color and romance. In the eighteenth century, it was famous for the "conspicuous consumption" that Peter Burke has noted of seventeenth-century Venice.[2]

Western sinologists have often compared particular Chinese cities to supposed counterparts in Europe. Such comparisons are not inappropriate: they render the strange familiar and comprehensible. Despite marked contrasts in the scale of the two cities, the reader could do worse than to imagine Yangzhou as a Chinese Venice. In the eighteenth century, at the height of its prosperity, it was a wealthy, beautiful, and historic city, ringed and divided by canals—not as many as in Venice but numerous enough to make boats a part of everyday life. Like Venice it had a Grand Canal and if not gondolas then certainly *huafang*—the "painted barques" that inspired the title of Li Dou's (d. 1817) famous guide book, *Chronicle of the Painted Barques of Yangzhou*.[3] And like Venice, Yangzhou continued to attract visitors long after its heyday.

In Chinese literature, Yangzhou was constructed as a site of dreams and illusions. "I dreamed of Yangzhou," wrote Zheng Banqiao (1694–1765), "then awoke and wondered whether Yangzhou was dreaming of me."[4] The most obvious allusion in these lines is to a passage by the philosopher Zhuangzi (369?–286? B.C.), who dreamt that he was a butterfly and then awoke and wondered whether he was a figment of the butterfly's dream.[5] Since Zhuangzi's time, a dream of Yangzhou had been fully elaborated in essay, story, and drama.[6] Its origins lie in a poem by Du Mu (803–52):

> My restless spirit roamed drunkenly over river and lake;
> Slender-waisted girls of Chu took my heart at leisure.
> After ten years from a Yangzhou dream I woke,
> Known only for living in houses of pleasure.[7]

A millennium later, Zhou Sheng's *Dream of Yangzhou* portrayed Yangzhou society and its "houses of pleasure" in a period of transition in the middle of the nineteenth century, a still lively consumer society that the author recalled nostalgically after the Taiping Rebellion, when the city was devastated.[8]

In the twentieth century the dream faded. Yu Dafu's (1869–1945) "Old Dream of Yangzhou" contrasted the romantic literary construction of Yangzhou with the realities of the dilapidated city of the Republican era.[9] Feng Zikai's "Dream of Yangzhou," published in 1958, was a revolutionary dream in which his instinct to express the regrets for the past felt by Yu Dafu was countered by his self-conscious realization that the glories of old Yangzhou had been created at the expense of working people.[10] These later writings were products of a time when China and its cities were being reconfigured under the twin imperatives of nationalism and modernization, and viewed through different eyes. In the 1890s Geraldine Guinness found Yangzhou's pagan temples horrifically numerous, its inhabitants exotic rather than familiar in appearance.[11]

Not only had the glories faded; the era in which their traces might have been appreciated had passed.[12] The empire that had nurtured Yangzhou was gone, and the newly established Nationalist government in Nanjing was bent on making a nation. Emily Honig's study of Subei people in Shanghai has shown that in this context, Yangzhou migrants in Shanghai were derided as people from a backward place and even accused of collaboration with the Japanese.[13] The city was in essence the product of a past that in the Nationalist era was seen as a problem. The leisurely lives of its urban elite, past and present, became signs of the decadence to which China's malaise in the first half of the twentieth century could be attributed.[14]

The journalist Fan Changjiang (1909–70)[15] summed up the resulting tensions in images of Yangzhou:

> There are two Yangzhous
> One decadent, one heroic.
> Yangdi had harlots as his guests,
> The salt merchants were also of this make.
> Lord Shi was upright and brave,
> The eight eccentrics' works were also great.
> Right now the trunk is sturdy,
> And blossoms waft about.[16]

In brief, Yangzhou had some worthwhile cultural heroes: the valiant Shi Kefa (1602–45), who defended the city against the Manchu onslaught in 1645, and the "eight eccentric" painters of the eighteenth century, significant contributors to one of China's

great art forms. But the city's courtesans had been recast as
whores, and the salt merchants, together with Emperor Sui Yangdi (r. 605–18), were tainted by association with them.

⚬ℬ

Fan's poem draws attention to Yangzhou's place in two important narratives of Ming-Qing history: the seventeenth-century
story of the Qing conquest, which features Shi Kefa in a major
role; and the eighteenth-century story of rising merchant wealth
and social change, which is dominated by the figure of the upwardly mobile salt merchant. The fame of Yangzhou's place in
the Ming-Qing transition is attributable in some measure to the
widespread availability in translation of Wang Xiuchu's "Ten-
Day Diary of Yangzhou," an eyewitness account of the massacre
of 1645.[17] As a showcase of merchant wealth and power in the
eighteenth century, its place was firmly established by Ho Ping-
ti's article on the salt merchants of Yangzhou, published in 1954
and for a long time virtually the only English-language historical
study of commercial activities and merchant culture in the Qing
period.[18]

Most historians of the Qing, from fields as diverse as literature,
philosophy, painting, and theater studies, have sooner or later
found reason to cite Ho's pioneering study.[19] Together with
subsequent works by the same author, it has powerfully informed two assumptions generally held about Chinese society in
the late Ming and the Qing. One is that the social distinctions between merchants and scholar-officials began to blur in the late
imperial period. The second is that native-place particularism, on
which might be based an idea of China as at best a fragmented
society, was giving way to increased interregional social and
economic integration.[20] Yangzhou, with its large populations of
merchants and sojourners—overlapping categories, to say the
least—features in broad-brush portrayals of Chinese society as
the standard illustration of these processes.

The significance of Ho's research is evident in its continuing
relevance despite nearly half a century of profound change in the
field of Chinese history. Yangzhou, with its degree-winning merchants and consumer economy, has proved to be an eminently

citable site in a revisionist historiography that emphasizes the monetization of the economy, the growth of interregional trade, the establishment of (even if peripheral) foreign trade, the expansion of cities and changes in rural-urban relations, the rise of powerful merchants and the blurring of social boundaries, the proliferation of urban institutions, the growth of print culture and literacy, and a shift from metaphysics to evidential scholarship.[21]

As a cornerstone for this historiographical edifice, Yangzhou embodies the problem of the relationship between local history and the history of China. When a city is cited as an example of what was happening within Chinese society, it is in effect deemed to be a microcosm of China. Ho Ping-ti inadvertently pointed to the difficulties attendant on this assumption when he failed to find incontrovertible evidence of a plethora of native-place associations in eighteenth-century Yangzhou.[22] Native-place associations, as Ho showed, are an index of commercial activity, particularly of engagement in long-distance trade. If Yangzhou lacked such associations and moreover failed to be represented in other great cities by its own native-place associations, in what sense was this city a representative example of urban development and social change in late imperial China?

As a "part of the whole," eighteenth-century Yangzhou presents in reverse the problem of Shanghai identified by Linda Cooke Johnson. Johnson criticized the notion of pre–Treaty Port Shanghai as a "sleepy fishing village," waiting to be urbanized by the West. Throughout the late imperial period, she argues, Shanghai was actively developing roles in the economic and social organization of its own hinterland.[23] Yangzhou in the eighteenth century is not without similarities to Shanghai in the late nineteenth century. It is viewed as having been pre-eminently a center of commercial activity, its pivotal position in the salt trade bestowing on it a position of self-evident importance in the long-distance trading network on which the economic integration of China depended. Its place and agency in a teleology of historical change are often implied, and sometimes directly stated.[24]

The problems of typicality and representativeness implicit in this historical construction can be resolved only by close comparative study of cities within particular urban systems, a task beyond the scope of the present book. Here it can be said that as the administrative center of that richest of all government monopolies, the Lianghuai salt trade, Yangzhou enjoyed a close relationship with the imperial capital, Beijing, itself not an atypical Chinese city in the view of Susan Naquin.[25] Its geographical proximity to the great cities of Jiangnan further encourages a view of the city as emblematic of urban development in the empire's most highly urbanized region. Its name is often coupled with that of Suzhou, with which it shared a reputation for beautiful women and gracious gardens, among other cultural products. Yet as the administrative center of the salt monopoly, it was certainly as distinctive a city in the eighteenth century as was Canton, the dedicated port for foreign trade, in the early nineteenth. Yangzhou's uniqueness does not signify that this particular urban society has nothing to tell us about Chinese society at large, but it does warn against too facile a use of Yangzhou as exemplifying trends or phenomena in other places and other local societies.

Something Yangzhou had in common with other considerable cities of the eighteenth century was the high social profile of Huizhou merchants in the city, the subject of a pioneering study by Wang Zhenzhong.[26] The merchants of Huizhou, some 250 miles distant from Yangzhou, were active in important centers of trade throughout the Middle and Lower Yangzi valleys. In Hankou, one of the few cities of Qing China to have been studied in detail, salt merchants from Huizhou also played a key role in urban social organization.[27] But the particular character of the locality in which Huizhou merchants operated meant that the effects of their activities on local society differed considerably from place to place. Shiba Yoshinobu remarks on the presence of merchants from Fujian, Huizhou, and Shanxi in eighteenth-century Ningbo.[28] One of the differences between Yangzhou and Ningbo was that in Ningbo the influence of these "guest merchants" (*keshang*) was manifest in the rise of strong local merchant groups. In the early nineteenth century, Ningbo merchants "gradually rose to prominence." It is likely, argues Shiba, that "much of their commercial skill and financial acumen was learned from their rivals," principally merchants from Huizhou and Fujian.[29] No such development was observable in Yangzhou, where merchants from Shanxi and Shaanxi were also active. In late nineteenth-century

Shanghai, Ningbo was known as the home of its merchant-bankers.[30] Natives of Yangzhou were the proletariat.[31]

The most enduring legacy of Huizhou merchants in Yangzhou is the rich corpus of artworks produced through their patronage, and it is not surprising that art historians have produced some of the most interesting historical work on this city. In a recently published study of eighteenth-century painting in Yangzhou, Ginger Hsü considers the changing social and economic relationships behind the exchange of paintings and provides a more finely articulated account of social and economic relations within the city than has hitherto been available.[32] Hsü properly emphasizes the depth and complexity of the commercial society of Yangzhou. In his study of the late paintings of Shitao (1642–1707), Jonathan Hay takes a further theoretical step by arguing that Yangzhou "was in many ways a modern city" and moreover that its "more modern characteristics . . . were not unique to this one city."[33]

As Hay further suggests, there are limitations to considering a city as bounded by the specificities of the locality. Yangzhou's "social, economic and cultural character," he writes, was "defined not by localized relationships inherent to that place and space but instead, to a large degree, by its relationship to other places."[34] From this perspective, considering urban societies in parallel is less satisfactory than considering them as organically linked, and the Huizhou merchants themselves were important agents in effecting this linkage. Viewed as a site within an urban system, Yangzhou becomes less an example than a product of what was happening in the system as a whole (or in China at large). The fact that most of its famous painters came from elsewhere illustrates the point.

On the other hand, what was different about and particular to Yangzhou helps shed light on the contradictory images of the city in twentieth-century texts. There were other painters in Yangzhou, often described as producers of pictures (*tu*) rather than works of art (*hua*), to utilize a distinction made by Gong Xian (ca. 1619–89), and they came mostly from Yangzhou and its dependencies.[35] Likewise, there was another Yangzhou, one that was poor rather than rich. In bureaucratic records of the late imperial period lies the hint of an explanation for the counter-myth of Yangzhou that emerged in the twentieth century. In brief, they make apparent the relative weakness of local society, the indebtedness of the city to an infrastructure devised for the service and maintenance of the late empire, the environmental problems of the greater Yangzhou region, and the absence of an economic diversity that might have helped sustain the city through the period of change initiated by, or at least made apparent in, the Opium War of 1840–42.

Official and semi-official records, ranging from local gazetteers to archival documents, make an instructive contrast to the literary record of the same period. An official memorial or an imperial edict is likely to mention Yangzhou in connection with a flood in the region, the need for famine relief, or a malfunction in the salt monopoly, which dominated the regional economy. In poems and essays, by contrast, references to Yangzhou are always more often to the city itself, which was by and large sheltered from the immediate effects of natural disasters and which as often as not profited from irregularities in the salt trade. The Yangzhou of the bureaucratic record lies buried in archives or weighty compilations of official documents. The Yangzhou of letters continues to circulate freely throughout the Chinese literary world.

Yangzhou invites attention because of its place in cultural iconography; its shifting place tells not only of its own changing fortunes but also of changes in the geography of the Chinese imagination. The tarnishing of this particular icon was attributable at least in part to changes in the Chinese experience of cities and to corresponding shifts in the Chinese worldview. The name of the city, which was also that of the prefecture of which it was the capital, draws attention to other, actual landscapes: to the walled city, within which a complex, culturally distinctive urban society took shape in the late imperial period; to the twin counties for which it also served as a capital and from which it drew its native population; and to the larger region in which the city was located both geographically and administratively, the warp and weft of bureaucratic and economic ties linking it to other greater and lesser administrative centers, market towns and villages, rice fields, fish ponds, and salt pans. Were the observer to stand further back, a fourth landscape would emerge, because Yangzhou was an offspring of the empire, born of the conjunction of north and south and sustained by communications between the two. Even if viewed from a great distance allowing a sweeping view of its provinces, waterways, and outlying territories, the landscape of China does not overwhelm the observer's impression of late imperial Yangzhou. Rather, the eye is drawn to it by lines of communication: the Yangzi River running west to east, intersecting just south of Yangzhou with the Grand Canal.

The history of Yangzhou considered in these different landscapes is full of paradoxes. It was a wealthy city in a poor region. The city produced many talented scholars and artists, but few members of the bureaucracy. It was strategically located at the junction of two great commercial routes, and buying and selling were activities as common as eating and drinking. But in a period of flourishing interregional trade, Yangzhou people had no obvious presence in other great commercial cities. Such paradoxes lay at the heart of Yangzhou society in the late imperial period, and they constitute its history.

The elucidation of these paradoxes is facilitated by a particular corpus of writings that permits us to reflect on the city as a local place. In the late seventeenth century, Tobie Meyer-Fong has shown, early Qing scholar-officials set about reinvesting Yangzhou with meaning through a range of literary projects that added to the pre-existing iconography.[36] In the eighteenth century, men in Yangzhou busied themselves documenting aspects of the city as it visibly flowered under the impact of the extraordinary wealth of the salt merchants. Li Dou, writing in 1795, listed some of the important sources available to him in his ambitious project to document city and society during its heyday: Wang Yinggeng's *Gazetteer of the Eminent Sites of Pingshan Hall*, Cheng Mengxing's *Little Gazetteer of Pingshan Hall*, Zhao Zhibi's *Illustrated Gazetteer of Pingshan Hall*, and Wang Zhong's *Complete Records of Guangling*.[37] These works were the product of the period of the Huizhou ascendancy in Yangzhou. Wang Yinggeng and Cheng Mengxing, both active in the first half of the eighteenth century, were among the most prominent of the Huizhou merchants in Yangzhou at that time. Wang Zhong (1745–94) was also of Huizhou ancestry, although as we shall see, his position was somewhat anomalous in this respect.[38] Zhao Zhibi was salt controller at a time when this office was the most powerful of any in Yangzhou. With the exception of Wang Zhong's late and uncompleted work, all take Pingshan Hall, located in the hills northwest of the city and a place imbued with religious and historical meaning, as a central point of reference.

Li Dou's *Chronicle of the Painted Barques of Yangzhou* is the best-known and a much-cited work on the eighteenth-century city, and it is a rich source for the historian. Unlike the various gazetteers of Pingshan Hall, this "chronicle" pays close attention to the city proper. Li, a man of no prominence other than as author of this work, was registered as a native of Yizheng, the county directly west of Yangzhou. He shared this registration with many Yangzhou people and may safely be reckoned a local or native (*tuzhu*) in the terminology of the time.[39] (People of Huizhou descent, by contrast, were commonly although not invariably counted as "sojourners" [*liuyu*].) Compilation of the material for *The Painted Barques* commenced in 1764, two years after the third of the Qianlong emperor's Southern Tours. The emperor would make three more tours by the time Li Dou completed his project in 1795, the last year of the Qianlong reign. The impact of the tours on Yangzhou was extraordinary in terms of the city's expansion in the second half of the eighteenth century, and Li appropriately begins his work with reference to them. But in the same opening chapter, he mentions the teahouses, restaurants, and bathhouses around which much of the city's leisure life was constructed. The emperor's visits, although given pride of place, are incorporated into an urban panorama that features a large and diversified cast of townspeople.

This engaging and evocative account of the eighteenth-century city suggests in its construction a developing idea of the unique qualities of the locality. Li Dou could have organized his chronicle in the form of a local gazetteer, with different sections on the past, geography, notable buildings, scenic sights, biographies, and writings. Instead, he sensed the organic nature of the city. Past and present, people and places, writings and writers, are densely intertwined to produce a dramatically interactive account of urban society, the singularity of the literary performance matching that of the society itself in late imperial times. This acute sensitivity to urban life was evident subsequently in the lively interest in commodities and local culture evinced by Li Dou's contemporary Lin Sumen (1749–after 1809), who in 1808 published a collection of poems with commentaries on Yangzhou, and also by the self-described "Fool of Hanshang" (Hanshang mengren), who in the 1840s produced what Patrick Hanan regards as China's "first city novel," *Dreams of Wind and Moon*. Li Dou arguably established a way of imagining the city on which his successors could draw. Certainly he had been read by the Fool of Hanshang.[40]

The *Painted Barques* substantiated and developed the "dream of Yangzhou." Li Dou had an intimate knowledge of local society and paid particular attention to the achievements of otherwise unknown artists and scholars, but his chronicle noticeably avoided engagement with the workaday city of local administration. His was a celebration of the city, much like the *civis laus* of early Renaissance writings. In retrospect, his chronicle also marked the end of an era in which the dream of Yangzhou was recorded and the beginning of an age when it would be remembered. In works ranging from poetry collections to treatises on water control, Yangzhou scholars writing after Li Dou historicized their native place by documenting what had passed in the years since 1645. Again a heightened sense of the locality is evident. All the various authors were connected with one another in different ways, with noted scholar-official Ruan Yuan (1764–1849) serving as a pivotal figure in the social and literary relationships. The diversity of topics covered in their writings provided a more complex past for the city and also a richer regional context.[41]

Despite the many artists active in Qing Yangzhou, visual representations of the built environment are relatively few. They consist mainly of woodblock prints of famous sites and allusive, highly referenced paintings of its gardens, temples, and extramural scenery. In the urban landscapes of Shitao, the city looms tantalizingly through mists and vapors—a case of visual form as "suspended reality," writes Hay.[42] The demolition of the city walls in 1953 removed one of the strongest visual impressions of the city, evident in a number of Shitao's paintings,[43] and it was

Fig. 3   A city gateway and wharf, before the city
walls were demolished. Crayon drawing by
William Henry Scott, ca. 1946.

only through benign neglect that other vestiges of the past survived.

A rare pictorial record of the city before the dismantling of the walls can be found in a collection of crayon drawings by William Henry Scott, who served as a missionary in Yangzhou immediately after World War II.[44] The sketches show an old, walled city full of shadows, rubble in streets, the odd washing pole stretched across a narrow laneway with clothes hanging out to dry. Simply drawn human figures show men wearing long blue robes, a fashion not to last much longer. Telegraph poles in one drawing are a reminder of recent technological changes, although access to electricity was still a rarity. One wall of the city bears a sign in large red characters reading *wu mai ri huo* (Boycott Japanese goods!), a rather belated war cry for a place occupied by Japanese troops for eight years. The city's architectural marvels are also evident. Scott drew the Tianning Temple, the Lotus Flower Bridge, the White Stupa, and the Literary Peak Pagoda (Wenfengta), famous and emblematic sites, along with the city gates and the massive walls. Sections of the wall are shown as overgrown with vegetation, giving substance to the city's cognomen, the "Overgrown City" (Wucheng).

These sketches link the mid-twentieth century to an earlier visual history. The Literary Peak Pagoda was depicted in a watercolor by William Alexander in the late eighteenth century and appears in an album leaf by Shitao.[45] And the high-roofed city gate that Shitao painted, opposite the pagoda and fronted by the Grand Canal, may have been the same that Scott drew two and a half centuries later, viewed from closer at hand (see Fig. 3). This was the South, or River Pacification Gate (Jiang'an men), which was sheltered behind a triple *enceinte* (*zicheng*) or walled gate enclosure. A distinguishing feature of Scott's drawing is its attention to economic life in progress at the wharf, to which Shitao also alludes with his hazy depiction of a long row of boats along the barely suggested route of the Grand Canal. In the eighteenth century, reed sellers from Guazhou, southwest of Yangzhou, regularly came and went from the wharf outside South Gate. They stayed in small hostels in South Gate Street and feasted on bowls of dried bean-curd strips (*doufugan* or *gansi*), a local specialty, served at food stalls near the wharf.[46]

♋

Scott's naïve, charming drawings, their references to the grandeur of the past nonchalantly incorporated into representations of everyday life in the late 1940s, draw attention to Yangzhou as a local place with a local history, one in which sojourners and even emperors played an important part without ever changing the specific advantages and disadvantages of the city's location. The present book is an essay on this history. In the following chapters, I examine the making of the city in the late Ming and Qing periods with reference both to the historiographical problem of the city's place in history and to the semiotic problem of the meanings accumulating around Yangzhou over time. These problems are interrelated: the actual construction of the city—its academies of learning, its philanthropic institutions, its gardens, its teahouses, its brothels—underpinned the construction of a certain idea of Yangzhou, even if this idea incorporated and was enriched by earlier historical references.

As noted previously, the book is divided into four parts. Part I, consisting of the present and the following chapter, introduces the city both as a historical problem and as a place that is historically known through the records of its past. Part II historicizes the High Qing city through attention to developments in the late Ming and early Qing, traces the rise of the Huizhou merchants in Yangzhou from the late sixteenth century, and problematizes the history of the Ming-Qing transition with reference to the Yangzhou massacre of 1645 and the issue of loyalism in the second half of the seventeenth century. The eighteenth-century city proves by and large to have been recognizable in small before the end of the Ming.

Part III examines the relationship between the city and the greater Yangzhou region during the eighteenth century, focusing in Chapters 6 and 7 on the administration of the salt trade and the problems of water control in the city's hinterland. Although salt signified wealth, floods meant poverty. These two factors helped ensure Yangzhou's regional dominance. Chapter 8 shows what was happening in the city itself during the same time span: flourishing, as might be expected, but in different ways at different periods of the long eighteenth century.

Part IV explores the city from the perspective of gender and native place in the social hierarchy and simultaneously queries the conventional description of Yangzhou as the pre-eminent site of the blurring of social boundaries held to characterize Chinese society in the eighteenth century. The chapters in this part show that the particular patterns of spatial and social differentiation observable within the eighteenth-century city were in significant part an expression of relations between Huizhou and Yangzhou people. Finally, the Postscript lays the foundation for a nineteenth-century history of the city, a history that allows us to see the historically specific characteristics of earlier times clearly. A short foreword to each part introduces the themes of that section of the book.

Despite the cross-cultural naming practices noted above, it was long the practice in western historical sociology to differentiate sharply between the cities of China and those of Europe.[47] In terms of political organization and functions, they were certainly not the same as each other, but the experiences of pre-industrial city life across cultures probably have more in common than is often recognized. This introduction began with a reference to Venice but it might conclude with a reference to Amsterdam. The chronicles of Amsterdam, like those of Yangzhou, are full of tales of upward and downward social mobility of immigrants, who excited the hostility of locals while contributing mightily to the city through their patronage of architecture, the arts, and learning. But around 1680, according to Peter Burke, "Amsterdam stopped growing." The Baltic grain trade went into decline, and after 1672, "only one first-generation immigrant entered the Amsterdam elite, which naturally became predominantly rentier in composition."[48] Much the same occurred in Yangzhou around one and a quarter centuries later, when the salt trade went into decline, and the Huizhou wellspring of Yangzhou immigrant society dried up. Around 1800, Yangzhou stopped growing.

CR

TWO

A Name and a Place

Yangzhou is an inland city. Although situated in the coastal province of Jiangsu, it lies more than a hundred miles from the nearest coastal town and even further from the nearest seaport, Shanghai. It was not always so far from the ocean: sixty miles to its east lies Taizhou, which in the earliest records is referred to as Hailing, "hill by the sea." Over the centuries the sea retreated, leaving in its wake mudflats that added to the miles between Yangzhou and the coast. The sea yielded the salt that was basis of Yangzhou's wealth in late imperial times, but it otherwise entered little into the lives of Yangzhou people. Theirs was a world of rivers, lakes, and canals, which before the mid-nineteenth century were crowded with vessels coming and going. When the Macartney embassy passed down the Grand Canal in November 1793, John Barrow saw "a thousand vessels, at least, of all descriptions" at anchor by Yangzhou.[1]

The word "Yangzhou" is not self-explanatory as the name of a city. *Yang* means "vast" or "spreading" and *zhou* means "region," "province," or "state." In antiquity this name denoted a vast region south of the Yangzi River, far from the center of the burgeoning political society that would one day subsume it. Thousands of years ago, so the story goes, Chinese civilization began to emerge from the loess plains of the middle reaches of the Yellow River, gradually spreading west, east, and south from its fertile core. Territorial emblems changed as the territory "beneath heaven" (*tianxia*) grew. The Yangzi, once far away, became near, and what lay beyond it, Jiangnan, "south of the river," became "a glorious and beautiful land."[2]

Between the Yangzi and the Yellow rivers lay the Huai. Rising in the Tongbai Mountains, the Huai flowed through what are now the provinces of Henan, Anhui, and Jiangsu before reaching the sea. The Huai and Yellow rivers were linked in their lower reaches by the Si River,[3] but no corresponding waterway connected the Huai with the Yangzi. Rather, boats from *Yang zhou* "followed the course of the Yangzi and the sea, and so reached the Huai and the Si."[4] For waterborne traffic in very early times, the coastal route offered the only passage between north and south China. The founding of Yangzhou was dependent on this fact. The distant southern province of Yang zhou gave its name to the city of later times, but it was long before there was such a city. It began as a rough wall providing defense for a few soldiers keeping watch on top of a hill. The year was 486 B.C., and a war was being waged by the rival kingdoms of Wu and Qi. Fu Chai (r. 495–476 B.C.), king of the southern state of Wu, led his men north across the Yangzi River and looked around for a favorable site for a garrison.[5]

The land to the east of the present city of Yangzhou is flat, but to the northwest rises a range of low hills. On top of a prominence later known as Shugang, which offered a good view over the flatlands, Fu Chai's men erected the garrison wall. To the east, they then embarked on that most famous of engineering activities in China, the construction of a canal. This linked the Yangzi River with the Huai and was no doubt meant to facilitate the transport of troops and provisions to the borders of Qi.[6] Fu Chai lost the war but won for himself a name in the annals of Yangzhou. The garrison he built, apparently unoccupied after his retreat, holds a place of honor as the ancestor of the present city of Yangzhou.[7]

The garrison was called Hancheng, and the waterway was called the Han Gou (Han canal). Hardly any artifacts have been recovered from this early time, or for a period of around three centuries thereafter. During these centuries China became a different place. The quarreling states had been welded into an empire, ruled first and briefly by the Qin dynasty (221–207 B.C.) and then for a greater duration by the Han (202 B.C.–A.D. 220). In the year 195 B.C., the first Han emperor presented his nephew Liu Bi the southeastern fief of Wu. The Wu region covered much of the area that in antiquity had been called "Yang zhou." Liu Bi, lord of Wu, built his capital on the site of Hancheng and named it Guangling, which means "broad tumulus."[8] The city wall measured some fourteen and a half *li* around (a little over five miles), surrounding an area large enough to include small plots of cultivated land in addition to residences, barracks, and market-places.[9]

The Han dynasty lasted over four centuries. During this time canals were dug to and from Guangling, dikes were built, mulberry groves were planted, and salt was extracted from the sea. Most of the initiatives in water control—including the rerouting of the Huai-Yangzi canal—took place in the Later Han (A.D. 25–200), allowing us to imagine a gradual thickening of population and extension of cultivation, with irrigation and transport needs increasing over time.[10] Until the end of the dynasty, Guangling continued to be a frontier town, the place where the north met the south, but the meetings were more frequent now, and the city's geopolitical foundations more firmly established. From this time, there would always be a city by the canal that ran north from the Yangzi River to the Huai.[11]

The Han was followed by a long period of disunity. Guangling's strategic significance, foreshadowed when Fu Chai built his garrison centuries before, was repeatedly demonstrated in sackings of the city during wars between rival kingdoms. In the late fifth century, the poet Bao Zhao (414?–66) reflected in verse on the destruction of the city at the end of the Han, but the lines were appropriate to his own era, which was marked by massacres at Guangling in 451 and 459:

> The winds of war arise,
> It is cold on the city walls.
> Wells and roads are destroyed,
> Fields and paths in ruins.
> A thousand years, ten thousand eras,
> All finished: how can one speak of it?[12]

Bao Zhao called his poem "Ballad of the Ruined City," and Guangling was thereafter to be known also as "Wucheng," the ruined
or "overgrown" city, a poignant reference to the transience of its
fortunes.

The city was not fully restored until the empire was reunified
under the Sui dynasty (581–618). The heir apparent to the Sui
throne, Yang Guang, better known to history as Emperor Sui
Yangdi, was stationed at Guangling in 589, with jurisdiction over
the vast southern province of Yangzhou. He redeveloped the site,
creating a large and prosperous city that he called Jiangdu, or
"Metropolis-on-the-Yangzi."[13] This term survived the centuries as
a name for either the city or its immediate administrative area,
but it is most famously associated with the short-lived glories of
the Sui. It was to Jiangdu that Yangdi fled during the rebellion
that led to the fall of the Sui, less than thirty years after its rise.
Murdered in 618, he lies buried not far from the present city.
More than a thousand years later, Shitao painted his tomb into an
autumn landscape, in a sorrowful reflection on the passing of a
dynasty.[14]

From the Sui city of Jiangdu emerged the city of Tang times
(617–907): a city on the plain rather than on the hill. Its growth
was uncertain at first, judging by many changes of name and
administrative status. Changes in dynasty commonly entailed
changes in place-names, not only because the administrative hi-
erarchies, borders, and spans of territorial authority were altered
but also because renaming stamped the empire with the sign of a
new dynasty. The Sui city of Jiangdu had been a city of considerable administrative importance, governing sixteen districts north
and south of the Yangzi, in the present provinces of Jiangsu and
Anhui.[15] Its significance was reaffirmed in the early Tang, when
the city was made the seat of a governor-general in 624. At this
time, it acquired the name Yangzhou, although Jiangdu denoted
the local jurisdiction.[16]

This was not the last change in name that the city was to un-
dergo — Guangling was revived for a brief period in the middle of
the eighth century — but as Yangzhou it passed through its golden
age, and in practice the name endured. Poets imprinted it in the
literary corpus, which was central to the creation and mainte-

nance of a historical tradition. Li Bai (701–62) immortalized it in the eighth century when he saw Meng Haoran (689–740) off from Yellow Crane Tower—"in the third month, amidst the mists and blossoms, bound for Yangzhou."[17]

During the Tang dynasty, Yangzhou flourished. In the words of E. H. Schafer, it was "the jewel of China in the eighth century . . . a bustling, bourgeois city where money flowed easily . . . a city of well-dressed people, a city where the best entertainment was always available."[18] It owed its prosperity to the growing volume of traffic between north and south. As the number of trade shipments from the south to the court in Chang'an (Xi'an) increased during the seventh and early eighth centuries, the inland waterway system expanded to accommodate it.[19] Yangzhou was well placed to benefit from the improvements in communication and gained a name as a place where commerce reigned at the expense of agriculture.[20] It became a hub of interregional and even foreign trade and was host to a large population of merchants from inner and western Asia.[21] After the outbreak of the An Lushan rebellion (755–63), it became the seat of the Huainan military governor and the administrative center of the newly established government salt monopoly.[22] The urban population swelled because of the influx of soldiers, refugees, and salt merchants.[23]

This mixture of economic, bureaucratic, and strategic functions meant that Yangzhou remained a city of considerable importance until the Huang Chao rebellion (874–84), when it was reduced to a pawn in the hands of rival hegemons.[24] The city wall, and much within it, was probably destroyed during the Southern Tang retreat from Later Zhou armies in 957.[25] Li Yu (936–78), the last of the Southern Tang emperors, mourned the city's ruin in verses reminiscent of Bao Zhao's ballad of five centuries earlier:

> South of the river, north of the river, lies my native land.
> For thirty years my life passed in a dream.
> Henceforward the palaces and pleasances of Wu shall
>   be desolate,
> Terraces and halls of Guangling shall be wild and forlorn.[26]

The history of the city walls from the tenth through the eighteenth century says much about the subsequent vicissitudes in the city's fortunes. During the Northern Song (960–1126), Shen Kuo commented on Yangzhou's former "extraordinary prosperity" (*zui you fusheng*), which he clearly felt to be expressed by the impressive dimensions of the Tang city wall. Measuring "15 *li* and 110 paces from north to south, and seven *li* and thirty paces east to west," it was more than three times as long as the Guangling wall had been.[27] The wall constructed in the Later Zhou (951–60), by contrast, surrounded only the southeastern quarter of the Tang city. During the Northern Song no further building of walls was undertaken, a fact consistent with the city's diminished status in this period.[28] But under the Southern Song (1127–1278), when the capital lay south rather than north of the Yangzi, Yangzhou became a bastion of defense for the dynasty, and its total walled area was increased by two further enclosures, one to protect a garrison on Shugang and the other, a smaller enclosure, to link the garrison with the main city. This complex triple city (*san cheng*) was conquered by the Mongols only with some difficulty in August 1276. The gripping tale of the siege has the people of the city starving within the walls, and the Southern Song defender Li Tingzhi (1217?–76) declaiming to the enemy from the battlements: "I received an order to defend the city. . . . I have not heard a command to the effect that I surrender."[29]

The Mongols had faced staunch resistance from the walled cities of the Song and were not much in favor of building walls.[30] They may have let those around Yangzhou quietly crumble. The Yuan dynasty (1271–1368) relied for its defense not on fortresses but on the deployment of large bodies of troops, especially in the Huai and Yangzi valleys.[31] It is believably stated in *The Book of Ser Marco Polo* that "the people live by trade and manufactures, for a great amount of harness for knights and men-at-arms is made there. And in the city and its neighborhood a large number of troops are stationed by the Kaan's orders."[32] In 1357, however, Yangzhou was occupied by the rebel forces of Zhu Yuanzhang (1328–98), destined to become the first emperor of the Ming dynasty (1368–1644). Zhu was at this time fighting rival rebel Zhang Shicheng (1321–67), and he set his men to building a wall as a defense against Zhang's troops.[33] From this enclosure developed the city observed by Barrow in 1793, rectangular in shape, bounded by a waterway and surrounded by walls, "ancient and covered with moss," although their antiquity was of relatively recent date.[34] A second city wall was constructed in 1557 to protect residents who had settled outside of what would thenceforth be known as the Old City.[35]

Under the Ming, Yangzhou was made the capital of a large prefecture encompassing ten districts north of the Yangzi, and it underwent a last change of name when the prefecture was dubbed Weiyang. This name was etymologically related to Yangzhou, its origins being a line in the *Book of Documents*, "Huai, Hai wei Yang zhou." This could plausibly be read as "between the Huai and the sea is only *Yang zhou*," although the meaning of the word *wei* is indeterminate in this context.[36] The name "Yangzhou" was retained in popular usage, and after the Ming "Weiyang" was elevated to poetic or literary usage.[37] Jiangdu, the city's name under the Sui, survived as the name of the city's home county.[38]

City and Empire

By the late Ming, certain features had emerged as constants of Yangzhou's history. The city was basically a product of empire. Geopolitical factors had been critical determinants of the city's historical development, and the state of relations between north and south China at any particular time had either encouraged or inhibited its growth. Originating as a garrison to serve a southern king in his advance against a northern one, its military importance was confirmed in the succession of battles fought over it. Yao Wentian (1758–1827) commented on this feature when he began his history of Yangzhou with the words: "Weiyang [Yangzhou] is an important crossing between north and south. From the Qin and Han dynasties on, it has acted as a thoroughfare for military forces. The rise and fall of its departments and counties, the changes in its boundaries, outnumber those of other jurisdictions."[39]

Military conflict over Yangzhou was an outcome of the strategic significance of the Huai–Yangzi belt. From quite early in the history of imperial China, a frontier tended to be created between the Huai and the Yangzi during periods of fragmentation. The border separating Wu and Wei during the Three Kingdoms period (220–65), Wei and Qi during the Northern and Southern

Dynasties (420–589), the Southern Tang and Later Zhou during the Five Dynasties (907–60), and the Song and Jin empires in the twelfth and thirteenth centuries were dictated by the courses of these two rivers. During the Southern Ming, in the year leading up to the calamitous siege of Yangzhou in 1645, the lower Huai-Yangzi region served as a buffer zone between the advancing Manchu forces and the Southern Ming capital in Nanjing.

Northern-based dynasties such as the Tang and the Yuan were naturally interested in controlling or having access to the resources of the south. Dynasties such as the Southern Song or Southern Ming were equally interested in protecting their territories from northern invaders. Yangzhou, located on the main north-south transport route, was thus a likely military target in interdynastic conflicts. Possession of the city signified easy access to the Yangzi for a northern power or a viable defense of the Yangzi by a southern power. In times of peace and stability, the city was equally likely to be the headquarters for significant administrative functions. Since peace was fortunately the more frequent condition, Yangzhou was on the whole well served by its location.

When the empire was united and the political situation stable, Yangzhou usually benefited from two major institutions: the Grand Canal and the salt monopoly. The Grand Canal was less important during the Southern Song, when the capital was located in Hangzhou, than it had been during the Sui and Tang dynasties; but it was revitalized and extended far northward under the Yuan, when the capital was Dadu (modern Beijing).[40] The Ming dynasty, which initially ruled from Jinling (Nanjing), eventually followed the Yuan example and relocated to Beijing in 1428. The maintenance of the Grand Canal was thereafter one of the dynasty's major preoccupations. As a route for interregional communications and trade, especially for the yearly transport of the grain tribute from south to north, the canal conferred enormous importance on Yangzhou, which was one of its main ports.

The salt monopoly, which was to provide the foundations for Yangzhou's re-emergence as one of the premier cities of China in the sixteenth to eighteenth centuries, was already an important factor in the urban economy of the Tang city. Under the Song, Yangzhou did not have much of an urban economy to speak of, but the salt tax continued to be a significant source of government revenue and was critical to the financing of the Southern Song's defense.[41] During the Yuan, salt was probably the largest single source of state revenue, and the wealth of the Lianghuai merchants was already recognized as in effect a private bank on which officials could draw to fund public works such as canals.[42]

The canal and the monopoly were developed under the aegis of successive ruling houses and were intended to benefit and serve them, but they also underpinned the city's prosperity in times of peace. Yangzhou was the beneficiary, in sum, of an artificial infrastructure, the survival of which depended on imperial policy and the efficiency of which was ensured only by able administration. The city's logistical advantages were shared to some degree by the city of Huai'an, located further north on the Grand Canal, near the junction with the Huai. In the Ming and Qing, Huai'an had administrative roles in the salt trade, grain transport, and customs collection comparable to those of Yangzhou.

It was Yangzhou's closeness to the Yangzi that determined its historical ascendancy over its northern neighbor. During the Sui dynasty, Yangzhou became the ultimate outpost of the north. Beyond it, one was forced to negotiate the Yangzi—the great divider over which an early emperor once sighed, "Alas, it is indeed Heaven which separates south from north."[43] Having crossed the Yangzi, one entered into unmistakably southern terrain, where economic and demographic growth during the Tang and Southern Song dynasties would give the Wu region lasting advantages over the rest of the empire.[44]

Even given the relative advantages of its location near the Yangzi, Yangzhou was vulnerable to the mutability of administrative structures. During the Northern Song dynasty, the nearby Yangzi port of Yizheng, then known as Zhenzhou, became the seat of the grain transport commissioner and pre-empted Yangzhou as a center for the tea and salt trades.[45] Likewise in the Yuan, Yangzhou briefly gained heightened administrative importance when it became the provincial capital, before that role was returned to the southern city of Hangzhou.[46] Despite its attractively central position, then, Yangzhou's potential could be dramatically reduced by bureaucratic fiat.

City and Prefecture

Between the fourteenth and early twentieth centuries, the geo-political organization of China proper remained relatively stable, despite the enormous expansion of the empire under the Qing. During these centuries, Yangzhou was capital of a large prefecture. The custom in Chinese place-naming practices of not distinguishing the administrative center from its jurisdiction has sometimes led to a confusion of city and prefecture. References to the greater Yangzhou area in the prosperous eighteenth century, for example, reveal a general tendency to extrapolate from urban wealth to regional prosperity.[47] The fact is that even when the city was flourishing, the inhabitants of its vast agrarian hinterland were lucky to be making ends meet. Nonetheless, it was largely from its own dependencies that Yangzhou drew its great wealth, and if at certain times it seemed that the city was in the prefecture but not of it, at other times the links were in plain view.

Yangzhou prefecture occupied the greater part of the middle portion of Jiangsu province, covering more than 20,000 square kilometers (see Map 1). The middle or central part of the province was divided from the south by the Yangzi River and from the north by the Huai River, although the bed of the Huai was gradually being taken over by the Yellow River from the late twelfth century. The distinctions between the three sectors were apparent to observers. A late Qing commentary records the contrasts:

The northern region, around Xuzhou especially, is in no wise different from that of the North of China. It is even less rich and has but sparse clumps of bamboo, while the willow, poplar and a few acacias are the only trees that afford a little verdure to this impoverished tract. The mulberry is scarce, and the country has neither rice nor the tea-plant. There are a few fruit-trees and the fruit is excellent, especially the peaches. The Central region is not much superior to the northern, but the canals and lakes teem with fish, and the cotton which grows there is of excellent quality. The Southern region is the most favored, cotton, rice and the mulberry constituting with the ordinary cereals the staple products. The bamboo thrives well, but the tea-plant is backward. The hills are completely denuded. In the Yangzi river, as well as the canals and lakes, a great variety of fish is found.[48]

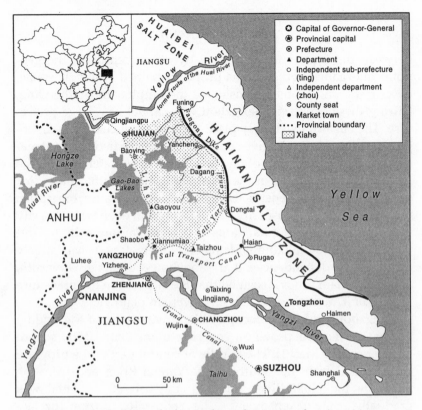

Map 1   Jiangbei in the late eighteenth century, showing cities,
salt yards, and major waterways.

Land and tax quotas in the eighteenth century bear out the
relatively poor conditions prevailing in the central region relative
to the southern. With less than half of the total registered land
area, southern Jiangsu yielded around three-quarters of the
commuted land-poll tax and a greater proportion of the rice tax.
In terms of tax levies, it exceeded the central and northern sectors
combined in all categories except wheat, the bulk of which came
from the north. Although the central part of the province was not
as poor as the north, the difference between these two regions
could hardly be compared to the enormous gap in prosperity between the south and the rest of the province.[49]

Central Jiangsu is an area in search of a name, as Jeffrey
Kinkley once remarked of west Hunan.[50] Jiangsu province was

created in 1667, when the province of Jiangnan, the Nan Zhili (southern metropolitan province) of Ming times, was divided in two.[51] Its name is a combination of the first syllables of the prefectures of Jiangning (Nanjing) and Suzhou, whose capitals were the centers of provincial authority. Nanjing was the seat of the governor-general of Liangjiang and Suzhou the seat of the governor of Jiangsu. The governor-general had authority over Anhui, the other half of the former province of Jiangnan, and the neighboring province of Jiangxi. In practice, the governance of Jiangsu province was divided between Suzhou, which controlled the Wu-speaking southeastern region of the province, and Jiangning, where the offices of the governor-general took responsibility for the remainder.[52]

The term Huainan, meaning "south of the Huai River," was sometimes used to denote central Jiangsu. Other terms used were Huai-Yang-Tong or Huai-Yang, both of which derived from the territorial scope of the circuit intendancy, and Jiang-Huai, which referred to the area between the Yangzi and Huai rivers.[53] Jiangbei and its later equivalent, Subei, were other possible terms. Both are translatable as "northern Jiangsu," but Jiangbei carried the additional and rather older meaning of "north of the Yangzi." The Ming geographer Xie Zhaozhe (1567–1624) was contrasting the northern and southern halves of China, not of Jiangsu, when he wrote: "There are no sluice gates in Jiangnan; there are no bridges in Jiangbei; there are no thatched houses in Jiangnan, there are no privy pits in Jiangbei."[54]

Jiangnan, meaning "south of the Yangzi," also had shifting meanings. Now used in reference to a loosely defined area of southern Jiangsu and northern Zhejiang provinces,[55] it was the name used in the early Qing to denote what became the two provinces of Anhui and Jiangsu. It survived into the eighteenth century within the river administration, but confusingly, the Jiangnan director-general of river administration (*Jiangnan hedao zongdu*) had nothing to do with hydraulic management south of the Yangzi. He was stationed well north of Yangzhou at Qingjiangpu, near Huai'an, and his jurisdiction ended at the northern bank of the Yangzi.[56] Clearly, this title derived from the old provincial name rather than from the literal meaning of the term "Jiangnan."

The naming of Jiangnan and Jiangbei was informed by a mix of sometimes contradictory geographical, cultural, and administrative factors. For the purposes of this book, the term Jiangbei, with all its ambiguities, suffices as a name for central Jiangsu. It has the advantage of an etymological relationship with its antonym, Jiangnan, and by the twentieth century, moreover, was used mainly to denote what lay north of the Yangzi and south of the Huai.[57] Qing Yangzhou is often thought of as being a Jiangnan city, quintessentially southern, belonging with Suzhou and Wuxi rather than with the definitively northern Huai'an.[58] In the eighteenth century, however, even Huai'an had something of a southern charm, evident in its southern-style garden quarter.[59] Indeed, before the breakdown of the inland waterway system and the boom in the coastal trade in the nineteenth century, links between the southern and central parts of Jiangsu were much closer than subsequently. In this period, Yangzhou could clearly be referred to in some sense as belonging to "Jiangnan," a term rich with historical connotations and at least as adequate as "Jiangsu" for conveying a general idea of the city's spatial context.

Properly speaking, however, Yangzhou belonged to Jiangbei, and the characteristics of even the urban population are best understood in this context. In the 1542 edition of the Yangzhou gazetteer, one contributor put a question about the composition of the city's population:

At the beginning of the [Ming] dynasty, it was found on inspection that native households in the prefectural city of Yangzhou numbered only eighteen, and then subsequently around forty. The rest were sojourners or people taking shelter from warfare and so on. Since then more than 170 years have passed. Although grain production is now fixed, the [native] population of the city has still not increased. Why is this?[60]

The question was rhetorical. The writer supplied his own answer:

It is due to injury from natural disasters, epidemics, and famines; to the rivers and seas running amok and emergencies in the grain tribute. A livelihood cannot be made, even without considering the heaviness of taxes and the burden of the corvée. Liu Yan said: "When the population grows, the revenue from taxation will increase of itself. This is an established principle." Now the territory of

Yangzhou is broad and far-reaching. The residents are many, but the villages and great households few. Why is this? The land has not been fully tilled; the irrigation ditches have not been fully developed; customs have not been fully reformed; wanderers have not fully returned. In the management of the fields, there is the problem of salination; in sericulture, things are not as they should be.[61]

In sum, agriculture was in a sorry state, industry was undeveloped, people were living close to subsistence levels, and evidence of prosperity was wanting. All these factors differentiated Jiangbei from Jiangnan in the middle of the Ming dynasty and did so still in the middle of the Qing. One consequence, as we shall see, was the relative weakness of the local gentry in Yangzhou; another was the peculiar bifurcation of the urban society into local and sojourner or immigrant communities.

On a map of late imperial China, the boundaries of Jiangbei are clear. To the south lay the Yangzi, to the north the Yellow River, occupying the bed of the Huai, which it had completely usurped by the late fifteenth century. To the west were the Grand Canal and a series of lakes, and to the east the sea. Only in the far southwest do aquatic borders fade from view, so that Yizheng county was divided from Luhe by a line on a map. This region incorporated all of Yangzhou prefecture, including the later independent jurisdiction of Tongzhou.

The number of county-level jurisdictions in Yangzhou prefecture changed over time as some counties were split off and others subdivided. In the early seventeenth century, the prefecture was composed of three departments and nine counties, but by the end of the eighteenth century the number of dependent jurisdictions had been reduced to seven.[62] One major change concerned the city's home county, Jiangdu, which in 1732 was split in two to form a second county centered on the city, Ganquan. The new county border ran right through the middle of the city. The land to the north and west of the city was allocated to Ganquan, the rest to Jiangdu. Jiangdu was the wealthier of the two counties, for to it fell a number of prosperous market towns. Ganquan, by contrast, incorporated some rather unprofitable farmlands. As a Ming gazetteer noted: "West of the city [later Ganquan], it is mostly hilly and the people are worn out. They put all their efforts into tilling the soil and have rather a name for being quiet and submissive. As for the people of Wanshou and Guiren [in Jiangdu], the land there being ample and the soil rich, they are all wealthy and proud."[63]

Southwest of Yangzhou lay Yizheng, known in the Ming as Yizhen and before that as Zhenzhou.[64] Yizheng was a twin town for Guazhou, a bustling riverine port within Jiangdu county, directly south of Yangzhou. Different branches of the Grand Canal connected both Yizheng and Guazhou to Yangzhou, and both places were lively mercantile centers. Locals and sojourners competed hotly for business, and fraud was reportedly common.[65]

North of Yangzhou lay Gaoyou and Baoying. These two jurisdictions spanned the Grand Canal. To the east were farmlands that in the middle of the nineteenth century, and probably for centuries before that, were devoted to the cultivation of "rice, vegetables, grains, beans, peas, and buckwheat" — unless the canal was in flood, which it was as often as not. To the west, people lived on the lakes, supporting themselves by "fishing, rearing ducks, cutting reeds, and cultivating water chestnuts and water lilies."[66] Boat people continue to the present day to be a feature of rural life in these counties. Further east lay the county seat of Xinghua, in the geographical center of Jiangbei, an urban island constantly under threat from the surrounding bodies of water. Here "the land is low, the soil poor, and the floods many. Products are basically very few."[67]

South of Xinghua and directly east of Yangzhou city lay Taizhou.[68] Taizhou was originally a large jurisdiction, stretching from the border of Jiangdu county in the west all the way to the sea and encompassing the most valuable salt yards in the empire. In the sixteenth century it could be claimed:

Of the six salt controllers in the empire, only the Lianghuai salt controller was powerful; of the three sub-controllers [in Lianghuai], the Taizhou sub-controller is foremost; and in Taizhou, Anfeng is the greatest salt yard. In this hideaway of merchants and salters, the profits from salt top the wealth of the southeast. Of the needs of the empire, what we depend on for frontier provisions, one-half is supplied from here.[69]

In 1765, Taizhou was divided to create the coastal county of Dongtai. A former dependent county, Rugao, was removed from

Yangzhou prefecture in 1724 as part of a major administrative re-structuring centered on the creation of the independent depart-ment of Tongzhou (present Nantong). Its jurisdiction stretched out along the north bank of the Yangzi River, southeast of Yang-zhou. It embraced Tongzhou, Taixing, and Rugao, all originally part of Yangzhou prefecture.[70] Haimen county, east of Tongzhou, disappeared from the administrative map in 1671 when the county seat was swallowed up by the sea, but it was resurrected as an independent subprefecture (*ting*) in 1775.[71]

Further north, outside Yangzhou prefecture but still within Jiangbei, lay the counties of Shanyang, Yancheng, and, as of 1731, Funing.[72] These counties belonged to Huai'an prefecture but were economically and culturally closer to Yangzhou than to the rest of Huai'an, which lay beyond the Yellow River. The heavily diked lower reaches of the Yellow River constituted a formidable im-passe between Jiangbei and Huaibei and divided Huai'an prefec-ture in two. Within the Skinnerian macroregional model, Huai'an prefecture could be considered doubly peripheral: its Huaibei counties were the periphery of the North China macroregion, and its Jiangbei counties were the periphery of the Lower Yangzi macroregion. Urbanization was correspondingly low. From the late fifteenth century, when the full stream of the Yellow River was flowing south, trade along what had been the lower reaches of the Huai valley suffered a long-term depression. According to the Funing county gazetteer: "Formerly, the waters of the Huai flowed peacefully along their course to the sea, and the southern [grain] transport went by sea from here. Maluo, Lupu, Yangzhai, and Yukou were all great towns. Qinggou was a thoroughfare for Shandong salt. Huankui was also flourishing. Now all are de-serted."[73] Huai'an city itself lay within Jiangbei, close to the junc-tion of the Grand Canal and the Yellow River. Yangzhou and Huai'an were connected by an unbroken stretch of the Grand Ca-nal and had shared interests in the salt trade and water control. Traffic between the two cities was constant.

Apart from the three jurisdictions of Yangzhou, Huai'an, and Tongzhou, Jiangbei included by environmental chance the little county of Jingjiang, which was sandwiched between Rugao and Taixing counties on the north bank of the Yangzi. Jingjiang be-longed to the southern prefecture of Changzhou. It had once been an island in the middle of the Yangzi, but shifting currents had altered the level of the river bed and by the end of the sixteenth century, Jingjiang was geographically part of Jiangbei. Culturally, however, it remained part of Jiangnan. Thus it was said: "Jingjiang belongs to Wu. Its rites, festivals, and customs are still broadly the same as in all the districts of Jiangnan."[74] At least until recent times, the people of Jingjiang crossed the Yangzi to conduct business rather than deal with their Jiangbei neighbors.[75]

Such was Jiangbei for nearly four hundred years between the end of the fifteenth century and the middle of the nineteenth. This periodization is determined by movements of the Yellow River, which was the most significant geographical feature of the region, although not quite within it. The shift of the Yellow River's waters southward between the twelfth and fifteenth centuries, culminating in the sealing of the northern stream in 1495, permanently altered the lie of the land in Jiangbei and introduced a highly unstable element into the natural environment. The result was that "injury from natural disasters, epidemics, and famines" became characteristics of Jiangbei at a time when Jiangnan was experiencing extensive commercialization and urbanization as its rural areas switched to cash cropping and cottage industry. In consequence, Yangzhou prefecture was quite different from, for example, the Jiangnan prefecture of Suzhou, even though Yangzhou and Suzhou are often spoken of in conjunction.

A Central Place?

The distribution of market places within Jiangbei was shaped by waterways, the most important of which was the Huai-Yang section of the Grand Canal. The Grand Canal served principally as a route for interregional trade, and the bulk of waterborne merchandise carried past Yangzhou was undoubtedly destined for markets outside Jiangbei. In the eighteenth century, tea, cloth, raw cotton, oil, and "miscellaneous goods" from Suzhou and Hangzhou accounted for much of the merchandise being transported northward; the north, in turn, sent wheat, beans, and dried and salted meat and fish south. The thousands of vessels in the official grain fleets were responsible for healthy volumes of interregional trade. Before 1729, northbound vessels were al lowed to carry up to 60 catties of local products in addition to their grain cargo, an amount increased to 100 catties that year. On the return journey, relieved of the grain, they carried wheat, beans, melons, and nuts.[76]

Yangzhou was the most important town on the Huai-Yang section of the Grand Canal, serving as a customs port and providing a large consumer market for goods from elsewhere. The only internal waterway linking north and south China connected it directly northward with some eighteen greater or lesser administrative cities in the provinces of Jiangsu, Shandong, and Zhili, as well as with a rather greater number of towns and market centers. To the south lay Zhenjiang and the mouth of the Nanhe—the southern branch of the Grand Canal—which led to the prosperous cities of the Lower Yangzi delta: Changzhou, Wuxi, Suzhou, Jiaxing, and Hangzhou.[77]

In the Tang dynasty, this location made Yangzhou arguably the greatest entrepôt in the empire. Demographic changes since the Tang had resulted in a diminution of Yangzhou's significance as center of commerce and port of trade. In the Ming and Qing dynasties, the south so exceeded the north in levels of population, primary and secondary production, urbanization, and cultural productivity that it was no longer plausible to view Yangzhou as the gateway to China proper. Except with respect to the salt trade, Yangzhou might be regarded simply as a station en route from one part of the empire to another.[78]

The tribute grain, the salt monopoly, and its status as a customs port remained factors in Yangzhou's favor, and while these factors still operated, it was a busy enough port and center of trade. Customs quotas set in 1749 show that with an expected return of over 200,000 taels, Yangzhou ranked seventh in the empire out of the 40 customs barriers for which individual quotas were stipulated. It was one of only nine customs stations to return more than 100,000 taels in 1735, the base year for the establishment of the quota.[79]

Within Yangzhou, there were no less than twelve markets in the late eighteenth century. (By contrast, Ningbo, another prosperous town in the southeast, had eight in the 1780s.)[80] Around the city, market towns proliferated, suggesting a typical case of central-place organization. The spatial balance was distorted, however, in favor of the southeast, where the greater number of market towns was concentrated. The canals passing through Jiangdu and the towns situated on them served to differentiate this county from Ganquan, where waterways were scarce, and the general level of prosperity lower.[81]

The implications of the Grand Canal for urbanization both south and north of Yangzhou are clear. North of Yangzhou there were, in addition to the cities of Gaoyou, Baoying, Huai'an, and Qingjiang (created 1760),[82] at least nine market towns before the confluence of the Grand Canal and the Huai/Yellow River. To the south lay the largest market town in Yangzhou's home counties, the walled town of Guazhou, where "residents and merchants gathered like spokes to the wheel hub" and "merchant vessels passed to and fro, many laying anchor here. All sorts of things could be purchased and trade flourished."[83] Within Yizheng county, there were only six settlements large enough to be termed *zhen* (towns), and three of them were located on the western branch of the Grand Canal, between Yizheng and Yangzhou. Two of these were no more than *shi* (periodic market towns) in the Ming dynasty but evidently developed substantially during the Qing.[84] Another market town, Yangziqiao in Jiangdu district, lay at the confluence of the Guazhou and Yizheng branches of the Grand Canal.

Urbanization in Jiangbei was otherwise shaped by the official salt transport routes. Apart from the administrative centers, the two most significant towns in Jiangbei east of Yangzhou were Hai'an and Xixi. Both derived their importance from the salt trade. Hai'an lay within Taizhou and was situated at the junction of a canal system that connected it with Taizhou city itself, Rugao, Tongzhou, and the Salt Yards Canal (Chuanchanghe, lit. "link the yards canal"). It was a substantial town capable of supporting an academy.[85] It was also the point of convergence for salt boats proceeding from the Tongzhou salt yards to Taizhou and evidently the center of a thriving trade in smuggled salt, for a salt detective was stationed here. Xixi, located five *li* southwest of Dongtai city, was a place where merchants congregated in great numbers. It was the gateway to trade in the markets of the salt yards.[86]

The location of lesser market towns confirms the formative influence of the official transport routes on urbanization and market activities in Jiangbei. Of the 50 towns (*zhen*) listed within Yangzhou prefecture in the Yongzheng period, around half were located either on the Grand Canal (including its two southern forks leading through Yizheng and Guazhou) or on one of the salt routes leading eastward from the Grand Canal to the salt yards. The salt yard centers themselves contained markets, all of which were located along the Salt Yards Canal, running parallel to Fangong Dike. In addition, more than a quarter of the towns were border towns, in some cases provincial border towns, where administrative, defense, and revenue functions were significant factors in the urbanization process (see Appendix A).

From central place theory, we would expect to find a hierarchy of market places distributed at regular distances from the major urban centers, with due allowance made for physiographical barriers. In fact, the marketing structure, although showing signs of centralization, appears to have been predominantly dendritic. The major market towns lay on the route of the grain and salt export routes — the Grand Canal and its southern branches, the Salt Transport Canal, and the Salt Yards Canal. Yangzhou itself, in the period under study, is less adequately explained by central place theory than by network system theory, which postulates that a given city may owe its existence or importance to its role as a gateway linking its hinterland to a network of long-distance trade. In central place theory, distance and the difficulties of transport are regarded as disincentives that limit the economic significance of long-distance trade and consequently its function as a stimulus for preindustrial urbanization. In network system theory, the profit motive is viewed as overcoming these disincentives.[87]

Network system theory helps make sense of the extensive system of trade across regions in the late imperial period, when regional systems were still quite distinct. Yangzhou was a node in this network. The Huainan salt-marketing area, which supplied Yangzhou with its wealth, stretched from southwestern Jiangsu through Anhui, Henan, Hubei, Hunan, and Jiangxi to the periphery of Guizhou. In other words, Yangzhou's important economic relations were with places outside the Lower Yangzi macroregion, contradicting the supposition that cities within a macroregion have stronger economic ties with one another than with cities in other macroregions. This explains why Yangzhou did not survive as a strong trading port. In the late nineteenth and twentieth centuries, administrative changes in the salt monopoly and infrastructural changes attendant on the rise of Shanghai, the advent of steamships, and the development of railways deprived the city of its distant trading partners. Forced back onto its own hinterland, Yangzhou was known in the 1930s primarily as an exporter of pickled vegetables, cosmetics, and toothbrushes. It attracted few visitors. Even the refugees who had once gathered at its gates in times of hardship bypassed it, heading south across the Yangzi River to greener pastures.

The situation in the eighteenth century was quite different. The city's location on the route to Beijing made it a natural port of call for travelers, and its wealth was a magnet for people from near and far. "Great scholars and high officials from every corner of the empire come here," wrote Li Dou. "I have not been able to note down all those whom I have seen or of whom I have heard."[88] Apart from the officials and scholars there were others, a motley crew according to Zhou Sheng, who in the fifth month of 1841 set off for Yangzhou on a passenger boat from Zhenjiang, on the other side of the Yangzi:

A red sun traveled the sky, and the river waters ran at the boil. The cabin was not fully seven feet long and there were twenty or thirty people sitting inside and out. The sour breath of the licentiate, the rancid breath of the village teacher, the monk's breath [smacking of] wine and meat, and the peddler's of garlic and onion, the stinking breath of the usurer, and the evil breath of the yamen runner kept pace with their sweat, collecting in the olfactories, making an indescribable odor. Then to the east there was a boat carrying a salt merchant and sounding the gong for crossing the Yangzi, and to the west a boat carrying an official.[89]

In these lines, Zhou neatly captured Yangzhou society in the Qing period, all the better for being situated in the middle of the Yangzi: an aspiring scholar, a teacher, a monk, a moneylender, a peddler, a yamen runner, all crowded in a boat with others of like kind, returning to or bound for Yangzhou to eke out a living or seize the main chance. Women were notably absent from the boat but not from the conversation. "Gentlemen," began one passenger, "you all know that there are beautiful women in Yangzhou. I don't know how these beauties compare with the delicacies of

Suzhou." At this, "the licentiate went into a frenzy, the village schoolmaster was no longer able to maintain his Daoist demeanor, and the monk regretted shaving his head."[90]

Separated from the hoi polloi by status, wealth, and power were the official traveling on one side and the salt merchant on the other, an appropriate synecdoche for the twin foundations of government and the salt trade on which the city rested in the Ming-Qing era. It is possible that the salt merchant observed by Zhou Sheng was a native, like himself, of Zhenjiang, for in the early nineteenth century Zhenjiang merchants began to invest in the Lianghuai salt trade.[91] But Yangzhou owed its prosperity in the immediately preceding centuries to men from more distant places, who made their homes in the city because their native places were so far away and who became a defining feature of the urban society they helped to create.

Cള

PART II

From Ming to Qing

C hina in the middle of the seventeenth century, like England
in the same period, was a "world turned upside down." The
change of dynasty, and especially the imposition of rule by a
non-Chinese people, was deeply felt by Chinese scholars, who groped for
appropriate intellectual and ethical positions from which they might
speak after the loss of the Ming. Although social historians have tended
to downplay the Ming-Qing divide in favor of a longer process of
change, recent studies of Manchu rule have again drawn attention to
the significance of the change of dynasty, particularly with respect to the
political culture of the ruling group.

In Yangzhou, the dynastic rupture was felt with unusual intensity.
The conquering forces took the city with great brutality in 1645, making
a martyr of the Ming defender Shi Kefa as the Mongols had of Li Tingzhi nearly four centuries before. Yangzhou was forever to carry the
signs of this trauma, showing its monument to Shi Kefa like a fading
battle scar. Yet an impressionistic view of the city a few decades after the
event suggests that it was not very different in its restored form from
what it had been before the siege. A century later, a late Ming resident
might have marveled at the changes in the built environment but would
not necessarily have been puzzled by them. Probably the most striking

novelty would have been men's hairstyles, since the mandatory queue had reduced the male half of the population to looking like northern barbarians.

The Manchu emperors featured prominently in the history of this provincial city, and the year 1645 could certainly serve as a point of departure for a history of the city that focused on the problem of Qing governorship. A longer durée is suggested by attention to socioeconomic developments. In this part of the book, Chapter 3 traces the growth of the salt merchant community in Yangzhou and shows that early in the seventeenth century Yangzhou society was already beginning to show the features for which it was known in the eighteenth century. Chapter 4 examines the sack of the city in 1645 through a close reading of Wang Xiuchu's famous "ten-day diary," one of the most famous records of the Manchu conquest. Chapter 5 explores the tensions between two sorts of history of the early Qing city: one a moral narrative centered on the issue of loyalism, the other a story of how the city and urban society were in practice reconstituted. Between the Ming and Qing dynasties there was certainly a rupture, but in Yangzhou the continuities were also marked. The flourishing urban society of the eighteenth century had deep historical roots.

ভ

THREE

City of Merchants

T he Chinese word for merchant, *shang*, supposedly dates
from the early Zhou dynasty, when the defeated people of
the Shang were forced to resort to trade for their livelihood. This etymological myth captures the notion of outsider associated with the merchant or trader. The standard expression for
merchant in late imperial times, *keshang* or "guest merchant," to
all intents and purposes meant "foreigner." Appropriately, the
god of wealth, based on a legendary Shang loyalist, is in some
legends given a Muslim identity and depicted as a swarthy foreigner, with curly beard and beetling eyebrows.[1]

In Yangzhou, this image has historical resonance. When Bagh-
dad was the greatest city in the western world and Chang'an
(later Xi'an or Sian) was the greatest in the east, Arab, Persian,
and sometimes even Jewish merchants traversed the distance between the two, traveling to India and China from as far away as
"Frank-land in the western Mediterranean sea . . . [bringing] back
musk, aloes, camphor, cinnamon, and other products of those
parts."[2] At this time, in the eighth and ninth centuries, Yangzhou
was at the hub of a communications network that linked
Chang'an with extensive maritime trade routes. Merchants from
distant parts stopped at Yangzhou on their journeys to and from

the capital or stayed there to buy and sell precious stones and rare medicines. Hence Du Fu's (712–70) verse: "The merchant from Araby leaves for Yangzhou."[3] When Yangzhou was sacked during the An Lushan rebellion, "several thousand Persian merchants were slaughtered."[4]

Merchants from these distant parts visited Yangzhou again in the thirteenth and fourteenth centuries, the era of the Pax Mongolica, when the descendants of Genghis Khan allowed safe passage across the silk route and welcomed visitors from across the seas. The Arab geographer Abulfeda (1273–1331) knew of Yangzhou, writing that "some who have seen [it] describe it as in a temperate part of the earth, with gardens and a ruined wall."[5] An Islamic missionary known in Chinese as Puhading, supposedly a descendant of Mohammed, was buried in Yangzhou around the time of Abulfeda's birth.[6] In the grounds of his elaborate tomb (see Fig. 4) are kept the gravestones of various of his compatriots, such as Erlüeding, who died in 1302 and was eulogized as "a great man," skilled in commerce, cultured, a merchant who cared for the common people.[7] Some years later, Catherine and Antonio Ilioni were buried in Yangzhou under tombstones with epitaphs in Latin beginning "hic iacet . . ." (here lies . . .).[8] They must surely have followed a trade route to China in search of wealth, in the mythic footsteps of Marco Polo who claimed to have governed Yangzhou for three years on behalf of the great khan.

This exotic history was buried with the Mongol empire. The tombstones with their incomprehensible scripts were used as building blocks for a new city wall in a new dynasty. Afterward, although there were still Moslems and perhaps for a while even Jews in Yangzhou,[9] they were not from across the seas; nor were they to our knowledge counted as men of substance. The merchants of later centuries were a different sort of outsider altogether, and they traded not in small, precious, easily transported commodities but in that basic ingredient in the diet of human beings: salt.

The salt trade in China has been extensively studied, not least with respect to the Lianghuai sector of the salt monopoly and the salt merchants of Yangzhou, who dominated the Lianghuai trade.[10] In general terms, no detailed argument is needed to

天方矩蒦

Fig. 4 Puhading's tomb, 1980. Constructed ca. 1275 on the banks of the Grand Canal, east of the city. Gravestones bearing inscriptions in Persian and Arabic are now kept in the grounds (photograph by the author).

document the importance of the salt monopoly to Yangzhou. Through most of the late imperial period, it was the major source of the city's wealth. Oderic of Pordenone, who traveled in China during the fourteenth century, commented in some awe that "the lord of this city hath from salt alone a revenue of five hundred tumans of balis and a balis being worth a florin and a half, thus a tuman maketh fifteen thousand florins!"[11] Around this time, the Lianghuai salt production quota was nearly twice as large as that of the second largest sector, Liangzhe, and in the late sixteenth century its annual revenues were nearly four times as much as the next most remunerative sector, Changlu (which incorporated Hebei and part of Henan).[12]

One reason for the wealth of the Lianghuai salt sector was the productiveness of the Huainan salt yards. The Lianghuai salt production zone stretched along the coast of Huainan and Huaibei (central and northern Jiangsu); the greater number of salt yards lay in Huainan, and more than four-fifths of the total Lianghuai

Map 2 The major salt monopoly sectors under the Qing (adapted from Saeki Tomi, *Shindai ensei no kenkyū*, p. 19).

salt quota was produced here. Another reason was the size of the designated salt-marketing area. In the Ming-Qing period, this encompassed all or part of seven provinces, from Henan in the north to the periphery of Guizhou in the far southwest (see Map 2). Many millions of customers distributed through a number of different provinces thus contributed to the liquid wealth accumulated in Yangzhou, which Ho Ping-ti has estimated exceeded that of any other city in China by the middle of the Qing dynasty.[13]

The salt trade had a formative influence on the structure of Yangzhou society and the character of urban culture, most obviously through the presence and agency of the salt merchants. These wealthy men and their activities have defined impressions of the eighteenth-century city. The present chapter traces their Ming origins, showing how deep were the foundations of an urban society that has often been described in terms specific to eighteenth-century social change.

The Lianghuai Salt Monopoly and Ming Yangzhou

Yangzhou was quite a modest place in the early Ming. A Korean visitor traveling up the Grand Canal in 1488 was struck by Suzhou and Hangzhou in the south and Linqing in the north but said nothing about Yangzhou, even though he passed by it on his journey.[14] Nanjing, as the first Ming capital, blossomed at this time, and so increasingly did Suzhou, located on the southern stretch of the Grand Canal in the heart of prosperous, fertile Jiangnan.[15] Yangzhou was eclipsed by its flourishing neighbors. Its military significance had been reduced, and its hinterland boasted none of the vitality of Suzhou's.

The transfer of rule from Mongol to Chinese hands had signified, however, a new set of strategic concerns in the empire: the northern frontiers, once secured, needed to be protected. The frontier lands were vast and inhospitable, and the supply of military provisions was expensive. Military colonies were established so that the army could feed itself, but poor soil and combat duties combined to frustrate the plan. These circumstances gave rise to a system whereby licenses to trade in salt were issued in return for supplying grain to the frontier garrisons. The system was suggested first in 1370 by the governor of Shanxi, who wanted it implemented in the districts of Datong and Taiyuan.[16] Entrepreneurs from these two districts thus had an early start in the race for control of the salt trade in Ming China.

The system was cryptically referred to as the *kaizhong fa*, a term that suggests "developing the border regions" (*kai bian*) and "obtaining the salt" (*zhong yan*), but the phrase may have originated in a 1395 memorial of the Board of Population and Revenue concerning the necessity to "embark on (*kai*) obtaining (*zhong*) the supply of grain" (*kai zhong na mi*).[17] Under the system, a merchant delivered grain to an official granary in the border region for distribution in case of need. In return, he received a license that allowed him to claim salt for the retail trade from a salt yard. The amount of salt was determined by the amount and the source of the grain supplied.

Merchants soon improved on this method by developing merchant agricultural colonies (*shangtun*) worked by displaced and landless people. This obviated the need for transporting grain over long distances. Variations in the system allowed for merchants to supply horses, iron, or cloth in return for salt. In 1438, a merchant from Yanzhou in Shaanxi could obtain the right to trade 100 lots of salt for a first-class horse or 80 lots for one of inferior grade, each lot (*yin*) being reckoned at 200 catties (*jin*). [18] In this way, the horse breeders of the northwest became caught up in a curious, attenuated relationship with the salt makers of the southeast.

This system worked well for the imperial government, but it was quite inconvenient for the merchants. From the border areas to the salt-production regions of Lianghuai was a long way, and the other seaboard salt zones were even more distant. Although the system did not lack participants, merchants on the whole preferred dealing in Lianghuai salt, which yielded great profits, unlike the salt produced in other regions, which tended to accumulate unsold:

Why is this? Lianghuai is where the Yangzi and Yellow rivers flow, easily reached from all directions. Transport by water is very easy. Zhejiang is rather more distant, and Shandong and Changlu are far to the east, hundreds of miles overland, a thousand miles by water. For this reason, merchants just apply for Lianghuai salt. For Zhejiang the numbers are much fewer, and no one at all applies for Changlu.[19]

Early efforts to resolve the problem of underparticipation in some salt zones took the form of restricting rights to Lianghuai salt and forcing merchants to take the remainder of their allotment elsewhere.[20]

As long as grain could be purchased or produced cheaply compared to the price of salt, this long-distance trade was a profitable undertaking for merchants, who grew rich at the expense of the government. Accordingly, during the Chenghua reign (1465–87), the government moved to commute the grain-for-salt trade into a monetary transaction. In its simplest form, commutation meant that merchants made a direct payment in silver to the salt controller for the trading rights to so much salt from the salt yards. The silver was transported to Beijing, and a portion of it was then allocated to the frontier areas for the purchase of grain.[21]

The grain supply system continued for a while alongside the commutation system, and in the middle of the sixteenth century there was even an attempt to restore it to a central position in the operation of the Lianghuai and Liangzhe salt trades.[22] From the late fifteenth century, however, merchants dealing in Lianghuai salt increasingly fell into one of three categories: frontier merchants (*bianshang*) procured rights to trade salt by providing grain or silver to the frontier; the inland merchants (*neishang*) could, independently of the frontier merchants, pay the salt comptroller the tax for so many lots of salt, which they then bought at the salt yards; the waterway merchants (*shuishang*) transported the bulk of the salt throughout the vast Lianghuai salt zone. The frontier merchants were "natives of the border areas" who dealt in grain, whereas the inland merchants were "from Shexian in Huizhou, Shanxi, and Shaanxi, resident [and/or] registered in Huai-Yang." Of the waterway merchants, the greater number were from Jiangxi and Hunan, men familiar with the transport routes of central-south China.[23] From the middle of the sixteenth century, the frontier merchants began withdrawing from direct involvement in the salt trade and merely sold their rights to the inland merchants.[24] Thereafter the link between Yangzhou and the northwest frontier was weak.[25]

The West Merchants

Although salt merchants were active in Yangzhou from early in the Ming dynasty, their steady growth in numbers in the city can be attributed to commutation. From the end of the fifteenth century, in the Hongzhi reign (1488–1505), "wealthy men from Shan-

xi and Shaanxi, many dealing in salt, [were] moving to live in the
Huai and Zhe areas, vacating the frontier."[26] Jingyang and Sanyuan were important sources of merchants from Shaanxi (see
Map 3). At the beginning of the Ming, Jingyang "was close to the
ways of antiquity; the people were still simple. In the cities they
rarely wore silk clothes or silk shoes; in the country old and
young alike went without stockings. Their tiled dwellings were
humble."[27] But this was a place where competition in scholarship
was keen, talent was abundant, and clerks were masters of the
law.[28] Men moved quickly to take advantage of the openings in
the *kaizhong* system, and the number of wealthy multiplied.
Weddings and funerals became lavish. And neighboring Sanyuan
"outdoes all other places [in the examinations]. Farmers work
hard. Handicrafts are not greatly developed, but the merchants
go far afield, not returning for years at a time. They are exhorted
to buy land, but many regard it as troublesome."[29]

In fact, the land in these districts was unproductive. The Confucian scholar Zhu Shi (1665–1736), who served as education
commissioner in Shaanxi at the beginning of the eighteenth century, wrote:

On examination I find that in the two provinces of Shanxi and
Shaanxi the land is poor and the people many. Even when the harvest is abundant, it is insufficient for food for these provinces. Grain
boats from all around the provinces of the southeast travel upriver
from the Yangzi and the Huai, gather at Henan and Huaiqing prefectures in Henan province. From the beginning of the Taihang Mountains at the township of Qinghua in Huaiqing, they travel through to
Shanxi, and from Stone Pillar at the Three Gates in Henan prefecture
they travel to the Tong Pass and Shaanxi. The people have long relied on this for something to put in their mouths.[30]

Shaanxi had been the metropolitan area in the Tang dynasty
and earlier, but with the shift of economic and political power to
the east, it had become isolated. The initiative in interregional
trade shown by the inhabitants of this province may be attributed
in part to this process of peripheralization. They were heirs to a
core culture and built on it to move back into a core economic region during the Ming. They did so against some odds. The Yellow River, to which Shaanxi was connected by the Wei River, was

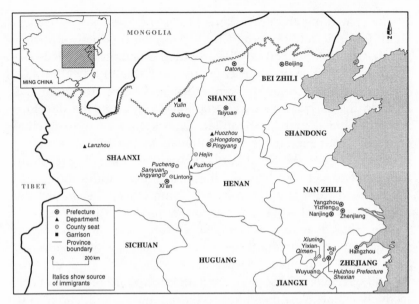

Map 3 Sources of Shan-Shaan and Huizhou migrants to Yangzhou. Among places mentioned frequently in sources are Shexian in Huizhou; Lintong, Sanyuan, and Jingyang in Shaanxi; Taiyuan and Datong in Shanxi. Ming borders are shown.

navigable but treacherous to the east, while to the north, between Shaanxi and Shanxi, river transport was rendered impossible by a series of waterfalls and rapids. The main communication routes leading out of the province were overland, one road leading northeast through Shanxi to the capital in Beijing, another southeast to the former capital, Nanjing (see Map 3).[31]

The Shanxi merchants were even more remote, isolated from surrounding provinces by hills of loess and rocky mountains. Colonel Mark Bell, passing through Shanxi in the late nineteenth century, saw Datong as full of potential: "Under an enterprising government, the neighborhood of Datong fu which possesses a large amount of fuel, is destined to become an important manufacturing centre, for Mongolia is a large producer of raw material such as camels' hair, wool and hides." The prosperity he envisaged for Datong was to depend on the Nankou mountain pass being made passable by carts, so as to allow easier communication between Beijing and the northern part of Shanxi province.[32]

The Shanxi merchants, reckoned to be the most powerful merchant group in late imperial China, were the products of a frontier that, as Colonel Bell implied, divided economic as well as political systems and was consequently favorable to a lively trade. The cross-frontier trade in oxen, hides, and horses for grain and cloth in the early Ming was a significant factor in the development of Shanxi merchant capital. Datong, one of the principal sources of salt merchants, was situated to the far north of the province, close to the Great Wall. Taiyuan and Xiangling (Linfen), also home to salt merchant families, lay in the middle and south, on the road from Datong to Xi'an. Communications shaped the native-place profiles of sojourner communities in Yangzhou.

The Shanxi merchants mixed closely with their Shaanxi neighbors. Empirewide there were at least twenty combined Shan–Shaan native-place associations in the Ming-Qing era.[33] In the salt gazetteers, these two merchant groups are referred to collectively as the West merchants (*xishang*) and in Yangzhou, as in many other places, eventually a single native-place association was established to serve the two.[34] Examination lists reveal the preponderance of Shaanxi over Shanxi merchants in the Lianghuai salt trade.[35] This was not due to their greater resources but to their more limited commercial interests. Shanxi merchants were involved in an extraordinary range of commercial activities, including the textile and grain trades as well as salt, and most important, banking. In the late Ming, they dominated commerce in the north as Huizhou merchants dominated it in the south, but according to Xie Zhaozhe, "their wealth was greater than that of Huizhou."[36]

On arriving in Yangzhou, men from the northwest found themselves in a place very different from their native provinces. During the Tang dynasty, when the Yangzi delta was developing rapidly, officials traveling to the southeast from Chang'an had marveled at just how different it was, and sometimes complained of it:

> Often fall the summer rains between the
>   Huai and the sea,
> Until at dawn the sky begins to clear.
> A cold wind blows from far away,
> And for a thousand *li* there spreads a single cloud.

Here it is damp and low, all noise and mud,
Till suddenly opens up the sky, and all is limpid.[37]

The dry, hilly provinces of the northwest shared few features with the low-lying, well-watered lands of the lower Yangzi. From a land of dusty roads and mountain tracks, the merchants moved to one of canals and lakes, where boats were more numerous than horses, and floods more common than droughts. The staple food was rice, not wheat, millet, or corn. The local patois, although belonging to the Mandarin family, can have been barely comprehensible to their ears. Here, indeed, they were far from home. Weeks or months must have passed before they heard about the earthquake that rocked the northwestern provinces in 1556.[38]

He Cheng, a native of Yulin in northern Shaanxi and the son of a salt merchant, was attending to problems in Yangzhou in the year the earthquake occurred.[39] A metropolitan graduate of 1532, he had served for a time as an official in Shanxi province, before retiring to live in Yangzhou. The city by this time had outgrown the walls erected in 1367: urban expansion probably began with the influx of merchants after changes to the salt system in the second half of the fifteenth century. Merchants built their houses east of the city, avoiding the domain of officialdom. Sojourning merchants in Suzhou followed the same practice, with the same consequences. In the middle of the sixteenth century each of these two cities had a substantial, wealthy suburb outside its walls.[40] In Yangzhou, the precincts of the extra-mural suburb included the sprawling offices of the salt controller and also the customs station.[41] Salt taxes, customs duties, and the collective private wealth of the salt merchants together made the Yangzhou suburb an attractive target for pirates who at the time were harassing the coastal provinces. One of the pirate leaders was a native of the Huizhou county of Shexian, whose compatriots were among the salt merchants of Yangzhou.[42]

In 1556, pirates penetrated the Yangzhou area. They attacked Guazhou, the Yangzi port just south of Yangzhou, setting fire to the grain tribute fleet. The following year they invaded Rugao, Haimen, and Tongzhou and then plundered the suburbs of Yangzhou and Gaoyou before setting up base further north in

Baoying. They remained a threat in the region for six years, ransacking the city of Xinghua, northeast of Yangzhou, in 1662.[43]

He Cheng, alive to the dangers posed to his compatriots, early urged the prefect to have a wall built around the unprotected settlement where most of the salt merchants resided. The suburb pressed hard on the left bank of the Grand Canal, and there was no room for the wall to be built without demolishing part of the residential area. The affected residents protested vigorously, but the raid in 1557 forced the issue. The new wall cost more than 46,000 taels of silver, 30,000 of which came from the pockets of the merchants themselves.[44]

In 1558, another Shaanxi merchant, Yan Jin, came to the city's defense:

In the thirty-seventh year [of the Jiajing reign period], pirates attacked Yangzhou. The prefect Shi Maohua called on the people to mount the walls, but they were all afraid. Jin led all the merchants from the northwest to ascend the parapets and shoot with strong bows. There was a man named Gao, skilled at archery, who killed off a number of leaders in succession. The brigands scattered in alarm. Moreover, when they heard so many Shaanxi accents among the men on the parapets, they reckoned that the fearless troops of the three frontiers must have arrived, and so they escaped under cover of nightfall.[45]

We are indebted to these pirates for a glimpse of the social composition of Yangzhou society in the mid-sixteenth century. At the beginning of the Ming dynasty, there were reportedly only a handful of households composed of natives of the prefecture in Yangzhou; everyone else had either fled or perished during the wars that ended the Mongol supremacy.[46] Over 170 years later population levels in the prefecture were still low, but judging by the extent of extramural development, this was not true of Yangzhou city. Where did all the residents of mid-Ming Yangzhou come from? The sixteenth-century Yangzhou gazetteer gives little hint of the diverse origins of people in the city, but the chronicle of the defense of the city in the 1550s suggests the numerical strength and social prominence of Shaanxi merchants.

The West merchants left few traces of their presence in Yangzhou. The salt merchants' contribution to urban culture was hinted at in a backhanded way by an official who arrived in Yang-zhou in 1529 and found it in a state of moral decay. Merchants, in his analysis, were responsible — their "ornate dwellings, beautiful clothes, sumptuous feasts" were leading people astray.[47] These were probably Huizhou merchants, judging by a contemporary complaint about the profligacy of Huizhou merchants in Yang-zhou and Huai'an.[48] West merchants had a name for frugality.

The style of local musical performances in Yangzhou may have been influenced by the Shan-Shaan sojourners. In the middle of the sixteenth century, singing and instrumental music in Yang-zhou were reportedly quite distinctive:[49] they differed from what could be heard in other places in the lower Yangzi valley, which in the Suzhou region at least was increasingly the Kun-style op-era (*Kunqu*). This might have been due simply to the fact that Yangzhou was separated by the Yangzi from the cities of Jiang-nan and thus likely to exhibit fairly distinctive cultural traits, but in the early Qing it was also the case that "there were great per-formances of Shaanxi Clapper Opera in one [Yangzhou] house af-ter another."[50] Shaanxi opera, probably introduced to Yangzhou by the West merchants, appears to have either provided the base for or at least strongly influenced the emergence of the local op-era (*luantan*) performed in Yangzhou dialect, which flourished in the eighteenth century well after Shaanxi opera itself had given way in Yangzhou to lavish operas in the Kun style.[51]

It is possible that footbinding practices in Yangzhou were also among the long-term consequences of the presence of the West merchants. The seventeenth-century writer Li Yu (1611–80?), who fancied himself a connoisseur of women, wrote in a well-known essay: "I have traveled all over the country and only in Lanzhou, Gansu province, and in Datong in Shanxi have I seen tiny feet showing their good aspects without their disadvan-tages."[52] Both Datong and Lanzhou were sources of salt mer-chants in Yangzhou. The northwestern provinces in general were noted for the ubiquity of footbinding, whereas in the Huai and Yangzi valleys country women often or usually went with feet unbound. But again according to Li Yu: "The practice of foot-binding in Yangzhou is more widespread than elsewhere. Even coolies, servants, seamstresses, the poor, the old, and the weak have tiny feet and cramped toes."[53] In this respect Yangzhou had much more in common with northern China than with its own region, and a connection with the influx of the West merchants is plausible.

The West merchants appear rarely in the record of the built city. A new wing on the Temple of Great Brightness, paid for by Shaanxi merchants in 1493; a gold statue inside the temple;[54] the New City wall, put up in 1557—these are things we know salt merchants from one place or another contributed to the material fabric of the city. They probably helped finance the renovation of the Yizheng Confucian School in 1528, the manufacture of a water clock in Yangzhou in 1529, and the establishment of the Weiyang Academy in 1535, but the local record notes only the names of salt officials in association with these undertakings.[55]

The position of the West merchants in Yangzhou was undermined in the late sixteenth century. A contributing factor was the weakening of the link between the supplying of grain to the frontier and the purchase of salt. The breaking point came in 1550. In that year, to raise revenues for military purposes, the imperial government allowed licensed merchants to carry two surplus lots of salt (*yu yin*) for every lot proper (*zheng yin*). The surplus lots had to be procured by payments in silver at Yangzhou and could not be secured at the frontier. The weight of the lot was also increased from 550 catties to 750. The frontier merchants were now virtually forced to trade an ever increasing amount of salt, rights to which had to be secured in a complex double transaction. In 1567 it was noted that they "had no time to wait" for the now lengthy process of salt inspection at the checkpoint in Yizheng and were "selling off their *yin*," presumably to the inland merchants.[56]

From this time on, Huizhou merchants, although active in the Lianghuai salt trade from much earlier, would always re-ceive first mention. Early in the seventeenth century, it could be boasted in Shexian, the capital county of Huizhou, that "among the so-called great traders (*da gu*) of the present time, none are greater than those of our county. Although there are [merchants] from Shannxi and Shanxi who also go to Huai'an and Yangzhou, they struggle to compete, and there are not many of them."[57]

The Huizhou Merchants

The prefecture of Huizhou belongs to Anhui province, which like Jiangsu, spreads across the Yangzi and Huai river valleys. During the Ming dynasty, the two provinces were part of Nan Zhili—the Southern Metropolitan province—centered on Nanjing, the first Ming capital, but, as noted in the preceding chapter, this province was divided in 1667 to form the two provinces of Jiangsu and Anhui.

Huizhou was a place of mountains and valleys pleasing to the artist's eye. The 72 peaks of the Yellow Mountains, soaring into the clouds, became a popular subject for paintings during the late imperial period, the sheer, sharp rock faces lending themselves well to the ink-laden brush. They were less amenable to the plow, and one of the constraints under which the people of Huizhou lived was a lack of arable land. Once the few plateaus and many narrow valleys had been brought under cultivation, farmers began to carve out terraces on the hillsides, which had to be laboriously watered and fertilized.

The terrain was both the curse and the blessing of Huizhou. Like the Venetians of medieval Europe, the inhabitants began to exploit what they had: forests of cypress and other conifers provided excellent wood for building and carpentry, and tea bushes yielded the leaf that was the basis for China's most popular beverage. By the ninth century, commercial farming in Huizhou was already well developed; sales of wood and tea to neighboring provinces provided the cash needed for grain purchases. Ink and paper later emerged as major products and enjoyed a guaranteed market. With the tools of the calligrapher and painter ready to hand, cloud-covered mountains all around, and merchants with mansions to adorn, the circumstances were ripe for painting to flourish. In the sixteenth and seventeenth centuries, Mount Huang became one of the most frequently portrayed scenes in Chinese art.[58]

Although Huizhou painters emerged from a merchant culture, Huizhou was not well positioned for interregional trade. The large interregional trading centers that developed in the late imperial period had good access to waterways: Hankou lay on the

Yangzi, Suzhou on the Grand Canal, Foshan in the Pearl River Delta, and Shanghai near to both the coast and the Yangzi. In contrast, the mountains of Huizhou—or Xin'an as it was earlier called—were a deterrent to easy travel. They provided a natural fortress within which Huizhou society, with its distinctive dialect, powerful lineages, and commercial orientation, gradually developed in the millennium from the third to thirteenth centuries. Huizhou's few waterways were improved to facilitate communications with neighboring and more distant provinces, and down these were borne the wood and tea on which the local people built their wealth.[59]

The limited amount of arable in the Huizhou area and its dispersal through narrow valleys, together with the growth of the labor- and capital-intensive timber and tea enterprises, was conducive to the growth of a trading culture and to its eventual spread. Large-scale emigration was facilitated by a strong clan system, to which is generally attributed the ability of Huizhou merchants to raise capital and invest in profitable enterprises abroad. They achieved such dominance of interregional and frontier trade in the late imperial period that their virtual disappearance in the second half of the nineteenth century never ceases to be surprising.

The prefecture of Huizhou consisted of the counties of Shexian, Xiuning, Wuyuan, Qimen, Yixian, and Jiqi[60] (see Map 3). A Huizhou merchant might theoretically come from any of these counties, but in Yangzhou most Huizhou merchants were from Shexian, the prefecture's home county. The neighboring county of Xiuning came a distant second. Although in Xiuning, as in Shexian, "the most powerful merchants were those who dealt in salt,"[61] Xiuning merchants were more numerous in the Liangzhe salt trade, in Zhejiang, than in Lianghuai.[62]

Of the remaining districts, Qimen and Wuyuan make only an occasional appearance in the Lianghuai salt records, and Yixian appears even more rarely. Within Huizhou prefecture, there was a tendency to regional specialization in commerce: Shexian's salt merchants had their counterparts in Xiuning's pawnbrokers and Qimen's tea merchants. Wuyuan, famous as the native place of the great Song philosopher Zhu Xi (1130–1200), was otherwise known for its tea and lumber.[63] Yixian and Jiqi were relatively poor in resources and inconsiderable as producers of commercial talent.[64] One consequence of this has been an almost total neglect of historical research into these counties. Shexian is almost definitively Huizhou and occupies centerstage in most works published on the Ming-Qing salt merchants or "Huizhou" merchants. Xiuning, Qimen, and even Wuyuan have also been topics of specialist research. Of Yixian, the prefectural gazetteer of 1566 simply notes: "There is little land, and the people are few. They are frugal in the extreme and are regarded by the people of Wuyuan and Qimen with great contumely." As for Jiqi, the most that could be said is that its "customs are similar to those of Yixian."[65] Jiqi was an out-of-the-way place, exploited in the early Ming by tax avoidance schemers from other counties and bypassed by the major overland route to Beijing.[66] When in the middle of the nineteenth century Wang Shiduo (1802–89) lamented this region's poverty, he was speaking of the poorest part of Huizhou.[67]

Commutation of the grain supply system greatly increased the competitiveness of the Huizhou merchants in the salt trade. Living closer than the West merchants to the Lianghuai salt-production and marketing areas, they were more conveniently situated to engage in the trade. In the late Ming, the standard merchant route from Yangzhou to Huizhou was around 700 *li* (approximately 250 miles) in length; that from Yangzhou to Xiangling was nearly three times longer.[68]

In Zhejiang, Huizhou merchants had an even more marked advantage. Although frontier merchants had at one point been forced to deal in Zhejiang salt as well as Huai salt, West merchants hardly feature in the Liangzhe records.[69] Zhejiang was farther from Shaanxi and Shanxi and closer to Huizhou than was Lianghuai, as well as being less profitable and thus less attractive to the West merchants. By contrast, the Huizhou county of Xiuning, which supplied the greater number of merchants to Liangzhe, was linked by the Xin'an River to the Qiantang River, which leads straight to Hangzhou.[70]

The relative position of Huizhou merchants in Yangzhou in the early sixteenth century does not compare with their later prominence. Harriet Zurndorfer explains why. Until the middle of the sixteenth century commodity farming continued to be profitable for Huizhou people. Thereafter, they were increasingly frustrated by taxes and levies. They turned to the salt trade "as a liberation from the drudgery of farming, [and thereby] also gained prestige, power, and above all wealth,"[71] abandoning the fields of Huizhou for life in cities far from home.

The great outpouring of Huizhou merchants to Yangzhou in the later sixteenth century appears to have promoted a strong sense of Shexian local identity. Timothy Brook has puzzled over the relatively late appearance of the first Shexian gazetteer, published only in 1609.[72] For Shexian elites, such a gazetteer might earlier have seemed in some sense *de trop*, since Shexian was the dominant county in Huizhou, and there was a strong tradition of Huizhou gazetteers. With the proliferation of Huizhou merchant colonies through the lower Yangzi Valley, Shexian suddenly assumed new significance on the cultural map of China: it was the native place of the majority of salt merchants in Yangzhou.

For salt merchants at this time, Yangzhou city was not necessarily the most convenient place to live within Yangzhou prefecture. Prior to the change in regulations in the middle of the sixteenth century, an inland merchant might pay the salt controller directly for a salt license and receive a ticket entitling him to salt in the salt yards. In this case, his presence was required in Yangzhou, and given the significance of connections in business, it might be advantageous for him to live there. On the other hand, he could purchase *yin* licenses from the frontier merchants. In that case he could go to the salt yards with license in hand, procure the salt, and bypass Yangzhou for the checking station at Yizheng. Further, the *yin* license was issued at Nanjing, not at Yangzhou,[73] and Yizheng was closer to Nanjing. It is noteworthy that in the Ming dynasty route book, *Shishang yaolan*, the route between Yangzhou prefecture and Huizhou takes Yizheng as its point of departure rather than the prefectural city.[74]

Yizheng was a riverine county. It marked the western limit of Yangzhou prefecture. In the Tang dynasty, when Yangzhou was already a great and wealthy city, Yizheng emerged as a small town called Baisha, but in the eleventh century it blossomed to prominence, pre-empting Yangzhou's former place as distribution and taxation center for the tea and salt trades. Under the Northern Song, Yangzhou had strictly local importance, governing only its own district of Jiangdu, while Yizheng (then Zhenzhou) was the major center of commercial activity on the north bank of the Yangzi.[75] It could play this role because it was situated at the junction of a branch of the Grand Canal and the Yangzi. It was more truly a Yangzi port than was Yangzhou. In the Ming dynasty, Yizheng was again subordinated to Yangzhou, but it was the location of the salt inspectorate, where the salt had to be weighed and transferred from salt lighters used on the Huai-Yang canals to larger vessels capable of navigating the Yangzi.

Many salt families from Huizhou registered in Yizheng, and it may well be that this was initially the more logical place for a transfer of registration for households engaged in the salt trade. The standard description of change in registration is: "so-and-so changed registration to such-and-such a place under the salt policy" (*yi yan gai ji, yi yance zhan ji*).[76] A 1699 listing of places of residence of Huizhou emigrants includes Yizheng, Yangzhou, and Huai'an. Huai'an, situated on the Grand Canal near its junction with the Yellow and Huai rivers, was the distribution port for Huaibei salt and hence the northern equivalent of Yizheng. But Yizheng itself has pride of place in the list: "At present all the rich people of Huizhou have made their homes in Yizheng, Yangzhou, Suzhou, Songjiang, Huai'an, Wuhu, and Huzhou, as well as Nanchang in Jiangxi, Hankou in Huguang, and even as far away as Beijing."[77] The Huizhou presence in Yangzhou was incontrovertible by this time, but the number of salt merchant families registered in Yizheng suggests at least a division of "Yangzhou" merchants between the two cities.

The rise of the Huizhou merchants in Yangzhou — or Yizheng — corresponds with the changing circumstances in Huizhou prefecture described by Zurndorfer.[78] The two prefectures had a symbiotic relationship, one reflected in the prominence of Huizhou merchants during the Ming and the correspondingly high profile of the salt merchants of Yangzhou in the Qing. Not until the second half of the sixteenth century, when an unfavorable tax regime began driving Huizhou people from the land, did Yangzhou gradually assume the features by which it would be known for the greater part of the Qing dynasty — the city itself enlarged by the second wall, increasing numbers of Huizhou sojourners in residence, and a flourishing demimonde.[79]

The rise of the Huizhou merchants in Yangzhou correlates not only with a transition in Huizhou society but also with the relative decline of the West merchants in Yangzhou. This might have been due to the scope of the West merchants' business activities, which embraced not only salt but finance and the interregional commodity trade, including the textile trade in Jiangnan.[80] Such breadth made Shanxi merchants in particular the undisputed merchant kings of the late imperial period, but it also meant that they were rather thinly distributed compared to the Huizhou merchants, who tended to cluster in family or lineage groups and in the salt, tea, and timber trades.[81] In any event, the West merchants were outnumbered and outspent by their Huizhou colleagues in Qing Yangzhou.

It appears probable that one further change in the salt monopoly in the late Ming advantaged the Huizhou merchants. In 1617, the dysfunctional frontier supply system was formally abandoned. Sales of official salt had slowed considerably, not least because merchants were required to pay advance taxes on salt shipments for future years and had responded by withdrawing from active involvement in the trade. Their places were taken by salt smugglers. To revive merchant interest, the imperial government now offered hereditary licenses to manage shipments of salt (*gang*) to registered areas within the Lianghuai salt supply region. This was known as the shipment system (*gangfa*).[82] Its immediate success explains the prosperity of Yangzhou in the last decades of the Ming, when the city's rural hinterland was suffering from floods and famine. The system lasted for two centuries, during which the Huizhou merchants were the dominant social group in Yangzhou.

A Huizhou Family in Late Ming Yangzhou

By the early seventeenth century, the Huizhou merchant families of Yangzhou were beginning to look like their eighteenth-century descendants in terms of social mobility and social roles. An examination of the family of Zheng Jinglian (fl. late 16th c.) shows that Yangzhou society in the late Ming foreshadowed that of the Qing. The Zheng family, probably farmers in Shexian, had fallen prey to local magnates and lost its wealth. They may have belonged to that well-documented social group, independent peasant smallholders, who were losing out to more powerful lineage members in the late Ming,[83] but they boasted (or were attributed with) some illustrious ancestors. Jinglian's great-great-grandfather and another family member were honored as martyrs to the cause of the Jianwen emperor, after both perished in the civil war that brought the Yongle emperor to power at the beginning of the fifteenth century.[84]

Zheng Jinglian was married and had at least one son, a five-year-old boy named Zhiyan, when he decided to seek his fortune further afield. Leaving the child in the care of his grandmother, he and his wife left Shexian to seek their fortunes in the booming, bustling, changing world of the late Ming era, although no one then knew, of course, quite how late it was. Five years later, the family made its way to Yangzhou. The move to Yangzhou probably took place in the last quarter of the sixteenth century, for Zhiyan's second son, Zheng Yuanxun, was born there in 1598. Exactly how the Zheng family subsequently made its dizzying ascent to the pinnacle of Yangzhou society cannot be ascertained. Zheng Jinglian may have muscled his way up, like many others, by buying *yin* licenses cheaply from the beleaguered frontier merchants and selling the salt dear. Zhiyan, who joined his parents in Yangzhou at some point, was reputed to have a sound grasp of "commerce of benefit to the state"; perhaps his grasp of commerce that benefited the family was even better.[85]

It took the Zhengs very little time to make good. Their passage to local power was not untypical of that described by Ho Ping-ti in his study of the salt merchants of Yangzhou: that is, wealth from the salt trade was used to finance entry into the ranks of the gentry. Zhiyan gained the licentiate degree and secured a good education for his sons, Yuansi, Yuanxun (1598–1644), Yuanhua (d. after 1655), and Xiaru (1610–73).[86] In the last two decades of the Ming dynasty, Zheng Yuanxun was to emerge as one of the most influential figures in Yangzhou society.

The confusion between Yizheng and Yangzhou in ascriptions of native place to the Huizhou merchants is well illustrated by this family; Yangzhou, Jiangdu, Yizheng, and Shexian all had claims to being the definitive native place of the Zhengs.[87] In terms of residence, the question can be resolved, because it is known that Zheng Yuanxun and his brothers lived in Yangzhou itself. At the same time, a Yizheng registration, descending from generation to generation, carried with it a history linking this family to the neighboring county. Furthermore, a branch of the family appears to have resided in Yizheng.[88] This might have been due to a division of labor within the extended family.

Zheng Yuanxun, the leading light in this family, topped the provincial exam in 1627 and acquired a high profile in and beyond the locality well before he gained his *jinshi* degree in 1643. His local biography attributes to him all the qualities of a Confucian gentleman, and his philanthropic activities in particular mark him as typical of leading Huizhou men in the seventeenth and eighteenth centuries. He showed his benevolence in the famine year of 1640 when he organized clan members to contribute more than 1,000 bushels of grain to establish a soup kitchen in Yangzhou. He promoted the registration of faithful wives and filial children, typical undertakings of the Neo-Confucian men of Huizhou. To friends from near and far, he was attentive and caring; he arranged for one visitor from afar to receive medical attention and for the corpse of another to be taken home for burial.[89]

He also participated in the literary activities characteristic of the disaffected literati of the late Ming. Despite his youth, he was reportedly consulted for advice by high officials who passed through Yangzhou and was repeatedly urged by them to take office; but after being appointed to a post in the Ministry of War, he retired on grounds of having to care for his aged mother.[90] He turned his energies to the Restoration Society, the successor to the ill-fated Donglin movement. The Restoration Society was itself a federation of local literary societies, which probably included Yangzhou's Zhuxi xushe—the "ongoing Society of West-of-the-Bamboo."[91] Zheng Yuanxun and his fellow *jinshi* graduate of 1643 Liang Yusi (d. 1645) were both members of this society, as was Zheng Yuanxun's nephew Zheng Weihong (d. 1645), another graduate of that year.[92]

Zheng's other undertakings in Yangzhou prefigure the patterns of cultural activity by which the Yangzhou salt merchants were to become known in the eighteenth century. First, he built a garden, or at least paid to have it built. There were gardens in Yangzhou before Zheng's time, but his Garden of Reflections

(Yingyuan) marks the proximate beginning of salt merchant gardens in the city.[93] It was established in 1634, at a time when other gardens were being built in Yangzhou and nearby towns.[94] All the Zheng brothers were prominent then or later as garden owners. According to Li Dou, they "competed with one another in building gardens."[95]

Zheng Yuanxun's garden was distinguished by its association not only with himself, the most prominent of the brothers, but also with its designer, Ji Cheng (b. 1582). Ji Cheng's fame is due principally to the publication of his book *Yuan ye*, commonly translated as "The Craft of Gardens," published in 1635 with a preface by Zheng, who probably sponsored the printing of the work.[96] The garden was located outside Yangzhou, apparently on the south side of the city by the western wall, with its main gate facing east toward the city.[97] The modern scholar of gardens Zhu Jiang, on the basis of textual evidence, describes the design of the Garden of Reflections as typical of a "large-scale garden," but in terms of acreage, the garden hardly seem to belong to this category.[98] Modestly described by Zheng as "only a few paces wide," the garden proper has been estimated at covering somewhat over an acre, not including an associated vegetable garden and flower plantation.[99] With its complex of halls, belvederes, and pavilions divided by rivulets and ponds and connected by winding covered walkways and bridges, it was no doubt large enough for his visitors to enjoy the pleasures of becoming lost along its paths.[100]

A surviving print of the Garden of Retirement (Xiuyuan), built by Yuan's younger brother, Xiaru, in the early years of the Qing, provides a good idea of some of the design features that were probably typical of large Yangzhou gardens in the seventeenth century (see Fig. 5).[101] Xiaru's garden, occupying around 50 *mu* within the eastern wall of the New City, incorporated two earlier gardens.[102] With its gentle undulations and curving boundaries, internal and external, it presented some classic characteristics of southern gardens: the winding corridors; the demarcation of separate areas of the garden, to form gardens within the garden; and the combination of "mountains and water," to create a sensation like that of gazing on a landscape painting. Just such a garden was built in Beijing for a homesick consort of the last Ming

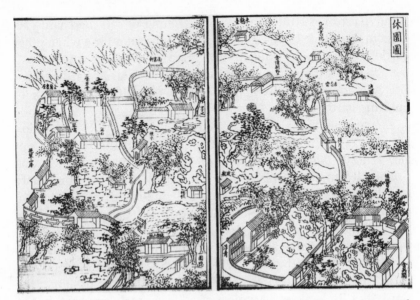

Fig. 5 Layout of the Garden of Retirement as depicted in twin wood-block prints in Zheng Qingyou's commemorative work, *Yangzhou Xiuyuan zhi* (1773).

emperor, who pined for her native Yangzhou.[103] The Garden of Reflections probably had all these features on a smaller scale.

Zheng Yuanxun justified his garden on the grounds that it would be a place where he could "care for his mother,"[104] but like the Garden of Retirement and other famous gardens of the succeeding dynasty, it was a venue for literary salons, which "brought together the most noted scholars of the empire."[105] Local Restoration Society gatherings were probably held here, since Yuanxun and at least one frequent guest at the garden, fellow-graduate Liang Yusi, were society members.[106] The literary activities conducted in the grounds were similar to those conducted in the eighteenth-century gardens. A memorable party held in the garden in 1643 marked "the blooming of the yellow peonies." Ballads and poems were composed, the guests competing with one another in elegance of phrasing and originality of conception; the collective product was printed and bound. Both the winner and the runner-up received a cash prize.[107] The scene must have resembled those described by Li Dou in the following century.[108]

Fig. 6   Zheng Yuanxun, *Lin Shen Shitian bi* (after Shen Zhou), 1631. In emulating the definitive literati artist Shen Zhou (1427–1509), Zheng confirmed his own literati tastes, which were elsewhere evident in the inspiration for and design of his garden (Suzhou Museum collection; published Cahill, *Shadows of Mount Huang*, p. 75, fig. 15).

Again like these later gardens, the Garden of Reflections was associated with painting. Zheng Yuanxun himself was a painter of minor note. James Cahill has described two extant landscapes as offering "not much" in pictorial terms but showing an interesting interaction between the Songjiang school of Dong Qichang (1555–1658) and the nascent Anhui (Huizhou) school[109] (see Fig. 6). As it happened, Dong Qichang was among Zheng's guests.[110] According to Zheng, Dong discussed the Garden of Reflections with him while it was being planned and supplied its name.[111] Dong's close associate Chen Jiru (1558–1639) inscribed a title board in the garden with the words "Pavilion of Delightful Seclusion," a phrase later used by Zheng as the title for the collection of essays in which Dong's influential treatise on the northern and southern schools of painting was published.[112]

The interaction between host and guest in this garden setting perfectly illustrates a social process commonly associated with Yangzhou at a later date: the gentrification of a salt merchant family.[113] A distinction is sometimes made between merchant and scholar gardens of the late imperial period, especially where the gardens of salt merchants were concerned, but the distinction is blurred at best.[114] The garden tradition in Yangzhou was continuous with that of Suzhou, the major center of scholar-garden development in the Ming, and if this was obvious in certain vulgar features such as ostentatious artificial mountains, it was also evident in the elegance of a garden designed by Ji Cheng, a native of Suzhou.[115] Zheng Yuanxun, a scholar-official from a merchant family, espoused the literati line when he wrote that "the owner (of a garden) must be sure of having the hills and valleys already there in his heart, and then the completed work may be either elaborate or simple, as he wishes."[116]

The roles of host, patron, scholar, and philanthropist Zheng Yuanxun performed were played out in Yangzhou by later representatives of the Huizhou diaspora. Leading members of this group owned gardens, held degrees, and distinguished themselves by contributions to good causes. The Manchu conquest robbed Zheng and his peers of the opportunity to achieve the collective historical prominence won by later Huizhou offspring in Yangzhou, but they imprinted a certain way of life on the fabric of the city. By the end of the Ming, Yangzhou had become in many respects a Huizhou city. Not even the fall of the dynasty changed this fact.

☙

FOUR

Yangzhou's Ten Days

arge-scale migration from Huizhou to Yangzhou commenced in the second half of the sixteenth century and
ended around the turn of the nineteenth. In the social history of Yangzhou, this was a relatively discrete period, during
which Shexian merchants dominated local society. It was punctu-
ated, however, by a momentous event: the interdynastic war that
raged through the empire in the 1640s and brought to the throne
the first in a long line of Manchu emperors.

Manchu emperors and officials were among the obvious new
actors in Yangzhou's post-Ming history. As agents in Chinese history more broadly, they share a timeline with the Huizhou merchants. Originally hunters and gatherers, the Manchus turned in
the sixteenth century to sedentary agriculture and animal hus-
bandry.[1] This was true particularly in the Liao River valley,
which lay closer to China than the other main areas of population
concentration and included a substantial Chinese population.
During the sixteenth century, the people here grew rich from selling ginseng, hides, and horses to the Chinese and Koreans. Under
the leadership of the intrepid Nurhaci (1559–1626), they gradually became powerful as well. In 1587, Nurhaci, now overlord,
built his first city.[2] In 1618 he formally established the Jin state

and simultaneously declared war on the Ming.[3] Less than three decades later, his fifteenth son, Duoduo (1614–49), known in Chinese as Prince Yu, was strolling the ruined streets of Yangzhou, inspecting the damage wrought during one of the worst incidents in the Manchu-Chinese war.[4]

In the annals of Chinese history, Yangzhou is an obvious point of reference for the dynastic transition. Its fall presaged the defeat of hopes for a Ming restoration, and the massacre of the populace by the conquering forces came to be seen as the tragic symbol of the end of an era. In the structural narrative supplied by the moral tale of the rise and fall of dynasties, the moment of its subjection marks a moment of historical rupture.[5] Wang Xiuchu's "Ten-Day Diary of Yangzhou," a memoir of the fall of the city in 1645, occupies a central place in the canon supplying this narrative.[6] A close reading of the "Ten-Day Diary" shows, however, that it supports alternative tales of the past and enables a view of Yangzhou's local history as continuous from the late Ming in some important respects.

The Manchus and the Fall of the Ming

The political instability of the later years of the Ming was apparent throughout the lower Yangzi delta. In 1604, the Donglin Academy was established in Wuxi, across the Yangzi and down the canal from Yangzhou. The academy provided a focus for the development of a reformist party dominated by a powerful alliance of officials from Jiangnan and the northwest provinces of Shanxi and Shaanxi—precisely the provinces that supplied Yangzhou with many of its salt merchants. In the space of twenty years, the Donglin party rose to a position of considerable influence before falling prey, in 1625, to a purge instigated by the eunuch Wei Zhongxian (1568–1627), one of the great villains of Chinese history. The purge was accompanied by the development of a personality cult centered on Wei, marked by the erection of temples and the development of rituals in his honor. In Yangzhou, Police Commissioner Wang Zheng (1571–1644) was one of the "two stalwarts" (er jing), both from Shaanxi, who refused to pay obeisance at the temple.[7] It was probably around the same time that Zheng Yuanxun, at great personal risk, provided a hiding place for a friend who had run afoul of a powerful eunuch.[8]

Political difficulties were accompanied by social and economic crises. In the 1620s, the northwest provinces were ravaged by drought, famine, and brigandry, misfortunes that gave rise to open rebellion.[9] The families of the West merchants in Yangzhou were no doubt affected by these developments. Things were not much better in Yangzhou prefecture. A bandit uprising in Xinghua in the late 1620s succeeded in isolating the county capital until the disturbances were suppressed by the valiant local magistrate.[10] In the 1630s, military officials in both Yangzhou and Yizheng had to defend these localities from rebels moving south from the Huai valley.[11] Shaanxi rebel Zhang Xianzhong (1605–47), who had emerged as one of the greatest internal threats to the Ming, was by now roving almost at will across north China. In 1637 he joined up with bandits in Jiangbei and launched an attack on Yizheng before retreating upriver to Hubei.[12] In 1635, starving refugees were dying on the roads around Yangzhou. In 1636, as the famine continued, men and women hanged themselves from trees or drowned themselves. In 1637, the Yangzi flooded northward, and countless people died in Yizheng.[13]

The empire was slowly disintegrating. The rivers flooded, the rain fell too often or too rarely, there was not enough to eat, gentrymen refused to take office, the Ming armies lost one battle after another. Harassed by peasant uprisings over two decades, threatened at its borders by the rising might of the Manchus, the dynasty fell in 1644. On April 24, just as Beijing was about to fall to the Shaanxi rebel Li Zicheng (1605?–45), the last Ming emperor began killing off his consorts, and the following day he took his own life. In Nanjing, loyal officials set up a rump court, the Southern Ming, hoping to save the south of China. Yangzhou was destined to be transformed from a city of leisure and pleasure into a garrison town, not for the first time in its history.

The loyalist regime in Nanjing was alive to historical precedent. Shi Kefa, the minister of war for the Southern Ming, insisted that the security of Nanjing depended on defending the lower Huai valley. The task was entrusted to four military careerists of dubious reputation, who, among other actions detrimental to the survival of the Southern Ming, quarreled over possession of Yangzhou. With the defense of the city politically and militarily divided, the residents of Yangzhou awaited the final outcome of the interdynastic conflict, which for them came brutally in the fourth month of 1645, in late May by the western calendar.

"Yangzhou shiri ji": The First Six Days

The fall of Yangzhou occupies a central place in the historiography of seventeenth-century China. The story has often been told. Wang Xiuchu, a resident of the city at the time, survived the sacking of the city and wrote a graphic account of the days of gruesome slaughter and destruction that followed the collapse of the Ming defense. His "Ten-Day Diary," a work proscribed throughout the eighteenth century, was circulated as anti-Manchu propaganda in the late nineteenth and early twentieth centuries. Subsequently translated into French and English, it is probably the best known of any of the accounts of the Manchu conquest of China.[14]

Nothing is known of Wang Xiuchu save what emerges from his account. His family appears to have been well established in the city: when the hour came for him to take refuge from the occupying troops, he did so in the company of his pregnant wife and son, three brothers, a sister-in-law, a nephew, and three members of his wife's family. Nonetheless, this was clearly a sojourning family. Wang consistently refers to people of Yangzhou (*Yang ren*) in a way that suggests he was not one of them. Two Yangzhou women are described in morally compromising circumstances that do no honor to the locality.[15] He referred to one neighbor, Qiao Chengwang, as a "West merchant," a phrase that distinguishes him from the Shan-Shaan sojourners. (The Qiao lineage, from Xiangling prefecture in Shanxi, was prominent in Yangzhou.)[16]

It is highly likely, however, that he belonged to the Huizhou community, even if, as one passage suggests, he and his brothers were not "rich merchants."[17] He lived in the southern part of the New City, directly next to the city wall, where the dwellings of the salt merchants were concentrated.[18] All his neighbors were merchants, which in Yangzhou meant people from elsewhere, and he himself was a man of some means, with a quantity of silver at his disposal. His wife had jewelry of gold and pearls. His surname is written with the common character for Wang, meaning "king"; the surname Wang written with the water radical is more often used in Huizhou, but the ordinary Wang is also to be found. Moreover, his brother had a relation by marriage named Hong, which is a common Huizhou name.[19] This relative is one of two women who are named in the diary. The other, a stranger to Wang, bore the surname Zheng, again a Huizhou name.[20] In contrast to his description of numerous other people in the diary as "Yangzhou people" or "dressed in the Yangzhou fashion"[21]—or indeed as "West merchants"—Wang does not distinguish these women by native place.

Why Wang Xiuchu and his family, along with thousands of others, were in Yangzhou when the city fell is a question worth pondering. There is a saying in Chinese: "xiao nan bi xiang; da nan bi cheng" (minor disturbance: flee the countryside; major disturbance: flee the city).[22] This principle was often put into practice. Before the Manchus arrived in Nanjing, "men and women, like a swarm of insects, pushed and shoved to get out of the city gates. Impossible to count, they helped the elderly along and carried their children in their arms."[23] Further south, in Pucheng in Fujian province, the Yangzhou-born Zheng Weihong, nephew of Zheng Yuanxun, died valiantly defending a city that was all but deserted by the time the Manchus arrived.[24]

These flights bespeak people's instinctive understanding of a fundamental element of military strategy in China: much of the fighting centered on cities.[25] In the case of Yangzhou, however, flight was rendered difficult by developments in the twelve months before the fall of the city. Beijing fell to the Shaanxi rebel and self-declared emperor Li Zicheng on April 25, 1644, but at that time the idea of a Manchu occupation of Yangzhou was remote. Zhang Xianzhong, so recently in the vicinity and now rampaging in the western province of Sichuan, seemed a more likely threat to the lower Yangzi region.[26] Given the choices, in which direction should one flee? Then in June, the city faced a threat from a different quarter when it was placed under siege by the Southern Ming general Gao Jie (d. 1645).[27] According to Dai Mingshi (1653–1713), numbers of people did flee the city at this point to seek refuge in the lake area to the north, but "many fell victim to bandits; whole families were wiped out."[28]

Shi Kefa was dispatched to Yangzhou to relieve the siege, but Gao withdrew only to the nearby riverside town of Guazhou. His threatening presence there effectively blocked the shortest escape route from Yangzhou to the south. In January 1645 Gao was murdered, but he left his army behind him. Moreover, with Gao out of the way, his rival, Huang Degong (d. 1645), set his sights on the city. These disturbances led to an influx of refugees, which boosted rather than diminished the size of the population.[29] Wang Xiuchu's wife and in-laws, for example, fled to Yangzhou from nearby Guazhou on the eve of the siege of Yangzhou, seeking refuge from the disorderly troops of the late and unlamented Gao.[30]

Traveling the open road during the last years of the Ming was hazardous for the wealthy. Residents of Yangzhou included numbers of salt and timber merchants who had fortunes stashed away in their homes, enough so that later they were able to offer the conquering forces thousands of taels of silver at a time in return for their lives.[31] Faced with a choice between losing their wealth to the soldiers and bandits roaming the countryside and using it to help defend the city, they chose the latter.[32] Shi Kefa's arrival gave them cause for optimism. Shi brought an army with him from Nanjing, and loyalists from elsewhere came to join his entourage. These tokens of Ming strength would in the end pale alongside the swelling ranks of Qing forces and collaborators, but in the meantime they were visible and the enemy was far away.

According to Wang Xiuchu, the residents of Yangzhou had very little notice of the battle for Yangzhou, but the letters of Shi Kefa show that he knew the city was doomed almost a week in advance of the attack. Shi had recently been leading an army in defense of territory further north but had been forced to retreat to the shelter of the city walls, along with his troops.[33] On the nineteenth of the fourth lunar month (May 14), he calculated that "the Yangzhou city wall will fall within a day and a night." Two days later, in his last letter home, he wrote: "The northern troops surrounded the Yangzhou city wall on the eighteenth but have not yet attacked. In any case, the people have already lost heart, and the situation cannot be saved. Sooner or later I must die, and I wonder whether my wife is willing to follow me?"[34]

Wang Xiuchu, by contrast, had a most confused understanding of the strategic balance, taking momentary hope from the appearance of orderly soldiers even when the attack was under way. The diary begins just before the attack takes place. At that time, the Ming soldiers were billeted around Yangzhou, to the great annoyance of the householders. Wang Xiuchu sought to buy himself out of this inconvenience by throwing a party for the captain in charge of his quarter. The night before the city's fall, this same captain issued a return invitation for a party that was to be attended by a well-known singsong girl. The captain was something of a musician, and he intended to accompany her singing on his lute. The company had assembled in anticipation of an evening of dissipation when a message from Shi Kefa arrived. "On reading it," recorded Wang, "the captain's face changed color. He immediately climbed onto the city wall, while all the guests dispersed."[35]

The next morning all was chaos. The narrow streets suddenly echoed to the sound of horse hooves. Ming defenders were fleeing the enemy, and the enemy cavalry was close behind, entering the city from the north. Roof tiles clattered under the feet of soldiers descending from the city walls, victors with bows and swords pursuing the vanquished. Courtyards, living quarters, and then the sleeping chambers of private dwellings were invaded with impunity. Soon soldiers of the occupying force could be seen going from door to door, demanding silver, and before nightfall a great massacre had commenced. It was raining steadily, but fires were burning as houses were put to the torch, and the night was lit up "as though by lightning or sunset." Next day "the ground was stained with blood and covered with mutilated and dismembered bodies. . . . Every gutter and pond was filled with corpses lying one upon the other." The slaughter lasted from the twenty-fifth of the fourth month till the first of the fifth. On this last day, troops who had been under the command of Gao Jie and then went over to the Manchus entered the city: "Every grain of rice, every inch of silk, was now gobbled up as if by a lion."[36]

According to Wang, the number of bodies cremated in the wake of the siege totaled 800,000, an impossibly large figure. Casualty figures for the Ming-Qing war are usually implausible, and for good reason: witnesses to battles and atrocities, undoubtedly overwhelmed by the extent of the horrors unfolding before their eyes, had no way of counting the bodies of the slain. The numbers are usually given in *wan* (ten thousand) and convey a sense of the uncountable rather than the counted.[37] One historian has estimated the population of Yangzhou at the time of the siege at 20,000 to 30,000, a figure far less than the number of fatalities given by Wang Xiuchu, but enough in combination with the number of Ming troops to account for thousands of bodies littering the streets in the wake of the massacre.[38] Frederic Wakeman considers this estimate for the city itself too low but gives credence to numbers that are too large, elsewhere referring to the "million or more citizens of Yangzhou."[39] Yangzhou at this time cannot have had a larger population than the 175,000 that G. William Skinner estimated for the early 1840s.

Wang Xiuchu's figure is best read simply as a statement of the horrific extent of the carnage. That it was horrific is better attested by the difficulties of disposing of all the bodies. The solution was to pile the corpses in great mounds and set fire to them,[40] but even after this the city remained littered with relics of death. When Zhou Lianggong (1612–72) arrived later in the year to take office as the first Qing salt controller in Yangzhou, he found a mountain of white bones outside the Guangchu Gate, north of the New City. He established a public cemetery for their burial.[41]

Not all the deaths were inflicted by the invaders. The local annals, in recording the righteous acts of the sons and daughters of Jiangdu district, tell a sorry tale of mass suicide by hanging and drowning.[42] Special attention is given to the many female martyrs. By recording these women, local literati were able to combat the less edifying examples of Yangzhou womanhood, who might serve as a sad metaphor for the inconstancy of Ming survivors or even as an explanation for the dynasty's collapse. Wang Xiuchu's theory on the reasons behind the fall of the Ming was prompted by the sight of a Yangzhou woman consorting with the conquering forces in the midst of the carnage:

The woman was a native of the city. Her face was heavily powdered, she was wearing splendid and gaudy clothes, and she laughed and flirted and dallied, seeming to be in high spirits. Whenever she came across any fine things the soldiers had, she begged and took from them with no sense of shame. The soldiers often said among themselves: "We conquered Korea and robbed tens of thousands of women, and not one of them lost her chastity." How could our women be so shameless? Alas, that is why China is in chaos![43]

The women no doubt had a hard time of it. According to Wang Xiuchu, some sought safety in temples but died there of starvation or fear. Others, it is known, were abducted and taken to places far away, living out their lives as the wives or concubines of men who spoke another tongue.[44] The women honored in the local records, however, are not these unhappy creatures but those who committed suicide. Zhang Guohua's wife, née Shi, piled up kindling, sat on top of the heap, set fire to it, and perished in the flames.[45] Likewise, the wife of Zhang Sixiang, née Lu, together with her daughter-in-law, née Sang, stopped up the doorway with kindling and lit a fire within the house. Forty-seven members of the household perished.[46]

Shi Kefa, too, perished. His body was never found, but his clothes were entombed on Plum Blossom Hill, on the banks of the city moat north of the city. This was the site of a garden created in 1592 by the then-prefect of Yangzhou and named for the hundred plum trees planted on it.[47] Quan Zuwang (1705–55), who was a frequent visitor to Yangzhou a century later, wrote a short piece in honor of the site, recording the gallantry of the loyal officials who died by Shi Kefa's side:

On the twenty-fifth, the city fell. The Loyal Martyr [Shi Kefa] took a sword to stab himself, but all the others struggled to restrain him. The Loyal Martyr called out loudly to [Shi] Dewei. Dewei wept but would not hand over the sword. Then, with a group supporting him, they went. On reaching Little East Gate, they encountered a host of soldiers. Assistant Commander Ma Wulu, Prefect Ren Minyu, and Generals Liu Zhaoji and others all died. The Loyal Martyr stated: "I am Minister Shi." He was pulled to the South Gate, where the eminent Prince Yu [Duoduo] called on him to surrender. The Loyal Martyr died cursing him. Earlier, His Honor had left word: "When I am dead, bury me on Plum Blossom Hill." In the end, since Dewei could not find the corpse, he buried his clothes and cap.[48]

There were, as Quan noted, other versions of what happened to Shi Kefa. Some said that they had themselves seen him riding away on a white horse; that he had left Yangzhou by Tianning Gate on the north side of city, and that he had drowned himself in the Yangzi. Others said he had not died at all, and many later uprisings featured his name.[49] So many uprisings, so many stories, meant that this Ming hero would haunt the Manchus until they themselves laid claim to him.

Alternative Histories 1: Zheng Yuanxun

The historical significance of Yangzhou's ten days is considerable but complex. In the early and mid-Qing, the massacre described by Wang Xiuchu was at least covert historical knowledge. Wang's diary was circulating, perhaps in manuscript form, at least until the time that it was explicitly proscribed in the 1770s. Someone bravely secreted a copy somewhere, for toward the end of the nineteenth century it came to the attention of the late Qing nationalists, who made it generally available through the pages of the *National Essence Newspaper* (*Guocui xuebao*).[50] By this time Yangzhou had ceased to be a very important city, but the nationalist cry of "Yangzhou's ten days" effectively installed it in a central place in nationalist historiography. This historiography was mutable, like the content of nationalism itself. By the 1930s, the Japanese had succeeded the Manchus as enemies of the nation. In 1934 a revisionist interpretation of Wang Xiuchu's diary attempted to show that natives of Yangzhou, accused of collaborating with the Japanese during the attack on Shanghai in 1932, had also collaborated with the Manchus in 1645.[51]

A second famous version of the fall of Yangzhou was penned by the loyalist historian Dai Mingshi,[52] who focused on the defense of the city—and of the dynasty—mounted by Shi Kefa. Quan Zuwang's description of Shi's death, to judge from the sequence of events and similarity of vocabulary, would appear to have been based on Dai's account.[53] Despite the proscription of Dai's writings under the Manchus, the defense of Yangzhou by Shi Kefa was a historical event that could be and was publicly commemorated during the Qing. It was recorded in the *Ming History*, a project fostered by the new regime and one to which a number of Yangzhou scholars contributed.[54] Prominent martyrs of the Ming were commonly recognized by the Qing emperors and endowed with posthumous honors for their righteousness.

As the site of the resistance and death of Shi Kefa, Yangzhou has acquired a significance in the history of the fall of the Ming well beyond that of the fall of the city itself. In words from a recent Yangzhou publication, "Shi Kefa belongs to Yangzhou and also to the entire Chinese people."[55] Shi's tomb has long been a major tourist site. In the seventeenth century both the loyalist poet Wu Jiaji (1618–84) and the Qing official Wang Shizhen (1634–1711) paid it homage; in the eighteenth Yuan Mei (1716–97) wept over it, and the Qianlong emperor (r. 1736–95) saluted it. In the twentieth century Yu Dafu, Yi Junzuo (1898–1976), Guo Moruo (1892–1978), and Tian Han (1898–1968) paid it homage in verse.[56] During the Cultural Revolution, Shi Kefa came under attack as a feudal element and an impediment to historical progress,[57] but this was a passing phase. After the fall of the Gang of Four, Plum Blossom Hill was restored and is now one of the most publicized tourist attractions of the city.

Zhu Ziqing found occasion to comment on the prominence of Shi Kefa's tomb. "Both inside and outside the city," he wrote, "there are many ancient monuments, including Wenxuan Mansion, the Tianbao Wall, Thunder Mountain Bund, the Four-and-Twenty Bridge and so on; but people often go just to Shi Kefa's Plum Blossom Hill and no further." There was, as Zhu notes, another Yangzhou: "If you are more or less at leisure, invite two or three friends along, seek out secluded spots, visit these old sites, which are yet full of interest. Naturally, take along some ground peanuts, a bit of five spice beef, and a little white wine."[58]

This was a local Yangzhou, full of meaning to someone such as Zhu who had grown up in the city, and it has its parallel in a local history, fragmented as this is through family records, obituaries, and the occasional writings of late Ming and early Qing scholars. With its focus on the moral stature of local heroes, the local history simultaneously echoes and diverges from the larger narratives centering on the "ten days" and Plum Blossom Hill. Those who wrote on the fall of the Ming were above all preoccupied with ethics. The historians and hagiographers of Yangzhou shared this preoccupation with their more famous contemporaries such as Dai Mingshi, but their accounts differed sharply in details that, although ultimately immaterial to the simple nationalist myth invoked at the end of the nineteenth century, were highly important from the point of view of Yangzhou.

The conflicting accounts of the death of Zheng Yuanxun, perhaps the most prominent gentryman of late Ming Yangzhou, encapsulate the tension between the two histories. Zheng receives brief mention in Wakeman's sweeping narrative of the Manchu conquest, making his appearance in a footnote; his name crops up just once in the *Dictionary of Ming Biography* as the owner of a garden designed by the famous landscape artist Ji Cheng.[59] But he has a central place in narratives of the fall of Yangzhou and is given due attention by both Dai Mingshi and local chroniclers.

A prelude to Zheng's role in the events of 1645 is provided by tumultuous events centered on the city six years earlier. In 1639, Yuan Jixian (1598–1646), subsequently to acquire fame as a Ming loyalist, was appointed assistant commissioner for military affairs in Yangzhou, probably on the recommendation of his friend Wu Sheng (1589–after 1644), who was a native of the Yangzhou county of Xinghua and was then serving as junior vice minister of war.[60] At that time, it is said, the metropolitan official Yang Xianming held sway over the Lianghuai salt monopoly, and "from censor and salt controller down, all kowtowed to him." When Yuan Jixian failed to show the same compliance, Yang had him impeached and removed from his post. This caused an uproar among the local people who forced a closure of the city gates to prevent Yuan's expulsion. For ten days all communications with the city were at a standstill. The impasse was broken when Zheng Yuanxun and his youngest brother, Zheng Xiaru, "alone went forth, talked frankly [to Yuan Jixian] about local affairs, and then persuaded the people to open the city gates."[61] This anecdote shows the Zheng brothers acting out the time-honored role of gentry leaders in a local crisis.

The Zheng family's fortunes reached their zenith in 1643, when Yuanxun and his nephew Weihong both secured the coveted metropolitan degree, incontrovertibly establishing the credentials of the Yangzhou branch of the Zheng clan. There were four other Yangzhou graduates in this year: Liang Yusi and Wang Yuzao, both from the rural Beihu (North Lakes) district of Yangzhou;[62] Gong Weiliu, a man of Taizhou but originally from Zhili; and Zong Hao, registered in the Yangzhou district of

Xinghua but resident either in Yangzhou or in the nearby market town of Yiling.[63]

Five of these successful candidates took office. At the time of the fall of the Ming, then, the class of 1643 was scattered over the face of the empire. Zheng Yuanxun, who declined office, remained in Yangzhou. Zong Hao was stationed as an intendant in Pingliang, in the devastated province of Shaanxi;[64] Wang Yuzao held the district magistracy of Cixi, in Zhejiang,[65] Liang Yusi was magistrate of Wan'an in Hubei,[66] and Zheng Weihong magistrate of Pucheng, Fujian province, where he remained after his subsequent promotion to censor of Huguang.[67] Gong Weiliu held a position in the Hanlin Academy in Beijing and was on the spot when the capital fell. He retired to Taizhou and steadily refused to take office under the new dynasty.[68]

The news of the emperor's death and the fall of Beijing sent a frisson of horror through Yangzhou. When it came to the ears of Zheng Yuanxun,

he put on garments of hemp and wept at the Sacred Temple. The men of Yangzhou heard that there were bandits on their way down from Shandong and prepared to flee with their wives and children. Yuanxun expended his fortune summoning volunteers, whom he called on to observe the principles of loyalty and filiality. Then the people became firm in purpose.[69]

The immediate problem for Yangzhou was, as the standard accounts relate, not bandits but the client armies of the Southern Ming, which competed for access to Yangzhou and its wealth. When Gao Jie placed the city under siege in the summer of 1644, Zheng Yuanxun's hour had come. According to the loyalist historian Dai Mingshi, he behaved in a craven fashion:

[When Gao Jie arrived at Yangzhou,] his soldiers failed to put away their weapons. The people had the greatest antipathy for them and ascended the walls to defend the city. In the countryside all around, countless numbers were slaughtered.

Zheng Yuanxun, metropolitan graduate of Jiangdu, a man full of pride and self-importance, went forth to negotiate. He entered Jie's camp, partook of wine, and talked with Jie in a thoroughly congenial fashion. Jie presented him with a pearl disc. Yuanxun returned to the city, even more swollen with conceit, and spoke to the people, saying: "General Jie's arrival is by virtue of a summons in letters of credential. If he goes to Nanjing, he is heeded. Why not in Yangzhou?" The crowd went into uproar, accusing Yuanxun of selling out Yangzhou in the name of virtue. They killed him and consumed his flesh on the spot.[70]

The local annals provide a different account. His biography in the Yizheng county gazetteer commences with a record of his examination successes and continues with attention to the social features of the Confucian gentleman. These acts of private and public piety establish him in the local record as a man of unwavering righteousness. The account of his death does not belie the record:

The metropolitan graduate Zheng Yuanxun knew Jie of old and, fearing that the city could not be defended, went to his camp to speak with him. Jie was pleased, and to show his good intentions, he lifted the cordon and withdrew his troops to a distance of five *li*. On the city walls, an assembly subsequently deliberated. Yuanxun said: "Jie comes here by imperial command. He cannot be denied. He should be suffered to enter the city without stirring up turmoil." The gentry and commoners then clamored together: "Do you not see the slain, all piled up beneath the city walls?" Yuanxun said: "These include those killed by Yang Cheng. They were not all killed by Commissioner Gao." Yang Cheng was a general of the Yangzhou garrison [under Gao's command]. Since many of his troops were lawless, Yuanxun mentioned them. Those assembled mistakenly thought he said "Yangcheng" [Yangzhou city]. They accused him of betraying the city and tore him limb from limb.[71]

A more detailed version of this tale in the prefectural gazetteer reveals a more complex set of relations among the main actors in the story. According to this version, local officials and gentry were considerably at odds. The assistant commissioner for defense, Ma Wulu, a native of Shaanxi, bore a grudge against the police commissioner, Tang Laihe. This Ma, as noted above, was later to die by Shi Kefa's side. Yuanxun got along well with Laihe, whom he knew as the son of a fellow metropolitan graduate. Ma Wulu on this account set himself against Yuanxun and declared that the city had to be defended, that no compromise should be made with Gao. Subsequently some of Zheng's own militia captured a number of Gao Jie's troops who were out gathering kindling, and they strung them from the walls of the city. Gao Jie, in a rage, then laid waste to a local village. But Gao Jie owed Zheng a debt. When he was in Shandong as commander under Governor Wang Yongji, he had committed crimes that led Wang to impose the death penalty on him. Wang was a native of the Yangzhou department of Gaoyou and had passed the provincial exams at the same time as Yuanxun. Yuanxun had successfully utilized this connection to have Gao exonerated, perhaps in the interests of military defense of the Yangzi–Huai region. It was for this reason that he felt confident of approaching him after Gao cordoned off the city.

The lifting of the cordon allowed the city gates on the northwest side to be opened to bring in grain and fuel. Gao Jie withdrew his troops at Yuanxun's request, promised to punish his subordinate Yang Cheng, and presented Yuanxun with several hundred passes allowing traders to go in and out of the city. Yuanxun distributed these on demand, but the supply was rapidly exhausted, leaving latecomers disappointed and resentful. It was then that rumors began circulating: "Some voiced their suspicions, telling people: 'Gao Jie avoided the death penalty because of some fellow called Zheng; this can't but be a relative [of Zheng Yuanxun]; he is certainly receiving a payoff; he must die.' In the course of an evening these words swept through the city."

That night, Ma Wulu unleashed a barrage of stones at Gao Jie's troops. Enraged, they again pressed around the city with a great hubbub, as though about to attack. At this juncture, Zheng Yuanxun sent to Gaoyou to seek Wang Yongji's intervention. Wang was able to pacify Gao, and the local gentry subsequently went out to see him. In the meantime, however, some soldiers plundered the market town of Xiannümiao east of Yangzhou. In the middle of the night, the city was in turmoil, and the rumors against Zheng gathered momentum. The reports circulating in the streets had it that someone called Zheng belonged to the party of bandits and that Gao Jie's promise to punish Yang Cheng meant retribution not against his own subordinate but against the city of Yangzhou (Yangcheng). An armed mob surrounded Zheng Yuanxun and fell upon him. His bondservant Yin Qi also died in the struggle.[72]

In place of an arrogant, self-serving Yuanxun selling out the city, the local accounts thus have a gentry leader of impeccable

credentials endeavoring to save the city, his actions fatally mis-
understood by the city mob. Both versions are essentially loyalist, Dai Mingshi's famously so, but the local chroniclers defended the name of a leading member of the local gentry and his
family. The different accounts, including Dai's, are premised on
popular lore. Dai Mingshi's account is consistent with popular
rumor at the time of Zheng Yuanxun's death and has held sway
because Dai was eventually celebrated as a Chinese martyr. The
Yangzhou literati, however, took issue with these rumors and
ensured in the wake of the fall of Yangzhou that Zheng Yuanxun would be honored. Yangzhou scholar Jiao Xun (1763–1820)
recorded that he had been killed by "evil people."[73] Li Dou,
whose potted biographies of Yangzhou people rarely amount to
more than a few lines, accorded the Zheng family and its gardens several pages and recounted the entire story of Zheng
Yuanxun's death, based on information in the local gazetteers.[74]
Yuanxun's biography in the early-nineteenth-century prefectural
gazetteer also runs to many pages.[75]

Alternative Histories 2: Zong Hao

The local histories relate other tales, not all of valor. Among
Zheng Yuanxun's fellow-graduates of 1643, four others were
deemed worthy of honorable mention in the local records. One
was his nephew Weihong, who died a martyr's death in distant
Fujian, valiantly defending to the last the all but deserted city of
Pucheng, where he was stationed.[76] In Wan'an in Hubei, Liang
Yusi attempted suicide when the city fell. Rescued by family
members, he was subsequently taken as a prisoner to Nanchang,
where he succeeded in putting an end to his life.[77] Both these men,
like Yuanxun, are recorded in the upright and loyal category of
local biographies.

Of the survivors, Wang Yuzao was serving as district magistrate in eastern Zhejiang when the dynasty fell. Like Liang Yusi,
he attempted suicide. Subsequently he unsuccessfully sought to
join the retreating forces of Ming prince Zhu Yihai (1618–62). He
then adopted the guise of a Daoist monk, wandering around for
eight years and enduring desperate hardships before returning

around 1653 to his home in the Beihu area, just north of Yangzhou, where he led a life of seclusion.[78]

Gong Weiliu survived the fall of Beijing and returned to
Taizhou. He was recommended for office under the new dynasty
but declined the honor of appointment, pleading the precedent of
"an only son resigning to support [his parents]," essentially the
same excuse as his father, Gong Jilan (js 1637), had used to avoid
office in the late Ming. In Taizhou, Weiliu built himself a studio
and followed scholarly pursuits, which included editing the 1673
edition of the Taizhou gazetteer.[79] Both Gong and Wang, in
avoiding service to the new dynasty, were true to the ethical imperative "A good official does not serve two emperors." They
were worthy subjects for the compilers of local records.

The case of Zong Hao, the last of the six graduates, is different.
Virtually nothing is known about Zong personally because the
local gazetteers mention little more than his name and his degree,
but it is probable that apart from the Zhengs, uncle and nephew,
he was the only one who lived in or near to Yangzhou.[80] The contrast between the pages devoted to Zheng Yuangxun and the few
characters accorded to Zong Hao suggest a different sort of historiographical tension at work, one possibly formed in the first instance by the network of relations between people in Yangzhou.

The Zong family was well known in late Ming Yangzhou and
provides an instructive contrast to the Zhengs. The fortunes of
both families were built on salt, but although this is clearly
stated in the case of the Zhengs, it is virtually hidden from sight
in the local biographies of the Zongs, who are described simply
as natives of the Yangzhou county of Xinghua. This might have
signified that they were landowners, except that the appearance
of this family in the Lianghuai salt gazetteer reveals that they
were by origin salters (zao). Zong, a rather rare surname, is not
uncommon in the Huainan salt production zone. Zong Bu and
Zong Jie of the Caoyan Salt Yard were followers of the views of
Wang Gen (1483–1541) of the sixteenth-century populist Taizhou
School and were salters from Caoyan Salt Yard.[81] Zong Hao's
grandfather, Zong Mingshi (js 1589), was from the neighboring
salt yard of Xiaohai, where the family probably made its fortune
as owners of a saltern.

Zong Mingshi moved to Yangzhou in his youth, probably with his family. He passed the provincial exams in 1588 and the metropolitan exams the following year, subsequently accepting a position in the Board of Works in Beijing. He was blessed with four sons, of whom the eldest, Wanhua, passed the provincial exam in 1609. Zong Hao was the son of the youngest, Wanguo.

The salter families, though capable of achievements in the examination arena, did not establish themselves in Yangzhou with quite the éclat of the merchant families. The Zong family achieved local notoriety by quite a different channel. According to Shen Defu (1578–1642),

[Mingshi's] eldest grandson was wild and neglected his studies. Zong invited a well-known scholar from Dantu, Chen Xiao, to tutor him for the examinations. Chen was overly severe in disciplining his pupil and would beat him unmercifully. The grandson resented this bitterly and complained. Chen, overhearing him, was furious and thrashed him even more severely. Subsequently, in deep dudgeon with his pupil, he bought some bad meat laced with arsenic and consumed it. By evening he was raving mad and calling for water but would let no one come in. Soon after, he died in the schoolroom.[82]

This must have been the cause célèbre of Yangzhou in 1610. Chen's son, either believing that his father had been poisoned by the youth or holding that a suicide was attributable to him, took the case to court. The local gentry closed ranks and would provide no evidence. Zong Mingshi eventually made an out-of-court settlement with the Chen family, at no small cost to himself, but the local magistrate was unsatisfied and sent the case to the governor's office. At the time of Shen's writing, despite widespread public support for the grandson and numerous representations to the powers on high, the boy was still languishing in jail.[83]

This piece of gossip does not, of course, appear in the local gazetteers, but the career of Zong Mingshi's eldest son, Wanhua, receives due attention. Wanhua was appointed magistrate of Jingzhou county in the 1620s. Here he successfully quelled an uprising of the White Lotus sect. More than 8,000 rebels are said to have lost their heads to the swords of his 3,000 soldiers, and he himself became a local hero, his valiant deeds inscribed in stone. Later, as assistant prefect in Chaozhou (Swatow), he came into conflict with Governor-General Xiong Wencan (d. 1640) over methods of suppressing pirates. Accused by Xiong of collusion with the pirates, he was flung into jail, where he died before his case could be heard.[84] He was posthumously vindicated when Xiong Wencan was executed for failing to suppress Zhang Xianzhong in Hubei.[85]

The detailed record of Zong Wanhua's career in both the local and the salt gazetteers makes the brevity of the entry on Zong Hao the more obvious, particularly since Zong Hao's academic success was greater. Something is known about other members of this family as well. Hao's cousin, the poet Zong Yuanding (1620–98), was one of the most prominent literary figures in early Qing Yangzhou, an eccentric figure who made money at Rainbow Bridge outside the city by selling flowers woven out of grass.[86] Another cousin, Zong Guan, was a well regarded poet in the locality.[87] It is impossible, then, that simply nothing was known about Hao's official career at the time by his younger contemporaries.

Some light is thrown on this issue by the appearance of Hao's name in the list of local officials in early Qing Jiangnan: the blood had hardly dried on the streets of Yangzhou when he crossed the Yangzi to become prefect of Changzhou.[88] The omission in the local chronicles of a proper biography for him looks like a moral judgment by local historians: he was the only one of the six graduates of 1643 to take office under the new dynasty. Perhaps he himself felt the burden of this decision. His son is recorded as having passed the provincial-level examination as a resident of Jiangning prefecture (Nanjing).[89] If this change of registration was made at Zong Hao's initiative, then he had effectively cut his ties with his ancestral place as well as with the Ming.

Zong Hao, unlike his uncle Wanhua and his cousins Yuanding and Guan, has no place in local histories of the Ming-Qing transition.[90] The central place of the Zheng Yuanxun, by contrast, was kept in popular view by a memorial built to honor him and his younger brother, Yuanhua. Yuanhua, the father of the Ming martyr Zheng Weihong, had, through Yuanxun's offices, served creditably under the Ming as a military subprefect.[91] The shrine was erected by local people on a plot of land adjoining Yuanxun's garden, on the southwestern side of the city. The garden went to rack and ruin in the early Qing, but the shrine appears to have survived and was described in detail by Li Dou in the late eighteenth century.[92] Its lateral position, to the west of the city facing eastward, contrasted appropriately with that of Shi Kefa on Plum Blossom Hill. Shi's tomb faced southward, the proper position for this symbol of the fallen Ming. Zheng's took the position of guest at the banquet, but a most honored guest in a city that, despite his Huizhou origins, claimed him as its own.

"Yangzhou shiri ji": The Last Four Days

The depiction of carnage in Wang Xiuchu's diary has for at least a century overshadowed other accounts of the fall of Yangzhou, and the impression it leaves of a city laid waste at the beginning of the Qing is such that a long period of convalescence might well be imagined as necessary to its recovery. Analyzing the diary with an eye to local detail rather than to national significance suggests otherwise. The siege and massacre occupy only the first six days of the diary: the last four days covered in the diary are also important. On the second day of the fifth month, the seventh day of Wang's record,

civil administration was established in all districts. Officials were instructed to issue proclamations to calm the people. . . . On the third day a proclamation for charity and relief was issued. . . . On the fourth day the weather was fine. The sun shone down hotly on the corpses from which issued an unbearable stench. Everywhere around us corpses were being burned. . . . Now the fifth day arrived, and people hidden in secluded places were beginning to reappear.

On this last day, Wang Xiuchu's brother died of wounds he had sustained during the siege. Moreover, Xiuchu himself was still afraid to return to his own house: with his wife and child he led for a time the life of a recluse, trying to avoid attracting the attention of the desperadoes roaming the city streets. Nonetheless, the return of law and order had been initiated with the appointment of officials. Efforts were being made to clean up the city and feed the survivors. The deaths of Prefect Ren Minyu, reportedly killed as he accompanied Shi Kefa out of the city, and of District Magistrate Luo Youlong[93] had sundered the city's ties to the Southern Ming, a regime that in any case collapsed with the flight of the Hongguang emperor from Nanjing and the surrender of the southern capital without bloodshed three weeks after the massacre in Yangzhou.[94]

By the time Wang came in later years to write his memoir, he was able to reflect on the contrast between those terrible days and the situation of the members of a "younger generation who are fortunate enough to lie in tranquility, enjoying the happiness of leisure" even if they were also, in his view, self-indulgent and neglectful of morality.[95] A close reading of his diary compels the conclusion that in his lifetime he must have witnessed the flowering of late Ming Yangzhou, its destruction during the interdynastic war, and then its quite rapid revival under a new dynasty, the Qing.

CB

FIVE

The Loyalist City

Around fifteen miles north of Yangzhou and west of the Grand Canal lay the North Lakes district, Beihu, a patchwork of lakes, fields, and ponds interlaced with rivulets and raised paths that enabled pedestrians to avoid getting their feet wet. Periodic markets summoned the peasantry and petty traders bearing in their panniers and on their donkeys the simple commodities necessary to life: grain, cloth, and livestock. The great families here in the late Ming dynasty were the Liang and Sun families. Liang Yusi, Zheng Yuanxun's fellow graduate of 1643, was a scion of these Liang.[1]

To this out-of-the way place, the widow of one Ruan Bingqian, fleeing the disturbances around Yangzhou, brought her family in 1644.[2] Six generations later, this family produced Ruan Yuan, Yangzhou's most eminent scholar-official in the early nineteenth century. In Ruan Yuan's generation, one of his cousins married Jiao Xun, another famous scholar of the period and a native of Beihu.[3] Ruan Yuan was born in Yangzhou city, but Jiao Xun grew up in Beihu and documented the locality in a short topographical and historical work on the model of a local gazetteer.

Intermarriage among the prominent families of Beihu was probably common. Jiao Xun was also related on the maternal side to the Wang family of Beihu. According to Jiao, scholarship in Beihu began to flourish with Wang Nalian (*js* 1607).[4] Wang Nalian was the father of Wang Yuzao, Zheng Yuanxun's fellow graduate of 1643. Jiao recalled visiting a maternal uncle when young and seeing in his house a portrait of Wang Yuzao with his head bare, wearing no emblems of office. The uncle, a great-great-grandson of Yuzao, explained that this was due to Yuzao's having "returned from Zhejiang [where he held office in the late Ming] to a life of seclusion, devoting himself to farming. For the rest of his life he went without [the official's] cap and hence was portrayed in this way."[5]

As noted in the preceding chapter, Wang Yuzao took the path of eremitism after the fall of the Ming. He was the quintessential Ming loyalist, or *yimin*, to use the Chinese term: one of the "avoiding people" whose bodies were ruled by the Qing but whose hearts remained with the Ming. The term "Ming loyalist" has been defined as applying "meaningfully to anyone who pointedly altered his or her life patterns and goals to demonstrate unalterable personal identification with the fallen order."[6] Wang Yuzao's avoidance of office (*yi*) under the Qing and his retirement to a life of seclusion were among the least ambiguous signs of loyalist sentiment.

Wang Yuzao had two sons, Fangqi and Fangwei, both of whom followed their father in avoiding office under the Qing. Ruan Yuan, in an anthology compiled in 1799, recorded that Wang Fangwei was once approached by some former students of his grandfather, who besought him to enter into public life. Fangwei responded with a poem:

> I have been fishing at the lakeside now for
> twenty years,
> Gradually frosting at the temples: I rue my
> hoary head.
> Yesterday from the jetty I watched for the
> new waters to rise,
> But there was only an autumn tide, going
> out and coming in.[7]

These lines allude clearly to the lost cause of the Ming. The poem would appear to have been written around 1664 or 1665, twenty years after the fall of the dynasty or of Yangzhou, respectively. In 1659, loyalists in the Lower Yangzi region had briefly taken heart when Southern Ming forces under Zheng Chenggong (1624–62) occupied the riverside town of Guazhou, just south of Yangzhou. In this context, Wang Fangwei's poem might be read as an expression of disappearing hopes.

In their attention to the Wangs of Beihu, Ruan Yuan and Jiao Xun bear witness to the commanding presence of the "avoiding people" on the stage of early Qing China as constructed by later Qing literati. The association of these *yimin* with Yangzhou was strong in the eyes of later chroniclers and was reinforced by their own historicization of their native place. The prefectural gazetteer of 1810, to which both Jiao Xun and Ruan Yuan contributed, includes a small section on sojourners in Yangzhou; around half of the Qing entries deal with early Qing *yimin* living in or near Yangzhou. Included are prominent artists and writers of the time, such as the Huizhou painter Cheng Sui (1605–91?) and the poets Sun Zhiwei (1620–87) from Shaanxi, Fei Mi (1625–1701) from Sichuan, and Sun Mo (1613–78) from Huizhou.[8] To these names can be added those of other *yimin* and fellow-travelers, some listed in this same gazetteer under different categories: the poet Du Jun (1611–87) from Hubei, resident in Nanjing but an occasional visitor to the city; Gong Xian, painter of brooding landscapes, who came and went between Nanjing and Yangzhou; Zha Shibiao (1615–98), from Huizhou, another painter, who first took up the brush in the 1650s; and, most famous of all, Shitao, a scion of the Ming dynasts.[9] The immortality of the names of these various sojourners was ensured by their works, which had a special poignancy and appeal due to the tragedy and romance associated with the *yimin*.

Also given due attention in this gazetteer are the "reclusive avoiders" (*yinyi*) of Yangzhou prefecture, those who lived in relative seclusion and refused to take office. The terms *yinyi* and *yimin* are not interchangeable: *yimin* appears to be specific to people who survived the change of dynasty, whereas *yinyi* applies to men who for one reason or another refused to take office under the Ming or in earlier dynasties.[10] In the early Qing, however, *yinyi* and *yimin* were synonymous. The *yinyi* entries include a number of Beihu loyalists, along with the poet Wu Jiaji, from

Anfeng Salt Yard on the coast,[11] and the cousins Zong Yuanyu and Zong Yuanding, cousins of the "twice-serving" official Zong Hao.[12]

Some of the entries in this section of the gazetteer are, as is common, merely reduplicated from earlier gazetteers. Others are new additions. The citation of sources for the new entries shows that considerable research had been invested in building a portrait of Yangzhou as a site of ethical practices. Among these sources are Ruan Yuan's own *Record of Heroic Spirits from Between the Huai and the Sea* (*Huaihai yingling ji*). The focus on loyalism in this, the largest of the Qing dynasty gazetteers from Yangzhou, is thus not evidence of a historical continuum in Ming loyalist sentiment. Rather, it should be understood in the context of a renewed interest in loyalism inspired by the Qianlong emperor, who had for his own ends fostered biographical projects centered on the issue of loyalty to the emperor.[13]

A spatial representation of loyalists in Yangzhou prefecture would reveal a surprising dispersion away from the prefectural capital. Prominent among Yangzhou *yimin* was former official Gong Weiliu, another of Zheng Yuanxun's fellow-graduates, who lived in Taizhou department, east of Yangzhou. A number of sojourners, including noted *yimin* such as Fei Mi, are recorded as having lived in or traveled to Taizhou for varying periods of time, visiting Gong in his Thatched Hall of Spring Rain.[14] The salt yards of Taizhou also boasted a number of loyalists, most famously Wu Jiaji.[15] In the neighboring jurisdiction of Rugao lived Mao Xiang (1611–93), a member of the Restoration Society prominent in the factional struggles in Nanjing in the dying days of the Southern Ming.[16] (Mao was connected by marriage to Gong, whose daughter had married Mao's brother.)[17] Among *yimin* from Yangzhou's home county, Jiangdu, Zong Yuanding is closely associated with the prefectural city, but he lived outside the walls; he and his cousins appear to have been domiciled in the market town of Yiling, east of the city.[18]

In the immediate Yangzhou area, Beihu appears to have been the locus of loyalist sentiment, although the prominence of Beihu scholars such as Jiao Xun and Ruan Yuan among chroniclers of Yangzhou may lend this district undue prominence in the written record. The loyalist community in Beihu was close, their interrelationships evoked with startling immediacy in the cryptic record of their lives. Nor were they only men. The wife of the recluse Fan Quan adopted the stance of a recluse, "wearing her hair in the mallet style and dressed in coarse cloth." A disciple of Wang Yuzao's elder son, Fangqi, she was fond of chanting lyrics (*ci*) in the company of fellow recluses in Beihu.[19] The appearance of several military graduates in Beihu genealogies otherwise confers a rather martial air on the record of this community. The brother of one Beihu loyalist was a military *jinshi* of 1640.[20] Ruan Yuan's great-great-grandfather served as military commander during the Wanli era, and his grandfather was a military *jinshi* of 1715. A string of other relatives had military degrees.[21] Jiao Xun, whose own family produced military graduates, wrote that "those who do not make sacrifices to [the God of War] are not of my clan."[22]

In a rural locality such as Beihu, loyalists could avoid engagement with official life more easily than in a city. The city in China self-evidently belonged to the dynasty. It was physically distinguished from sometimes larger, economically thriving market towns, by certain institutions: the Temple of the City God, the prefectural or county school, the yamen of the prefect or magistrate, and, of course, the city walls, although some strategically important towns were also walled. These features declared the city's relationship with the dynasty, as surely as the plaque on his robe declared an official's. To engage with the city was to engage with the bureaucracy and, ultimately, with the dynasty. Unsanctioned forms of social organization, such as secret societies, were more likely to be based in market towns than in the administrative centers of the empire.[23]

The weighty ritual significance of the city was well recognized. In describing Wang Fangwei's integrity of character and lofty adherence to loyalist ideals, Jiao Xun drew attention to his avoidance of the city: "He did not go to the city, he did not accept disciples, he did not travel, he did not go out eating and drinking. Utterly and profoundly, he devoted all his talents to the study of the principles of the *Book of Changes*."[24] Another Beihu loyalist, Xu Shiqi, refused to attend literary gatherings in Yangzhou, even those at which well-known *yimin* poets such as Zong Yuanding and Wu Jiaji would be present.[25] And in the nearby county of Xinghua dwelt the loyalist Lu Tinglun, who "for ten years did not enter the city."[26] This ethical imperative stemmed from a desire not to avoid social interaction per se but to avoid having to salute local manifestations of Qing authority.

Yet Yangzhou had an appeal for *yimin* scholars and artists. A combination of recent and remote historical events made the early Qing city a supremely appropriate context for reflecting on the past. Wu Jiaji, a frequent visitor to Yangzhou in the 1660s, wrote numerous poems about women sacrificing their lives in the interests of family, a number of which were set in the context of the fall of Yangzhou in 1645. Jonathan Chaves is undoubtedly correct in interpreting these as allegories of loyalty to the Ming. Sun Zhiwei, a close acquaintance of Wu Jiaji, wrote unambiguously: "A chaste wife is like a loyal minister."[27] Among artists, it is notable that the Yuan painter Ni Zan (1301–74), increasingly popular among artists and collectors in the late sixteenth and early seventeenth centuries, seemed especially relevant to *yimin* painters of the second half of the seventeenth century. The Yuan dynasty, with its Mongol overlords and loyalist literati, provided subjects of the early Qing with parallels for their own condition. Cheng Sui once wrote of his own dry, serious landscape painting: "It is easy for me to capture the spirit of desolation in Ni Zan's painting because we have had the same experience in our lives."[28]

Swapping cup and verse outside the city walls on a summer evening, painters and poets could together reflect on the downfall of the dynasty via traces of the past: Bao Zhao's *Ballad of the Ruined City*, penned after the destruction of this same city in the fifth century; Jiang Kui's lament on the sack of Yangzhou by the Jurchen in 1133: "Even the ruined lakes and trees / Are loath to say what happened then";[29] tales of Li Tingzhi gallantly defending the city walls against the Mongols until the bitter end, just as Shi Kefa was to do against the Manchus. Themselves from places devastated either during the rebellions of the late Ming or in the course of the Manchu conquest, sojourners and locals alike could hardly avoid finding reminders in Yangzhou of what had transpired in 1645. A place in the southeastern quarter of the city acquired the name Great Red Water (Hongshuiwang), according to Zhou Sheng, because it was "where soldiers had been stationed under the Ming. Men were cut down like so many stalks of hemp, and the blood ran in rivers."[30]

For those who stayed away from the city, a principled decision meant a punishing removal from the wellsprings of literati life. The exchange of manuscripts, the execution and appreciation of paintings, the convivial poetry parties, the learned discussions on points of textual criticism and principles of philosophy, were activities vital to intellectual life and literary productivity. Such activities flourished in Yangzhou in the second half of the seventeenth century, and *yimin* were prominent contributors. Images of Wang Yuzao attending to his fields in Beihu and his son Wang Fangwei laboring over the *Book of Changes* provide a reminder, however, that the city was a place of negotiation and compromise. A loyalist in Beihu could live out his years in a post-Ming twilight. Loyalists in Yangzhou were drawn inexorably into the time of the Qing. Their lives and social activities were shaped by the new order, which was to prove surprisingly familiar.

Reconstruction and Rehabilitation

A counterpoint to the romantic tale of early Qing Yangzhou as a city of *yimin* is the prosaic story of its restoration as a functioning node of imperial administration and trade. The theory in Yangzhou is that Shitao, among the most famous of *yimin*, chose to live in the city because the massacre provided fertile ground for the seeds of dissent.[31] As Jonathan Hay has shown, the circumstances under which Shitao came to live in Yangzhou reveal that the city had other attractions. As a small child, Shitao was hidden in a monastery in Guizhou by a servant who saved him from the massacre that obliterated the rest of his family. After a peripatetic existence, which included some time in Huizhou and an occasional visit to Yangzhou—the first in 1673—he spent eight years in Nanjing, largely within a reclusive loyalist community, before moving to Yangzhou in 1686. Yangzhou was a much smaller city than Nanjing, and he was not impressed with it, comparing living there to "turning round and round in a well." It had the further disadvantage in his eyes of lacking any interesting landscapes, an accurate enough observation.[32] But Nanjing had its own problems. In Nanjing in 1685 Shitao had painted an album for sale, "but no one was interested."[33] Yangzhou, where he lived intermittently from 1686 and permanently from 1697, offered him better means to support himself.

Among the writers and artists associated with loyalism in Yangzhou, Shitao was a latecomer. Gong Xian, who like Shitao had a base in Nanjing, was able to support himself in Yangzhou in the late 1640s and 1650s, returning frequently thereafter,[34] and so apparently was Zha Shibiao, the self-styled "Yangzhou traveler," who was at least sporadically in Yangzhou during the 1660s and probably resident there from 1670.[35] It would appear that within a relatively short period after the city's devastation, the urban economy had revived to the point that men could make money from teaching and the sale of their works.

The poet Sun Zhiwei provides the hint of an explanation for the early restoration of a cash economy in Yangzhou. Sun first came to the city with his father as a boy of twelve, around 1631, but returned to his native place, Sanyuan in Shaanxi, to pursue his studies. Sanyuan was laid waste by the rebel Li Zicheng in the late Ming. After the fall of the dynasty, Sun again repaired to Yangzhou, bringing his wife with him. They lived there in the garden residence that his father had established.[36] Sun went into business, undoubtedly in salt, and quickly amassed a fortune of 1,000 taels.[37] But his fortunes did not thrive. According to Wang Maolin (1640?–88): "The old business daily declined; so he sold the garden residence and went to live by the Temple of Dong Zhongshu [in the western part of the New City]. He called the place where he lived the Hall of Cleansing Waters."[38] The downturn in his fortunes, however, does not disguise the fact that he had resources in Yangzhou and was able to make money there. Wang Maolin, himself from a Huizhou merchant family from Xiuning, does not mention that the Sun family was in the salt business, but Sun Zhiwei's appearance in the Lianghuai salt gazetteer makes it evident that salt was the source of his money.[39]

The resumption of the Lianghuai salt trade received immediate attention from the new dynasty. In 1645, Li Fayuan, the first Qing salt censor in Lianghuai, licensed the issuance of 60,000 lots (*yin*) of salt to allow resumption of the trade according to the shipment system established nearly 30 years before.[40] There were numerous impediments to the smooth operation of the trade in the first few decades. Salters had fled the salt yards, markets were disrupted by war, and officials were imposing excess fees on the salt transport. The climate for investment was hardly encouraging. During Zheng Chenggong's incursion in 1659, the court attempted to collect 450,000 owing on the Lianghuai salt tax to cover the costs of defense. Several merchants were arrested and beaten to death for failing to produce the money.[41] Yet Huizhou men quickly reestablished themselves in the salt trade. When Li Fayuan reopened the salt register, Min Ding, a Huizhou merchant active in the trade before the fall of the dynasty, immediately acquired monopoly rights over 40,000 lots.[42]

The salt trade was a significant factor in the size of the sojourning *yimin* community in Yangzhou and in the corresponding flourishing of the arts in the second half of the seventeenth century. The salt merchants of Huizhou surrounded themselves with the cultural products of the literati tradition. The greater among them were art connoisseurs, the lesser dabblers in the market. When they emigrated, they took their cultural habits with them and very often their artists, too.[43] In brief, there were many artists in Yangzhou because the market for art products there flourished from at least early in the Kangxi reign. Wang Shimin (1592–1680) from Taicang, one of the "Four Wangs" of the early Qing orthodox school of painting, wrote in 1666 of people from Suzhou flocking to Yangzhou to sell their works of art to a collector from Yunnan. The collector, "unable to distinguish a pebble from a stone," used a middleman who had made no less than 20,000 taels by buying paintings cheaply and then selling them to his client at an enormous profit.[44]

To patronage of the arts, the Huizhou merchants added projects aimed at stabilizing the local society and restoring its cultural fabric, or, in other words, of creating a livable city. Min Shizhang (b. 1607), one of Shitao's many Huizhou patrons,[45] arrived in Yangzhou "empty-handed," and made a small fortune of "a thousand taels" by doing accounts for the local people. He then went into the salt trade, which made him really rich.[46] From the 1650s to the 1670s, he was active in a variety of philanthropic activities, such as restoring a bridge, founding an orphanage, reviving the poorhouse, and providing relief for refugees from Yangzhou's dependent counties, who "congregated [around] the prefectural city" when floods ruined seven harvests in succession.⁴⁷

Such activities took place within the framework of restored administrative structures. Yangzhou was formally inducted into the new dynasty by the first Qing prefect, Hu Qizhong, not long after the massacre. In 1645, Hu undertook the reconstruction of the prefectural school or "school temple" (xuegong), one of the most important emblems of the city's administrative status.⁴⁸ That same year, a provincial-level examination was conducted in Nanjing. Among the successful candidates were six from Jiangdu, Yangzhou's home county — the first of its gentry to make a pact with the new dynasty.⁴⁹

These various developments both symbolized the establishment of the authority of the new dynasty in Yangzhou and signaled to its inhabitants that the new order was in some respects to be a continuation of the old. Two years later, in 1647, the city walls were restored.⁵⁰ In the same year, the first metropolitan examination of the dynasty was conducted, and twelve candidates from Yangzhou prefecture — including four from Jiangdu county — gained the jinshi degree.⁵¹ This was as well as Yangzhou was ever to do in examinations under the Qing. Yangzhou's strong showing owed much to the lack of competition from unpacified areas in the south,⁵² but it was also indicative of a general readiness on the part of the elite to cooperate with the new regime. These graduates proceeded to hold office under the Qing. Such accommodation with the new dynasty was normal in the conquered areas, if not the norm, and it may be questioned whether Yangzhou would have shown much resistance to Qing forces if it had not been serving as a bastion of defense for the Southern Ming court in Nanjing⁵³

Huizhou families devastated by the war were among those who fielded sons in the examination arena. Two nephews of Zheng Yuanxun, a brother and a cousin of the Ming martyr Zheng Weihong, sat for the examinations early in the dynasty and took office in Beijing, the seat of imperial power.⁵⁴ The Huizhou diaspora accounted for the more prominent of Yangzhou scholar-officials under the new regime, including Wang Maolin, mentioned above; Wu Qi (1619–94), who won the Kangxi emperor's favor as a dramatist; and Wang Ji (1636–99), who made his name by serving as the Qing envoy to the Liuqiu islands (modern Okinawa).[55] Many of these families appear to have preserved fortunes made before the fall of the Ming, despite the looting and destruction of property during the sack of Yangzhou. As noted in Chapter 2, Zheng Xiaru, the brother of Zheng Yuanxun, purchased two adjacent properties in the eastern part of the New City early in the new dynasty and built a large and lovely garden there.

The Manchus, more closely than the Mongols in an earlier century, adopted the organizational principles of the dynasty they supplanted. The physical appearance of officials changed: Chinese men were ordered to shave their heads at the front, bind their hair into a queue at the back, and don the long robe (*changpao*) in accordance with Manchu practices. But the Manchu presence in Yangzhou was otherwise muted. The first Qing prefect was a native of Jiangnan, and the district magistrate a native of Henan. Over the next two decades these posts were dominated by northerners, mainly from Liaodong in the case of the prefectship, but until 1678 the field administration remained identifiably Chinese. The pattern was broken in that year by the appointment of a bannerman, although he was a Chinese bannerman, not a Manchu.[56] Yangzhou, perhaps surprisingly in view of its strategic position, did not become the site of a Manchu garrison.

The important offices of salt censor and salt controller were also held by Chinese during the very early years. One Manchu was appointed salt controller in 1652 and another as salt censor in 1658.[57] The lucrative post of customs superintendent was early held by Manchus, alternating or in conjunction with Chinese officials.[58] Other positions in Yangzhou were, and continued to be, held by ordinary Chinese officials. This was broadly consistent with the pattern of appointments in Jiangnan (the Nan Zhili of Ming times) as a whole, although in some places Manchus or Chinese bannermen did hold office as prefects in the early years of the dynasty. It is possible that in the case of Yangzhou, such appointments were studiously avoided in the years immediately following the massacre. In any case, it was under the guidance of Chinese officials that the early restoration of the city took place.

These officials provided part of the context within which Yangzhou's *yimin* culture flourished, but that understates their contribution to the making of this culture. Their patronage supported *yimin* endeavors, and their own literary activities dovetailed with those of *yimin* scholars. In the record of these activities can be discerned the process by which the *yimin* were accommodated by Qing society, and the city itself was rewritten into Qing times.

Rewriting Yangzhou

Nothing is as astonishing about the early decades of the Qing dynasty as the frenetic amount of writing undertaken across the spectrum of Ming-Qing loyalties. While wars were still being fought, rivers were running amok, and the food supply was still a problem, words were pouring forth on paper. In the very month that Yangzhou fell, an imperial edict commanded that work on the official Ming history, in effect the obituary of the Ming dynasty, commence. When the Ming history did get under way during the 1680s, some scholars known for their principled refusal to serve the Qing were drawn into association with it. The urge to record the past, and in this case in a form well established within the Chinese tradition—that of the dynastic history—proved irresistible to many.[59]

The inescapable impression is that Chinese literati, *yimin* and declared Qing subjects alike, were preoccupied in this period with writing and painting themselves back into the picture. Transcending the dynastic rupture by reference to textual, aesthetic, or ethical traditions, they placed themselves in a genealogy of Confucian scholars that could be considered unbroken, despite the rise and fall of other dynasties in other times. Simultaneously, they reinscribed the map of the empire, so that what had been Ming territory became Qing territory and places rich with Ming history acquired a Qing history as well.

A notable figure among the early Qing scholar-officials who helped redevelop Yangzhou as a Qing site was Zhou Lianggong, one of the first Qing officials to arrive in Yangzhou. Zhou was a metropolitan graduate of 1640 and had served under the Ming as magistrate of Weixian in eastern Shandong county, a position held a century later by the Yangzhou artist Zheng Banqiao. He was posthumously designated a "twice-serving official" during the Qianlong emperor's literary inquisition in the 1770s.[60] Appointed to Yangzhou as Lianghuai salt controller in 1645, he was moved within a year to the position of Huai'an-Yangzhou intendant and then transferred to Fujian, where he held a series of provincial-level positions between 1647 and 1654. In the 1660s, however, he held positions successively in Shandong and Nanjing, and travels on official business up and down the Grand Canal brought him back to Yangzhou frequently.

Zhou Lianggong's earliest literary enterprise under the Qing provides a good example of the compulsion of former Ming subjects to write. It is recorded of Zhou that when he was in Yangzhou "he spent much of his leisure time on a boat where he would take up [painting] albums at random."[61] These albums served as the inspiration for biographies of the painters, histories of the works themselves, and reflections by Zhou in the form of poems or colophons on the paintings.[62] In 1646, when he was either still in Yangzhou or further north in Huai'an, he commenced work on his famous Record of Reading Paintings, an important if cryptic record of seventeenth-century art and artists, including work produced in the early Qing.[63] The author of one preface to this work attributed its genesis simply to Zhou's interest in art,[64] but given its commencement date of 1646, it would seem that he was engaged in the profoundly cathartic act of recording what had, until 1644, been the present. In 1645, Zhou was busy organizing the burial of the corpses with which the streets of Yangzhou were strewn in the wake of the massacre. The boat on which he passed his leisure time must have taken him out onto Baozhang Lake, as it was then known—an easy walk from the city walls but out of sight of the devastation they contained. The image of this Qing scholar-official seated in a boat taking refuge from the cataclysmic present in the quiet contemplation of album leaves might be paired with an image of the Ming loyalist Wang Fangwei at his desk, laboring away at the Book of Changes. Divided by their declaration of loyalties, the scholar-official and the hermit were united in their attention to the past.

More intimately associated with Yangzhou over a longer period of time was Wang Shizhen (see Fig. 7), whose contribution to the making of early Qing Yangzhou has been sensitively explored by Tobie Meyer-Fong.[65] A metropolitan graduate of 1658,

Fig. 7 Yu Zhiding, *Wang Shizhen fang xian tujuan* (Wang Shizhen releases the quail), 1700, detail (Palace Museum Collection; published in Mayching Kao, *Paintings by Yangzhou Artists of the Qing Dynasty* [Beijing: The Palace Museum; Hong Kong: The Art Gallery, Chinese University of Hong Kong, 1985], p. 44, fig. 26).

Wang was appointed in 1660 to the post of police commissioner in Yangzhou, a position he held until 1665.[66] Wang's experience as an official gives some hint of the difficulties besetting Yangzhou in the early years of the Qing. Between 1645 and 1660, Yangzhou had accumulated tax arrears of 20,000 taels of silver. Prosecutions for tax evasion resulted. When Wang arrived in Yangzhou, he found the jails full of relatives of the accused, "wasted in body, pale of visage, chains clanking, weighed down by fetters," guilty only by association. One of his achievements in Yangzhou was to develop a plan through which, with the cooperation of the salt and civil officials, the arrears could be paid off.[67]

There were, moreover, still military disturbances in the region. The year before Wang arrived, the Yangzi ports of Guazhou and

Yizheng, both within striking distance of Yangzhou, were attacked by the forces of Zheng Chenggong in an alarming reminder that the interdynastic struggle was not yet over. The frightened residents of Yangzhou scattered inland, and if Wang Ji's family is any example, they took some time to return. Wang Ji moved his household east to Taizhou at this time and stayed there for at least three years.[68] Wang Shizhen arrived in Yangzhou in time to become involved in the criminal cases arising from the attacks and won a name for his fairness in dealing with the accused.[69] The trials were held in Nanjing in the last year of the Shunzhi reign, 1661, and were among the many official matters that took him across the Yangzi during his period of office in Yangzhou.

The record of Wang's travels within Jiangsu, both north and south of the river, provides a hint of the official's important role in effecting social integration of the gentry across administrative boundaries and over great distances. His journeys can be tracked through his literary activities. In the first month of 1661, he traveled to Suzhou via Wuxi, his first venture into this famous center of southern culture, home to the Donglin and Restoration literary circles of the late Ming. From Yuyang Mountain in the vicinity of Taihu, he took his literary name, "Man of Yuyang Mountain," first used in a work composed in Yizheng, where he spent part of the following autumn. In Nanjing that same year he stayed in the famous quarter by the Qinhuai Canal, celebrated for its courtesans and literati parties. In Jiangbei he also traveled widely: north to Gaoyou, Huai'an, and then across the Huai, east from Gaoyou to Xinghua, and from Yangzhou directly east to Taizhou and southeast to Rugao, where he met Mao Xiang.[70] All these places were celebrated in his verse.

Wang's literary activities served to reinscribe Yangzhou on the Chinese cultural landscape. His Yangzhou poems were typically topographical, taking their point of departure from a town, temple, waterway, or other site he was visiting. The evocative powers of such poems went well beyond their pictorial and sentimental elements. Replete with classical allusions (*dian*), in accordance with his theory of poetics, his poems allowed him to reflect on past and present through indirect references (*yuan*) to events of long ago, a tactic that reinvested the places themselves with a historical significance that the change of dynasty had threatened to undermine. "Purposefully, the east wind through willow and poplar blows; / The verdure extends to which bridge of the Ruined City?"; "Waves swirling round Thunder Pool — a single eddy; / Even now the waters murmur against Yangzhou"; "Three bearded gentlemen pass below Pingshan; / White-haired disciples long for the knowledge of yore"; "Green, green grasses half-cover Plum Blossom Hill; / At leisure I see the traveler off, before riding home": with verses such as these Wang remapped Yangzhou, variously quoting and paraphrasing the Tang poets Du Mu, Du Fu, and Wang Wei, returning again and again to major icons in Yangzhou's history: the city laid waste in Bao Zhao's fifth-century ballad; the Four-and-Twenty Bridge, first featured in a poem by Du Mu; the tomb of Sui Yangdi near Thunder Pool; Pingshan, the eminence northwest of Yangzhou where Ouyang Xiu (1007–72) had built a temple in the twelfth century; Plum Blossom Hill, where the clothes of Shi Kefa were entombed.[71]

In verse Wang also recorded his association with *yimin* literati in Yangzhou: a colophon on a portrait of the Shaanxi poet Sun Zhiwei, a reflection on reading the poetry of the Sichuanese poet Fei Mi, a dedication to the local poet Zong Yuanding; a farewell poem to the Huizhou poet Sun Mo, returning home to Huangshan.[72] These point more directly to Wang's major contribution to literary life in early Qing Yangzhou — his active cultivation of poets, which laid the basis for the city's emergence as the empire's pre-eminent center of poetry in the second half of the seventeenth century. This enterprise, too, had topographical expression. The two most famous literary events in Yangzhou during Wang's period of office were poetry meetings conducted by the Red Bridge (Hong qiao), located a short distance northwest of the city, at the southern end of Shouxi Lake.[73] Since the bridge had been constructed during the reign of the last Ming emperor (r. 1628–44), it had no great historical significance, but after Wang Shizhen's term of office in Yangzhou, "people passing through Guangling would often ask [the whereabouts of] the Red Bridge."[74] Poems Wang Shizhen himself composed at the Red Bridge affirmed the connection between past and present.

The circumstances under which a relationship between a Qing scholar-official and an *yimin* poet might develop are illustrated with unusual precision in the record of Wang Shizhen's activities in Yangzhou. One poet whom he sought out was a man called Li Yi, from the Yangzhou county of Xinghua. Li Yi had given up office when the Ming dynasty fell and, together with a cousin, established a poetry society in Xinghua. Wang Shizhen heard of Li Yi and went to call on him in Xinghua, some two to three days' journey from Yangzhou, arriving outside his door in all his official splendor. Li merely sent his apologies and declined to see him. Wang respected his position and did not force the issue.[75]

Wang had greater success in pursuing the acquaintance of Wu Jiaji. He wrote a memoir of how he came to meet him:

I had been living in Yangzhou for three years before I met Wu Jiaji of Hailing [Taizhou]. An impecunious scholar, Jiaji dwelt in the saline lands by the sea. His antiquated dwelling with its broken tiles lay hidden within a few *mu* of sorrowful bamboo. Here, where snakes and tigers lurked and rodents hissed, no guests from the four quarters came. In such surrounds, Jiaji intoned his bitter laments, seeking the acquaintance of none, and his fame did not extend beyond one hundred *li*.

Guangling lies one hundred *li* from Hailing, and from Hailing it is yet another hundred to where Jiaji lives. One might see his poetry and yet have no means of seeing the man. One evening as the snow fell deep and the wind whistled without remorse, [in company with] the lonely sound of the night watch, I was looking through some old writings beneath the lantern light when I received some of Jiaji's poems. Overjoyed on reading them, I went on to write a preface. Next day, in all haste I sent to Louxuan, Jiaji's abode. Jiaji was sensitive to my intent, and on my behalf poled a boat to the prefectural city, where we met to our mutual enjoyment.[76]

As Wang noted in his memoir, Zhou Lianggong knew Wu Jiaji before he did, and Wang Ji knew him even earlier. The manuscript of poems that Wang perused that snowy evening had in fact been given to him by Zhou Liangong in 1663 when the latter was passing through Yangzhou on his way to Shandong. Zhou Lianggong in turn had heard Wu Jiaji's name from Wang Ji. Zhou subsequently sponsored the publication of the manuscript, and both Wang Shizhen and Wang Ji provided prefaces for it, as did Zhou himself.[77] Wu Jiaji's fame, confirmed when his work was included in the imperial Four Treasuries collection compiled in the late eighteenth century, was thus firmly based on patronage from a circle of influential scholar-officials of the early Qing and dependent on communications within the Yangzhou area in the mid-seventeenth century.

The friendship and patronage of scholar-officials had an impact on Wu Jiaji's poetry. He was drawn intimately into literati society in Yangzhou, enjoying the company particularly of Sun Zhiwei, Sun Mo, Wang Ji, and Wang Maolin—a mixture of *yimin* and young Qing scholars, mostly of extra-local origins. Wang Maolin, like Wang Ji, was from a Huizhou family in Yangzhou, and, again like Wang Ji, served on the Ming history project in the capital. Wu was introduced into the family circles of both Wangs, especially of Wang Ji, offering poems on the wedding of a brother here, the birthday of a father there, the death of a mother else-where.[78] Farewell poems to a seemingly endless number of Huizhou people in Yangzhou attest to a wide circle of acquaintances, apparently developed in consequence of his friendship with Wang Ji. Wu's visits to Yangzhou also made him intimately acquainted with the city itself, as represented in its "famous spots and ancient sites." One consequence of this was a range of poems by him bearing on these sites—the Ruined City, the tomb of Sui Yangdi, Pingshan Hall, the tomb of Shi Kefa. In 1665, when Wang Shizhen was leaving Yangzhou, Wu was one of a company of poets who bade him farewell at a gathering at the Chanzhi Temple a couple of miles north of the city, a place already immortalized in poems by Du Mu and Su Shi, and hereafter to be remembered in association with Wang Shizhen himself.[79]

Yangzhou, then, provided a point of convergence for the *yimin* scholar and the Qing scholar-official in many senses. It was both a place where they could meet and a place that excited responses and reflections based on a common body of lore and learning and expressive of common sentiments. Wang Shizhen did the Qing dynasty a service in consorting with the *yimin*, for in so doing he brought Ming subjects firmly into the Qing domain. In effect, he did the same with the city itself, imbuing this symbol of the fall of the Ming with a living cultural significance. This achievement is most strikingly apparent in his poetry gatherings, which had the effect of inserting the Red Bridge into the lexicon of topographical references by which Yangzhou was identified as a cultural site.

Wang's activities in Yangzhou were virtually duplicated by Kong Shangren (1648–1718), his successor as scholar–official patron of poetry in Yangzhou. The two men met only in later life, when they were both in Beijing in the 1690s, but Kong had long been Wang's admirer. They were from the same province, Shandong, which made an association between them natural given their shared interest in poetry and the right circumstances for their meeting. As it was, Kong felt on his arrival in Yangzhou in 1686 as though he were walking on ground hallowed by the footprints of his illustrious predecessor. Kong's literary activities in Yangzhou closely resemble those of Wang Shizhen, a consequence, apparently, of conscious emulation.[80]

Kong, like Wang, had the sort of official position that might seem incongruent with sustained literary activities, were it not that the entire system of official appointments depended on scholarly attainments. Having won the attention of the Kangxi emperor during the emperor's first southern tour in 1684, this descendant of Confucius was subsequently appointed to a position in the river administration to help oversee the restoration of the hydraulic system in Jiangbei. This ensured that he would travel widely within the region, his official duties taking him from Yangzhou to Taizhou, Xinghua, Gaoyou, Baoying, and Yancheng, all lying fully or partly within the low-lying, flood-prone Xiahe area.[81] In consequence, he formed an even wider circle of local literati than had Wang Shizhen, although the two circles overlapped. A number of the *yimin* with whom Wang was associated were still alive, including Huizhou painters Zha Shibiao and Dai Benxiao, local poets Zong Yuanding and Zong Yuanyu, Rugao loyalist Mao Xiang, and the Nanjing painter Gong Xian.[82] Wu Jiaji died in 1684, but Kong cited him as one of his three exemplars in the art of poetry.[83]

Kong's social interaction with these *yimin*, including major literati gatherings held in 1686 and 1688, was reflected in another burst of poetic activity in the city that helped seal the city's name as the foremost center of poetry during the Kangxi period.[84] The poetry he himself composed during these years was edited by his *yimin* peers. Like Wang Shizhen, he recorded in verse the famous sites in Yangzhou: Thunder Pool, Plum Blossom Hill, Pingshan Hall, the Four-and-Twenty Bridge, the Ruined City.[85] But by this time, Qing Yangzhou itself had a history, and with a poem called "The Red Bridge Tavern," Kong took up a new cycle of historical references, commenced by Wang Shizhen:

[Red Bridge Tavern] is by the Little Qinhuai Canal. This is the place where Wang Shizhen enjoyed outings while in office. He wrote a poem called "Seductive Spring," now chanted by many.

> Of late the Red Bridge has won reknown;
> From the city men come here to slake their thirst.
> Fine are the verses the Commissioner wrote,
> And the local wine is far from the worst.[86]

In these few lines, Kong Shangren effectively historicized the Yangzhou of Wang's time. The immediate past was no longer the Ming, but the years of Kong's own childhood and youth. The more remote past he was able, with the clarity of distance, both to record and to lay to rest when he completed the work for which he is best known, the musical drama *Peach Blossom Fan*.

This dramatization of the last days of the Southern Ming court in Nanjing had been gestating in the author's mind since his youth and was finished in 1699. It was widely circulated and performed well before its publication in 1708. The Kangxi emperor himself is said to have enjoyed it, shaking his head over the plight of the Southern Ming emperor, whose cause was so badly served by conniving, morally bankrupt courtiers.[87] Yangzhou, bypassed by the emperor in his southern tour of 1684, makes a cameo appearance in the play as the site of resistance offered by Shi Kefa. Kong Shangren did not dwell on the massacre, but the steadfastness of the city in the Southern Ming cause is quietly affirmed in the concluding lines of Scene 35, uttered by Shi himself, who has been weeping tears of blood:

> Dust of battle everywhere,
> But here's a city that will not yield.
> Midnight tears from blurred old eyes
> Against a host will hold the field.[88]

Kong claimed that his interest in bringing this work to fruition had waned over time, and he completed it only on the urging of his colleagues.[89] The fact that he took up his brush again to pursue this task suggests the significance both of the lapse of time since the fall of the Ming in 1644 and the lengthening of his own years: he passed his fiftieth birthday in 1698. Facing old age, he perhaps felt free to reflect on the change of dynasty from the security of times well removed both by years and by circumstances from the chaos of the 1640s. The play opens with reflections on precisely this theme, as an *yimin* character recalls his former service in the Southern Ming's Board of Rites in Nanjing before noting how times have changed: "Now another cycle has dawned. Our ruler is supremely wise and virtuous, and his ministers are loyal and efficient. The people are quiet and contented after an uninterrupted succession of good harvests."[90]

Given the topic of the play, these lines might be interpreted as an opening gambit to neutralize possible political readings, but it was also the expression of a view widely held by Chinese scholar-officials in the late seventeenth century, who found in the restoration of order and the economy a justification for the change of dynasty. The year in which the prologue is set is 1684, the year that the Kangxi emperor made his first journey to the empire's southern realms. The preceding year, Qing forces had finally destroyed the dissident regime of Zheng Chenggong in Taiwan, thus bringing to a belated end the Southern Ming cause.

Transitions

The year 1684 is regarded as a turning point in Qing history, the year when things began to get better; the end of the period of mourning, as Hay suggests; the proximate beginning of China's "long eighteenth century."[91] This periodization places a question mark beside overt similarities between the experiences of Yangzhou by Wang Shizhen in the early 1660s and Kong Shangren in the late 1680s. Kong, surely, lived in a place more accustomed to Qing rule. In the 21 years between Wang's departure and Kong's arrival, an entire generation had come into being, and the Manchus had consolidated their hold on the empire.

In Yangzhou, the intervening decade—the 1670s—shows signs of a society in transition. Li Zongkong (*js* 1645), a local gentryman from a Shanxi family, wrote an evocative account of the city following particularly severe floods and a harsh winter in 1672. Yangzhou was the natural place of refuge for the victims.

Tens of thousands of starving people camped by the monasteries around Yangzhou, either erecting rush shelters or borrowing little boats and living along the canal. Although the governor-general, governor, and grain and salt officials all besought the merchants and people to provide rice gruel and clothing for relief, yet the snows continued deep and cold. Every day tens of people, at the very least, died of cold or starvation, and on the worst days it was more than a hundred. Countless died in the space of a month. The wasted children of the starving, small as turtle doves, pale as snow geese, dressed in rags and patches, went begging in and outside the city.[92]

When Li Zongkong set down these words, 45,000 refugees were camped outside the city in the wretched conditions he described. The salt censor ordered the salt merchants to organize relief measures. Four soup kitchens were set up outside the city walls to provide the indigents with a daily portion of rice gruel. Quantities of rice were sent to Taizhou, Gaoyou, and Xinghua, in Yangzhou's hinterland. In addition, over ten thousand lengths of cloth were to be provided. The total cost of the relief measures was estimated at 22,670 taels of silver.[93]

The first thing to be noted about the record of this disaster is that it shows the organizational capacities and financial resources of the salt administration and salt merchants being brought to bear on disaster relief, illustrating again the resilience of the salt trade and its significance in Yangzhou society. The second is that it brought home to Yangzhou the severity of the ongoing problems in its neglected hinterland, particularly the low-lying region of Xiahe. Within a few years, however, restoration of the Grand Canal–Yellow River complex commenced under the direction of Jin Fu (1633–92), who in 1677, his first year in office, initiated a major renovation of the Huai-Yang section of the Grand Canal.[94] The completion of this project by no means relieved the severe problems of water control in Yangzhou's hinterland: the opposite was alleged by one of Jin Fu's successors, who criticized the drainage system devised by Jin as responsible for continuing floods in Xiahe.[95] But it signified that the imperial government was attending to the restoration of infrastructure. In 1684, when the Kangxi emperor passed south on the first of his southern tours, he took a personal interest in the relief of Xiahe's problems and set in motion the development of a new hydraulic scheme.[96]

The 1670s was also the decade of the Sanfan (Three Feudato-
ries) rebellion, the suppression of which removed the southern
provinces from the semi-autonomous control of Chinese overlords and extended Manchu rule throughout the mainland. The
rebellion was marked by a significant cultural event in Yangzhou:
the restoration of Pingshan Hall on Shugang, both physically and
as a site of literary activity. Pingshan Hall, unlike Red Bridge,
was an established historical reference point well before the Qing.
It was built in 1048, when the great Song poet and official Ouyang Xiu was prefect of Yangzhou, and it was closely associated
with the Song poet Su Shi (1037–1101), because Su had visited
Ouyang Xiu there. Wang Maolin played a prominent role in
pressing for its reconstruction, but the Zhejiang scholar-official
Jin Zhen, who held various offices in Yangzhou, undertook the
organization of finances for the project.

According to Wei Xi (1624–81), Jin Zhen's fellow-provincial, Jin
had in mind "the moral education of the folk and the promotion
of virtuous customs."[97] Wei Xi either missed or diplomatically
overlooked the political significance of Jin Zhen's undertaking.
The Sanfan Rebellion erupted in 1673, and Jin Zhen, then acting-
prefect of Yangzhou, played a crucial role in keeping the city
calm as fighting broke out along the Yangzi.[98] The wine and poetry parties that subsequently took place at the restored premises
of Pingshan Hall might be viewed from this perspective as bread
and circuses. Nonetheless, Wei Xi's remarks on the significance of
poetry as a conservative and steadying social practice were fitting.
In a society populated by *yimin* of one sort or another, poetry was
certainly a safer option for the new dynasty than practical alternatives such as recourse to arms.

Wei Xi's record of the reconstruction of Pingshan Hall includes
by chance an impression of Yangzhou society as it was approximately a quarter of a century after the city's fall to the Manchus.
It will seem strangely familiar to anyone acquainted with the city
of the eighteenth century:

The customs of Yangzhou are, in fact, a melting pot for people from
every corner of the land. It is a concourse for the fish trade, for the
salt trade, and for money. Therefore its people have a great lust for
gain, love parties and excursions, procure singers and pursue cour-

tesans, wear fine clothes and live for the pleasure of the moment in order to show off their splendor to others.[99]

Wei Xi's record of Jin Zhen's project thus depicts a city caught between poetry and profit. We may reasonably doubt his description of the consequences of the project: "Every family and household was [soon] reciting poems, until the way of literature and of the *Classic of Poetry* gradually changed the atmosphere of money and horse trading." But poetry and profit together capture the tensions between past and future, as well as contradictions in the present; between the Yangzhou of Wang Shizhen, who served in the city when the massacre was still a living memory for all but the very young, and that of Kong Shangren, who described Yangzhou society in much the same terms as those used by Wei Xi.[100] Kong, like Wang, sponsored poets, and the literary circles of the two men overlapped. But Kong's direct contemporaries were born into a Qing world and like him encountered the Ming only through the memories and writings of their elders. In such a society, the aging *yimin* gradually faded from sight, and poetry itself lost the sharp edge evident in the writings of a Ming survivor such as Wu Jiaji. Kong sedulously cultivated the society of *yimin*, holding them in enormous esteem, but in *Peach Blossom Fan* he made a statement of hail and farewell that was appropriate to the condition of Yangzhou, and the empire, around the turn of the century.

☙

PART III

City and Hinterland

One of the great pleasures of life in Yangzhou was to take a boat out onto Baozhang Lake, or Slender West Lake (Shouxihu), as it came to be known. This popular leisure activity became a booming business in the course of the eighteenth century. The boats varied in size and design and became increasingly sophisticated over time as facilities were added for the preparation of meals. The most common was simply a converted salt boat, brought in from Taizhou after retirement from service and made over as a passenger craft.[1]

In the contrasting phases of the life cycle of a boat can be glimpsed something of the contrast between Yangzhou and much of the rest of Jiangbei. After years of service on the salt run between the coast and Yizheng, carrying great packs of salt and a few ill-clad porters, the Taizhou salt boat was often given a new lease of life on the waterways of Yangzhou. Smartly painted, its prow adorned with one or two giant characters declaring its name, it carried well-dressed gentlemen-at-leisure between the city and the lake, probably spending much time simply idling in the waters while its passengers enjoyed the great outdoors. It became an element in, as well as a symbol of, the local consumer economy that flourished on wealth from sales of the salt it had once carried.

Like the painted pleasure craft compared to the workaday salt lighter, Yangzhou appeared a prosperous and leisured place compared to its hin-

terland. It has often been assumed that Jiangbei as a whole shared in Yangzhou's eighteenth-century prosperity. Relative to rural Jiangnan, however, Jiangbei was poor and undeveloped; the salt-producing coastal areas in particular were plagued by economic and social problems. Further inland, problems in the regional water-control system adversely affected agriculture and were never resolved for long enough for anything resembling rural prosperity to develop.

As we shall see in the following chapters, the regional balance of resources created by exploitation of the salt pans was heavily in Yangzhou's favor because the salt monopoly funneled profits into the city. The jurisdictional powers of the salt administration were such that the prefectural capital even gained something from ongoing floods in the rural areas. Yangzhou's importance in the administration of water control grew as the imperial government struggled to find cost-efficient ways of keeping the floods at bay. These two aspects of the city's administrative roles in Jiangbei, the salt trade and water control, are examined in Chapters 6 and 7, respectively. In Chapter 8, we shall see that although floods and salt smugglers posed serious problems for the salt administration, the city itself was physically expanding in the second half of the eighteenth century, and the converted salt boats from Taizhou were in ever greater demand.

C3

SIX

Managing the Salt

A t one point in his famous guidebook to Yangzhou, Li
Dou takes the reader to North Willow Lane in the
northwestern corner of the New City, where the salt controller's yamen was situated:

In North Willow Lane . . . is Master Dong's Temple. Originally this
was the Zhengyi Academy, but in the Zhengde reign of the Ming
Dynasty, it was changed to the Zhengyi Temple, in honor of Dong
Zhongshu, chancellor during the Han dynasty. Here also is held a
copy of [Dong's] *The Spring and Autumn Collection*. During the present dynasty, Emperor Shengzu bestowed on it the phrase "The Illu-
minated Way of Zhengyi." Since then the shrine has been known as
Master Dong's Temple.[1]

Dong Zhongshu (ca. 179–ca. 104), official and Confucian philosopher of the Former Han dynasty, served briefly as chancellor
of the kingdom of Jiangdu, and the temple named in his honor
was constructed on the supposed site of his residence.[2] The imperial dedication to the shrine was made in 1705, on the occasion of
the Kangxi emperor's fifth visit to Yangzhou. Well might the emperor of a still young dynasty have thought to make this gesture:
Dong was the greatest contributor to a political philosophy that
explained the success or failure of particular reigns in terms of

the ethical conduct of sovereigns and reconciled the vicissitudes of the empire with the movements of heaven and earth.[3]

This monument to Confucian cosmology and ethics was adjacent to the yamen of the salt controller, which occupied several blocks east of North Willow Lane and was the largest and most powerful governmental institution in Yangzhou.[4] Yangzhou had a high ranking within the field administration: as a prefectural capital, it was ranked as a "most important" post, the highest of the four official post categories.[5] But the prefect was unimportant next to the salt controller (*yanyunshi*), who in practice managed the Lianghuai salt monopoly and increasingly other regional concerns as well.

As administrative center of the monopoly, Yangzhou was locked into a relationship with the salt supply and distribution areas of its own hinterland, an association that affected both city and region. In the second half of the seventeenth century, as an enlarged empire was gradually brought under the firm control of the dynasty, the reaffirmation of the monopoly structure established in the late Ming allowed the ruined city of the "ten days" to resume and even enhance its place among lower Yangzi cities as a center of commercial and cultural activity. In a demonstration of the city's relative independence of the rural economy, however, the rehabilitation of the region was much slower. The city grew great on salt, and the salt administration defined the city's most important relations with the region.

Monarchs and Merchants

Cui Hua, who was appointed salt controller in 1693, intimated the significance of the salt monopoly in the regional economy with an unabashed claim for the importance of the salt trade in his preface to the 1693 edition of the *Gazetteer of the Lianghuai Salt Regulations*:

Now among the profits from the mountains and marshes, the tax from salt is the greatest; and the Lianghuai salt tax amounts to half that from all the [salt] administrations in the empire. Through the Han, Jin, Tang, Song, Yuan, and Ming dynasties, although the salt levy has changed, it has never failed to supply the country, benefit the people, profit the merchants, and assist the salters.[6]

This unequivocal statement of the salt trade as central to the workings of the empire is consistent with the earlier claims of Wang Daokun (1525–93), who famously extolled the virtues of merchants.[7] Wang was a native of Shexian, the home of so many Lianghuai salt merchants, and his worldview was produced by the same socioeconomic changes that gave rise to the merchant-gentry families of late Ming Yangzhou in the sixteenth century. But whereas Wang's statements were made in his private writings, Cui's appeared in a semi-official publication. In the introduction, Cui Hua justified the compilation of the gazetteer by drawing a direct analogy with district gazetteers:

The glories of the empire can never adequately be recorded, but each province, prefecture, and so on has a written record of its geography, customs, population, tribute and taxation, establishment, and changes from beginning to end, so as to enable later generations to deliberate on the past and make adjustments for the present. . . . After the present dynasty came to the throne, the reforms were thorough, all the regulations were put in order, and only in a Lianghuai salt gazetteer did they fail to appear.[8]

The organization of the gazetteer echoed that of a local gazetteer with some fidelity.[9] Chapters on cosmology, territorial boundaries, officials, official buildings, examination successes, memorials, local customs, ancient sites, and biographies are arranged in the conventional way; the biographies are divided into sections featuring local exemplars of philosophy, filiality, administration, philanthropy, uprightness, and proper female behavior. The usual chapter on local products is devoted entirely to salt, and the final chapter on changes over time (*yan'ge*) is preceded, again conventionally, by four chapters of miscellaneous writings, one of which has poems by, among others, Zhou Lianggong, Wang Shizhen, Kong Shangren, and Wu Jiaji. With this gazetteer, the Lianghuai salt administration made the strongest of claims that the salt trade was equivalent to the land in its significance for the economic, cultural, and ethical workings of the empire. Its format suggests that the dynastic transition had encouraged or even accelerated the gentrification of the salt trade.

Yang Lien-sheng has noted the particular reliance by another northern dynasty, the Mongol rulers of the Yuan, "on the huge profits brought by Uighur and Chinese merchants."[10] Alien dynasties were likely to be reliant on merchants not only for funds but for active or tacit support of the dynasty itself. This dependency, S. N. Eisenstadt hypothesizes, arises because merchants are "free floating resources" on which rulers can draw, whereas landowners are "embedded resources," less amenable to recruitment by extra-local forces.[11] In keeping with this theory, merchants were arguably the most compliant social group in the Chinese empire.

One way the Mongols wooed salt merchants was to offer them easy entry into the examination system by permitting them to sit the first examination in the locality where they were in business. This was effected through the introduction of "transport registration" (*yunji*), subsequently known as "merchant registration" (*shangji*), for salt merchant examination candidates.[12] For a long time in Yangzhou, merchant registration benefited only the West merchants, because it did not apply to merchants operating within their home province. (Huizhou, the other source of Lianghuai salt merchants, belonged to the same province as Yangzhou until the division of Anhui and Jiangsu in 1667.) The West merchants clung to this advantage and objected furiously to the inclusion of the sons of Huizhou merchants when a school for the offspring of salt merchants and salters was established in Yangzhou in 1677.[13]

The examination system provided a means for the imperial government to bond the salt merchants to itself. This bond was maintained and strengthened under the Qing. Early in the dynasty, the court was especially dependent on the salt tax for revenue. In vast areas of China, including Yangzhou's hinterland, agriculture was in crisis, and extensive programs of rehabilitation would be required before the land tax generated substantial income. To finance its campaign against the Sanfan rebellion, the imperial government drew on the salt tax and contributions from the Lianghuai salt merchants.[14]

During the Qing's "long" eighteenth century, the relationship between dynasty and salt merchants was cemented. In 1689, Yangzhou experienced its first direct encounter with the new dynasty when the Kangxi emperor, on his second tour to the south, visited the city. Four imperial visits followed. The emperor was feted by the salt merchants on an increasingly elaborate scale: antiques and paintings were presented; operas were performed by troupes owned by the salt merchants; banquets were given.[15] In 1705, an entire palace was constructed on the grounds of the Gaomin temple, south of Yangzhou.[16] Half a century later, the Qianlong emperor emulated his grandfather, undertaking six imperial tours to the south and enjoying even more lavish displays of hospitality. These tours enabled the most powerful merchants to establish personal relationships with the emperor, whom some of them entertained in their lavish gardens. They were rewarded for their efforts with prestigious although empty titles such as "superintendent of the palace stud" and "superintendent of imperial gardens and hunting parks," and their gardens were graced with titles bestowed by the emperor.[17]

The imperial visits created a firm bond between the dynasty and the salt merchants and generated a system of mutual support that lasted into the early nineteenth century. The court was highly dependent on the salt trade both for taxes and for incidental funds and had to foster good relations with the merchants. Although under the Qing the central government's income from the land tax remained fairly steady despite the growth in population, the contribution of the salt tax climbed from 8.87 percent of total revenues in 1682 to 11.83 percent in 1766.[18] For extraordinary expenses involving military or hydraulic undertakings, the salt merchants were a ready resource and could be marshaled into "returning the imperial grace" (baoxiao) by making large contributions to the imperial coffers.[19] For their part, the merchants worked closely with the metropolitan government. The Imperial Household Department in Beijing, which was responsible for the costly task of provisioning the Forbidden City, lent out sums of between a few thousand and a million taels of silver to the merchants, which they invested with profit both to themselves and to their creditors, repaying the loans at 10 percent interest.[20]

Monopoly Merchants and Their Money

The wealth and power of salt merchants under the Qing were products in part of the reorganization of the salt monopoly in the late Ming. The old division of frontier, inland, and waterway merchants gave way to new categories under the shipment system introduced in 1617. The frontier merchants became redundant and disappeared. The waterway merchants, unable to compete in the restructured monopoly of 1617, were reduced to the status of small-scale distributors in Jiangxi and Huguang. The inland merchants ceased to be known as such and re-emerged to view as transport merchants (*yunshang*).

In the new shipment system, a quota of salt was set for each port, and the Board of Population and Revenue issued licenses for that amount of salt to merchants with sufficient capital to buy and transport the salt. To engage in the trade, a merchant had to provide annual evidence of his established place in the monopoly (*genwo*, lit. "basic nest") with a certificate (*wodan*, lit. "nest certificate"). (The word or perhaps the homophonically used character for "nest" [*wo*] in this context appears to have been derived from the Shanxi dialect; it is glossed by Wang Zhenzhong as "an official vacancy or position" [*gongchai kongque huo weizhi*].)[21] Each year, at the start of the transport season, the transport merchant reported to the salt controller in Yangzhou with his certificate and license in hand and was checked off the register; the controller then issued a license specifying the amount of salt and the destination of the shipment. This was known as *nianwo* (lit. "yearly nest").[22]

Brokerage was common in the salt trade, and some registered merchants were merchants in name only. They invested in the salt trade but leased their rights to "business merchants" (*yeshang*). Large numbers of unregistered salt merchants were involved in the day-to-day affairs of trade. These included the so-called leasehold merchants (*zushang*), who leased trading rights from the business merchants; and the proxy merchants (*daishang*), who were employed by licensed merchants to manage their business.[23]

The transport merchants were also distinguished from their close associates, the yard merchants (*changshang*). Yard merchants either operated independently or were employed by the transport merchants as the on-site buyers of salt at the salt yards. Some yard merchants owned salterns, originally the hereditary possession of salters (*zaohu*; see below).[24] A significant factor differentiating the yard merchants from the salters was native place. The merchants were of Huizhou or, more rarely, Shan-Shaan registration or descent. All the salters were natives of Tongzhou, Taizhou, Shanyang (Huai'an), or Yancheng, to name only the relevant jurisdictions within the Huainan salt-production sector.

The transport merchants were the great merchants of the time and were thought to possess individual fortunes amounting to millions of taels of silver.[25] Some were greater than others. In 1677, the salt controller appointed 24 head merchants (*zongshang*); the remainder, described as miscellaneous merchants (*sanshang*), were entered in the shipment registry (*gangce*) under the name of one of the 24 head merchants. This initiative was aimed at relieving the difficulties then being experienced by the imperial government in ensuring a flow of revenue to combat the Sanfan rebellion.[26] By making the most powerful merchants responsible for the rest, the salt administration passed on the burden of tax collection and lessened the number of merchants with whom it had to deal. The number of head merchants fluctuated over time: it increased to 30 during the Yongzheng reign but was reduced to 25 in the late eighteenth century and then to 16 in the early nineteenth.[27]

Throughout this period, the head merchants were the elite of the Yangzhou merchants. Among early holders of the post were Cheng Liangru and his son, Zhiying. The father demonstrated his abilities early in the 1660s by organizing the distribution of delayed salt sales over a number of years and, in 1665, by repairing the Fangong Dike. The son was a head merchant for twenty years and led the initiative to raise funds for the final suppression of the Sanfan Rebellion in 1681.[28] Zhiying's grandson, Cheng Mengxing, a *jinshi* of 1712 and sometime Hanlin compiler, retired from office to take over the family salt business. He created one of the great gardens of early eighteenth-century Yangzhou, the Bamboo Garden, and was a central figure in literary circles during the late Kangxi and Yongzheng periods.[29] Cheng Mengxing's contemporary, Wang Yinggeng, was the most prominent of the head merchants in the first half of the eighteenth century and took the lead in philanthropic and morally improving undertakings, such as disaster relief and the promotion of widow fidelity.[30] Both Cheng Mengxing and Wang Yinggeng produced anthologies on Yangzhou.[31]

Under the Qianlong emperor, further centralization of the salt merchant body was effected with the selection of one of the head merchants as principal merchant (*shouzong*), a position created some time before 1768 and apparently held for three years at a

time.[32] The most noted principal merchant was Jiang Chun (1721–
89), who probably held the position at various points in his long
career. Jiang was a poet, archer, and opera enthusiast, and a
dominant presence in Yangzhou society in the second half of the
eighteenth century. He took the initiative in developing regulations for the first visit by the Qianlong emperor in 1751[33] and
played host to the emperor in one or the other of his gardens on
four occasions between 1762 and 1784. In a career spanning fifty
years, he distinguished himself by spending a claimed one millions taels of silver on "disaster relief, waterworks, and military
needs."[34] In his old age, he was left impoverished, and the Imperial Household Department intervened to relieve his impecu-
nity.[35] Among later principal merchants were father and son Bao
Zhidao (1743–1801) and Bao Shufang, who were among the most
active members of Yangzhou society in the late eighteenth and
early nineteenth centuries, and Huang Zhiyun, owner of the famous and still extant Ge Garden, a controversial figure whose
abuse of his position attracted high-level criticism during an investigation of salt sales in 1822.[36]

This list of names includes many of the dominant figures in
Yangzhou society over a period of some 150 years. To them can
be added the merchant owners of private gardens known to have
been visited by the Qianlong emperor, all of whom must have
been head merchants.[37] Property owners, philanthropists, connoisseurs of the arts, and occasionally writers, these men, most of
whom were from Huizhou, assumed responsibility to varying
degrees for the upkeep and improvement of the urban infrastructure and of the regional water control and famine relief systems.
They stood at the junction between the Lianghuai salt merchant
body and the imperial government. One of the major roles of the
principal merchant was to organize merchant contributions to the
court, the "return of imperial grace," which in fact greatly increased in value after the institution of principal merchant was
created.[38]

The size of the merchant body is given only in very general
numbers. Tao Zhu (1779–1839) wrote in 1831 that there had once
been "several hundred" transport merchants, as did Li Cheng,
writing just a few years earlier.[39] The uncertainty of their numbers makes it difficult to gauge the wealth of individual mer-

chants and the profitability of the salt trade to any one of them. On the basis of such figures as are available, speculations can be offered. In 1726, total taxable salt from Lianghuai salt yards amounted to some 450 million catties. A rough estimate of the total sale price can be calculated from the wholesale price in Hankou in 1723, which was around 0.13 taels per packet of 8.25 catties.[40] This is equivalent to a wholesale gross of over seven million taels of silver. Taxes took some two million of this, and after business and other costs (wages, packaging, transport, levies), it is difficult to see how more than four million could have been retained as surplus. Wang Fangzhong estimates the actual profit (not including the retained value of the capital investment) as about half the sale price, which would be 3.5 million.[41]

If there were, generously, around three hundred transport merchants in Huaibei and Huainan at this time, the average profit from the 1726 sales would have been less than 12,000 taels per annum. The stipend for salt officials in Lianghuai, by way of comparison, ranged between 400 and 3,000 taels (see Appendix B). In 1748, the painter Zheng Banqiao earned around a thousand taels per year from the sale of his works.[42] Stipends for the 260 registered students at the two Yangzhou academies supported by the salt monopoly totaled a little over 6,000 taels per annum in the 1790s.[43] By these standards, an "average" profit would have provided a merchant with a substantial income, but one that would have required careful husbanding indeed to grow to the vaunted capital of millions of taels said to have been possessed by the wealthiest merchants.

Probably there was no such thing as an average profit. In the early nineteenth century, the "big merchants" from Yangzhou controlled 60–70 percent of the Hankou trade and the "small merchants" the remainder.[44] If these proportions held true for the preceding century, then in the 1720s perhaps upward of two million taels of profits was shared between the 30 head merchants (assuming these accounted for most of the "big merchants"), the remainder going to the more modest support of the *sanshang*. By this reasoning, the wealthiest merchants in the 1720s may have earned profits of between 50,000 and 100,000 taels per annum.

To their legal earnings from salt sales, merchants added profits from usury,[45] excess sales, and smuggling. Millions could be earned through a knowledgeable exploitation of the official salt trade. In 1768 a major scandal erupted in Yangzhou when a newly appointed salt controller found, on investigating the books, that ten million taels in profits over the preceding twenty years had gone unreported.[46] This case is well known because it resulted in death sentences for the former salt controller, Lu Jianzeng (1690–1768), and the former salt censor, Gao Heng (d. 1765) — who was the brother of an imperial concubine.[47] In all likelihood, however, the peculation was probably only a sample of the creative accounting employed to maximize profits more or less within the boundaries of the salt monopoly's regular workings.

Many active salt merchants (as opposed to the "business merchants") also traded in other commodities. A notification of a breach of customs duties in 1740 reveals that a vessel belonging to a Huizhou merchant from Xiuning had been forced back to the customs port in Jiangxi to make good on unpaid duties on a boatload of salted fish from Yancheng. The unfortunate merchant lost his entire cargo when a great storm blew up and sank the ship in port.[48] Yancheng was in Huainan salt production zone, and salted fish was a local product. The merchant was almost certainly running a sideline to his salt business.

A more significant sideline was grain. Hankou, the wholesale market for salt bound for Hunan and Hubei, was also the point of distribution for rice from Hunan and Sichuan, much of it bound for the cotton- and tea-producing, grain-deficit areas of the lower Yangzi region. The salt carriers did not return empty from Hankou. In 1731 the Huguang governor-general reported: "Investigating the Hankou area, I find that between the eleventh month of last year and early in the second month of this, rice boats of outside traders (*waifan michuan*) already number more than four hundred, and the number of great vessels of salt merchants engaged in transporting rice can hardly be calculated."[49]

Even taking into account all these ways of making money, those with great fortunes may have been relatively few in number. Whether very rich or moderately rich, however, salt merchants had a profound impact on the Yangzhou society and economy. It has been estimated that at the height of the Lianghuai salt monopoly's fortunes in the eighteenth century, those dependent on the salt trade in Yangzhou included "officials, merchants, clerks, gentry (*shidafu*), and boat haulers [hired] from around the city, and without doubt numbered in the tens of thousands."[50] The urban population, no doubt growing throughout this period, may have exceeded 100,000 by the end of the century, but probably not by much.[51] On this estimate, the salt sector was a pervasive presence indeed within the city.

To these "tens of thousands" must be added all those who made a living by selling services and commodities to them, from the artists whom they patronized to the fishers and farmers who supplied the city's food. Fisherman daily traveled from distances of up to 40 miles to supply the thriving fish markets, located at Huangjin embankment northeast of the city, and thence the restaurants that proliferated in the city during the eighteenth century. Salted and dried fish and a variety of other fresh- and saltwater products—water chestnuts, lotus root, shrimps, crabs, oysters—were brought to the markets from the coast and the Huai and the Yangzi rivers.[52] Wine and spirits were supplied to the city from numerous villages within Yangzhou's home districts by "an endless stream of donkeys and carts," as well as by boats from further afield in Tongzhou, Taizhou, Gaoyou, and Baoying.[53] When a salt merchant set foot in Yangzhou, the ripples from his wealth spread out in ever-abating circles before finally reaching the salters on the coast. The concentration of effects was, however, greater at the center than at the periphery.

Merchants and Salters

The other side to the story of Yangzhou and its merchants is to be found in the salt yards. The salt monopoly was the dominant form of wealth-generating economic activity in Yangzhou prefecture and adjacent coastal counties, and hundreds of thousands of people were involved in it in one way or another.[54] Huainan salt was produced along the coast of Jiangbei, the production zone being defined by borders that extended along either side of the Fangong Dike, which ran practically the full length of the Jiangbei coast. The zone passed through a number of county-level jurisdictions (four in the early eighteenth century, six by its end) and was internally divided into a number of yards (*chang*). There were 26 yards at the beginning of the eighteenth century but, because of amalgamations, only 20 by its end.[55] They varied in size but together covered an area of some 4,500 square *li*, over a thousand square miles.[56]

Much of the salt zone was given over to the production of the rushes that were used as fuel for the salt boilers. Both the rights to decoct salt and the rights to the rushlands were hereditary—although by the early Qing not inalienable—among salter people (*zaohu, zaoding*), who in the early Ming were recruited variously from the local populace, the criminal classes, and enemies of Zhu Yuanzhang, the first Ming emperor. In the fourteenth century, the Huainan coast had been among the first areas to raise the standard against the Mongols, a great rebellion being fomented under the leadership of salt worker Zhang Shicheng. Afterward Zhang came to a settlement with the Mongols. When Zhu Yuanzhang won the war and became emperor, he took punitive action against the Jiangnan families who had sided with Zhang: in a massive relocation program, they were uprooted from their homelands, transported to Shandong, given the hereditary registration of salter, and left to a lifetime of boiling salt. The same action was undertaken in Guangdong, where families were exiled to Huainan.[57]

It proved difficult to keep the salters at work. Unlike the peasants, who returned to their land after a crisis abated, salters took advantage of wars or floods or famines to run far away and never return. In the sixteenth century a sequence of floods, droughts, and epidemics reduced the salter population by more than half, either through death or flight. They were replaced by people who had broken the salt laws, along with criminals condemned to hard labor, exile, or death, these sentences being commuted to life in the salt industry.[58] Although salters—except, of course, for these condemned men—could climb the ladder of social success via the examination system, it seems probable that the odor of social marginality, buttressed by impressions of the poverty and lawlessness of the salt yards, clung to the image of the coast dwellers from generation to generation. In the words of one Shexian merchant, "The salters of times past were criminals; the salters of the present are rogues, sly in behavior with the strength of tigers, committing villainies without fear of reprimand from officials."[59]

Fig. 8 Boiling up the salt as depicted in 1693 (*left*) and 1806 (*right*) (SOURCES: prefatory illustrations in *KXLHYFZ, JQLHYFZ*).

Even the writer of these lines, however, could see that "sometimes [the salters] have no clothes or no food; sometimes their children have to be married off; sometimes they have to boil the salt with no fuel; sometimes they have to drop the price, and owe money to officials."[60] The illustrations in the 1693 salt gazetteer bear out the hardships of the salt makers, who are depicted as emaciated men, ribs in evidence; barely clad women and elders; children lingering around the work site. The unknown artist had a keen eye for social detail, in striking contrast with a successor, who, for the 1806 edition of the gazetteer, produced formulaic, schematic representations of the same stages of salt production (see Fig. 8). The contrast with depictions of salt merchants, painted in their gardens or studies — well-fed, well-clothed, at leisure — is even more startling (cf. Fig. 22, p. 254).

The salt merchant and the salt maker moved in different worlds. Plenty and penury, leisure and labor, divided them. The salters' work, as explained in detail in the salt gazetteers, involved many tasks: cutting the rushes for the fuel; walling off an area of coastal tideland rich in brine, which would come to maturity after a year; spreading ashes on the surface of the salt field and waiting for the salt crystals to appear (between the first and fifth lunar months); sweeping the whitened mass together with the ashes and trampling it to drain off the purified brine into wells; waiting for the salt to congeal, then shoveling it into the salt cauldrons after adding brine; boiling up over a period of twelve hours till good-quality white salt was produced; packing and transporting the salt by boat (or in Tongzhou by cart) to the yard depot; sifting to separate the coarse from the fine or edible salt; and finally packing the latter into rush bags, weighing it, and sending it on its way. The whole process was watched over closely and recorded by officials.[61]

The 450 million catties of salt that gave a few hundred salt merchants seven million taels or so before costs yielded only two or three cash per catty at the production site, or less than 1.5 million taels to be divided among the many thousands of salt workers.[62] Salt production alone was not sufficient to support them: fishing, farming, and salt smuggling were among their other means of support. Edicts prohibiting the cultivation of other crops on the reed lands suggest that salters were probably turning to agriculture to feed themselves.[63] Although from the middle of the eighteenth century the salters were able to mortgage or sell the grasslands, the exchange was limited to a transfer of land between salter groups, and the use of the land was still restricted to fuel production.[64] The number of salters in fact rapidly increased during the eighteenth century, outstripping the industry's need for labor. These people appear to have been condemned to residence on the coast and non-farming occupations in 1776 when the salt censor refused to re-register surplus *zao* (salters) as *min* (common people).[65]

This is not to suggest that the peasant's lot in Jiangbei was much superior to the salter's. Life east and west of the Fangong Dike in Dongtai county, or in other words the life of the *zao* and the *min*, was described in a local gazetteer as follows:

The people of the villages in the western part are all farmers, and with the incessant floods and droughts very few were able to put aside anything. They had no choice but to live simply. The salt workers in the yards draw their livelihood from boiling up the seawater. They have permanent employment but are not always working. They like to shirk and hate engaging in trade. They think nothing of wandering away elsewhere, drinking and eating and wearing out one another's hospitality.[66]

Apparently money flowed through the salt yards in a way unknown among the farming communities further inland. Of the salters actually engaged in the salt industry, however, only the salt masters, who owned the boilers, had any opportunity to amass wealth other than through smuggling. But they steadily lost their position under the Qing. During the dynastic transition, salt merchants moved in to buy up the salterns and began employing hired labor in place of the salters.[67] This was in part a consequence of the abandonment of the salt yards by the salters in the middle of the seventeenth century: in one yard, all the original salters are said to have disappeared by 1679.[68] By the early nineteenth century, fully half the salterns were owned by salt merchants.[69]

There were tensions between the salt masters — local people with a *zao* registration — and the merchants, who complained that they were being squeezed by the high prices charged by the *zaohu*. The countercharge, that the merchants forced the salt masters to borrow at high interest, appears a more likely scenario.[70] This was included among "seven bitternesses" of the salter's life. The others, as recorded in a local ballad, were the poor quality of their housing and food, the gathering of fuel, the straining of the brine, the boiling up of the salt, meeting of production quotas, and coastal floods.[71] The limited resources of the salt masters compared to those available to merchants became apparent when a restructuring of the salt monopoly was being discussed in the 1830s. The suggestion that all salt taxes should be levied at the site of production was rejected on the basis that the salters were "poor people."[72]

Salt Officials

Yangzhou, situated far from the coast, was intimately connected with it by administrative arrangements developed to protect the imperial monopoly on the salt trade. The Lianghuai salt administration was an extremely powerful sector of the bureaucracy, with ample resources, and incumbencies were much sought after. Shōjirō Takino has pointed out that the complex of high offices in

Jiangnan at the time of the Lu Jianzeng case shows Yangzhou in the middle of the eighteenth century to have been virtually a sinecure of the Gao family. The ill-fated Gao Heng was salt censor at the time of the scam. His father held the post of director-general of river conservancy for Jiangnan in 1748–53, and a cousin succeeded to the same office in 1761 before becoming governor-general of Liangjiang, a post that, as of 1731, included such duties as the director-generalship of the Lianghuai Salt Administration. Gao Heng was tried in 1768 under the authority of his own cousin, his administrative superior in two respects. Heng's brother held posts as inspector at the Shanghai Customs and as salt controller in Hedong. Another male relative of the same generation variously held office as salt controller in Hedong and Changlu and as supervisor of silk manufacture in Hangzhou.[73]

The authority of the senior salt officials was grounded not only in the office itself but in the position of the incumbent in the network of bureaucratic and social transactions that determined the hierarchy of power in Qing China. Lianghuai salt censors were drawn from the relatively select group of Manchu and Chinese bannermen, such as the Gao family. In keeping with the Manchus' dyarchic balance of bureaucratic power, a non-banner Chinese, such as Lu Jianzeng, was usually appointed to the position of salt controller. Lu was also well-connected, and family members and relatives by marriage occupied prominent places in the bureaucracy.[74]

The salt censor (*xunyan yushi*) was the senior salt official in Yangzhou, with general responsibilities for overseeing the operation of the Lianghuai monopoly. His office was at the interface between the salt administration and the field administration. He looked outward from Yangzhou, overseeing the distribution of salt to the six provinces within the Lianghuai salt-marketing area and reporting the completion of each salt transport. To him also fell the duty of negotiating arrangements for the control of salt smuggling with military and civil officials.[75] The location of his yamen in the Old City, where the offices of the field administration were also located, was consistent with this role. The post of salt censor was ungraded, and the appointee retained his previous official rank,[76] but since the salt controller's office was graded at 3b—above that of a prefect (4b) but below that of a governor

(2b) — the salt censor would normally have been drawn from a high-ranking office; otherwise he would have been in the embarrassing position of having to bow lower to the salt controller, who was technically his subordinate.

The imperial government played a balancing act between the salt censor and the salt controller: fluctuations in their stipends show that the office of salt controller was downgraded in the late eighteenth century relative to that of the salt censor (see Appendix B). For the greater part of the eighteenth century, however, the salt controller was in practice the most powerful official in city and prefecture alike. His yamen sprawled over several blocks of the New City and housed, in addition to himself, four subordinate ranked officials, a staff of ten head clerks, and a mass of supernumeraries. In the early Qing, there were ten bureaus in the salt administration, functionally divided more or less in imitation of the imperial government — appointments, population, rites, works, and so on. Over time their number increased as new bureaus were created to handle the overflow of work from the original offices, in particular from that of the treasury, and to manage new areas of responsibility. By the 1820s, there were twenty in all.[77]

The duties of the salt controller involved the close surveillance of salt workers and salt production, scrutiny of weights and sales of salt, the speedy remittance of the salt taxes, reporting on backlogs of salt, and, most important, the licensing and supervision of salt merchants.[78] In the course of the Qing dynasty, the controller came to play a central role in the management of both urban and rural affairs in Jiangbei, because his relationship with the merchants and his access to revenue from the salt trade allowed him to usurp positions normally occupied by senior officials in the field administration. This was particularly evident in hydraulic management, a subject considered in the following chapter. The medical dispensary adjacent to the salt-controller's yamen was both a sign of the benevolence of the bureaucracy and a reminder of the importance of the salt administration in Yangzhou.

Outside Yangzhou and Yizheng, the salt administration was most strongly concentrated in the salt production zone. Here spatial organization ran counter to the normal county, departmental, and prefectural divisions, and the salt administration impinged on the workings of the civil administration. Jurisdiction over the salt yards was divided in a rather uncertain fashion between civil and salt officials. Each yard had an administrative center distinct from the district-level capital, and in some cases — Yudong, Dongtai, Shigang — these were, or had been, contained within walls in the manner of a district capital.[79] The administrative center was physically defined by the presence of a salt receiver (*yankesi dashi*), one to each yard. The office of salt receiver was originally unranked, but in 1728 it was graded at 8a, equivalent in rank to the post of assistant county magistrate (*xiancheng*).[80]

Just as districts were grouped under the jurisdiction of a prefect or his equivalent, so, too, were the yards grouped under a sub-assistant salt controller (*yanyunsi yunpan* or *fensi*) who, with a rank of 6b, was equivalent in status to a first-class assistant department magistrate (*zhoutong*). The original 30 Lianghuai yards had been divided under the authority of the Huai'an, Taizhou, and Tongzhou subcontrollers. This division accorded fairly closely in spatial terms with existing prefectural, departmental, or district divisions.[81] In 1732, the administration at this level was reorganized by restricting the jurisdiction of the Huai'an subcontroller to the Huaibei salt yards and dividing the Huainan salt yards between the Taizhou and the Tongzhou subcontrollers.[82] From this time on, there was a considerable territorial overlap between salt and civil interests along the coast. Salt officials based in one district or prefecture now exercised a jurisdiction of sorts over salt workers living within the borders of another; civil jurisdiction over commoners living within the borders of salt yards that crossed the border of two district-level jurisdictions was either divided between the relevant local magistrates, as in the case of Caoyan, which "belonged to (*li*) the area of the two counties of Dongtai and Xinghua," or devolved wholly on one of the two, as in the case of Liuzhuang, which "lay between the two cities of Dongtai and Xinghua [but] was administered as one by Xinghua."[83] These changes had the effect of dispersing powers of jurisdiction and reducing the opportunities for collusion between salt and field officials.

The lowly status of the salt receiver, the only salt official in any salt yard, appears anomalous in the context of an industry bedeviled by the problem of smuggling. The official presence in the coastal districts was low relative to the vast size of the county ju-risdictions. Funing, Yancheng, and Dongtai counties were geo-graphically the largest in Jiangsu, being from two to ten times greater in area than most other counties in the province.[84] Before the establishment of Funing and Dongtai in the 1760s, the older ju-risdictions of Shanyang and Taizhou, from which they were cre-ated, were even larger, and the salt yards mostly lay at great dis-tances from the yamen of the local magistrate. Crime and disorder were rampant. "Among the salters in the yards," complained one official, "there are many instances of gambling, licentiousness, banditry, assault and battery, brawling, and the like."[85] In the early Qianlong reign, river and salt officials were deployed to compen-sate for the dearth of officials in the field administration. The first-class subprefect for water control appointed to the Dongtai salt yard in 1736 not only had to control the waters but also to "inves-tigate bandits and apprehend smugglers."[86] In 1746, salt receivers were accorded slightly lesser powers to "prohibit and restrain" criminals and send them to the local magistrate for trial.[87] Crime and disorder were again implicitly invoked in 1767 when Liang-jiang Governor-General Gao Jin (1707–79), arguing for the estab-lishment of a new county seat at Dongtai, pointed to the distance between Taizhou and the coast and used an old but telling apho-rism to describe the problem of administrative reach: "The whip is long, but it does not reach [the target]" (*bian chang mo ji*).[88] Salt smuggling was the foremost challenge to officials in these counties.

Smugglers

Despite the state's appropriation of the salt industry and its con-trol of salt marketing, salt was a significant item of local trade within Jiangbei. Most of the Jiangbei region was excluded from the official marketing area because it was too difficult to control smuggling in areas close to the salt production zone. This meant that salt was widely peddled in relatively small quantities throughout Jiangbei.[89] Larger quantities were smuggled out of the region for sale at competitive prices elsewhere.[90]

For salters, the private sale of salt was the one obvious means of making money. Officials faced difficulties in stemming illegal flows of salt from the salt yards for various reasons: the population was increasing while the number of ranked officials remained fairly static; large numbers of fishermen and boats both on the coast and in the interior ensured that mobility was high; the span of administrative control was too great to be really effective; returns from ordinary means of employment such as salt boiling, fishing, and farming were low; and the influence of or leadership provided by a local gentry was negligible. In the 1820s Li Cheng alleged that 30–40 percent of the illegal salt traded within the Lianghuai salt supply area came from the Lianghuai salt yards. The problem was not new to his time. In 1737, the salt censor requested permission to raise funds to purchase surplus salt from the salterns so as to limit the amount of salt available to smugglers.[91] In 1745, the production capacity of the Huianan yards was enhanced by the addition of 27 salt cauldrons, on the grounds that the amount of salt being produced was not enough to meet the quota.[92] At the same time, unsold official salt was piling up in nearby ports, suggesting that salt siphoned off at the site of production was reducing the demand for taxed salt and at the same time leaving not enough salt in the yards to fill later quotas.

The volume of salt controlled by licensed merchants in Jiangbei was actually quite small at the best of times, for in most counties salt was legally marketed in small quantities on a tax-free basis by peddlers. In the southern Huai'an, Yangzhou, and Tongzhou jurisdictions, quota salt was originally delivered only to Jiangdu, Ganquan, and Shanyang—the home counties of Yangzhou and Huai'an. The remaining departments and counties were considered as lying within the salt production area or too close to it for the sale of taxed salt to be considered a viable proposition. Yizheng was included in the peddled salt region for practical reasons: so much salt was spilled in weighing and repackaging that any attempt to impose quota salt on the county would simply be undermined by the private trade in waste salt.[93]

The peddler system allowed the poor and needy of the coastal districts to buy small quantities of salt from the salterns for sale to the untaxed counties in Jiangbei. In 1736 an attempt was made to streamline this system, and peddling rights were limited to old men over sixty, young boys under fifteen, young men with disabilities, old women, and orphans without means. On presenting

themselves to the relevant district yamen, applicants who satis-
fied the requirements were to be issued a stamped wooden tally
that licensed them to purchase up to 40 catties of salt a day from
the salt yards. This was known as the *laoshao* (old and young) system.[94] Almost from its inception, however, the age requirements
were abandoned, at least in the Lianghuai zone.[95]

Within the Huainan quota salt–supply area, Yangzhou's home
counties were regarded as the first line of defense against smuggled salt. They were categorized as local or "consumer" ports
(*shi'an*) and served as a barrier (*fanli*) or buffer zone to protect the
more distant "shipment salt" ports that received shipment salt
(*gang'an*). "Consumer" salt was cheaper than "shipment" salt,
since it was more lightly taxed, but it was more expensive than
salt in the neighboring peddler areas and difficult to sell if challenged by competition from the latter.[96] The vulnerability of these
two counties to the entry of smuggled salt was pronounced because they were virtually surrounded by peddler districts: Taixing and Taizhou to the east, Gaoyou to the north, and Yizheng
to the west. There was, moreover, a constant stream of traffic
passing along the Grand Canal, including the tribute grain vessels, which were notorious as carriers of smuggled salt.[97]

Despite this combination of unfavorable circumstances, the
quota of 26,710 *yin* for Jiangdu and Ganquan appears to have remained viable during the early part of the dynasty: if occasional
adjustments had to be made to dispose of unsold salt, this was in
keeping with the situation at other ports where a flood or a famine was likely to disrupt official salt sales in particular years. Not
until the Qianlong reign did there emerge serious, perennial difficulties in disposing of the quota. Efforts to control the trade in illicit salt in Jiangdu and Ganquan, together with the accumulation
of stocks of consumer salt in Yangzhou city, show that the private
salt trade expanded steadily over the eighteenth century.

Early in the dynasty, surveillance of transport routes within
Jiangbei appears to have functioned on an ad hoc basis; patrol
merchants (*xunshang*) and watchers (*xunyi*) in their service assumed the major responsibility for this. The patrol merchants were
probably Huizhou men of small means employed by their wealthier countrymen. The cost was divided between the salt transport
treasury and the merchants themselves.[98] Surveillance of the

Yangzi, which separated the Lianghuai and Liangzhe salt-marketing zones, was the responsibility of the Jiangning subprefect for Yangzi defense (*Jiangfang tongzhi*) and was carried out by troops from the Zhenjiang naval garrison, on the south side of the Yangzi. The subprefect was stationed at Nanjing, far from the problem area, and the number of troops involved in salt surveillance was small, factors that led in 1717 to the transfer of the subprefect to Sanjiangying in eastern Jiangdu county and to the subsequent detailing of 150 soldiers from Zhenjiang to the same area.[99] In the northwest, a tiny force of five soldiers monitored the movement of salt vessels passing across Hongze Lake, a vast body of water that provided access to Anhui from the Grand Canal.[100]

During the Yongzheng reign, a period marked by sweeping administrative and fiscal reforms, the salt monopoly came under close scrutiny. Lu Xun, grand secretary of the Board of War in 1723–25, estimated that the quantity of unlicensed salt sold both by smugglers and by registered merchants was several times the volume of licensed salt.[101] The Yongzheng emperor sought a solution via bureaucratic reforms. One of these reforms was the granting of official rank to salt receivers, noted above. From 1728 on, salt officials, in common with territorial officials, received a stipend to nourish honesty (*yanglianyin*; see Appendix C).

A second important reform in this period was the concurrent appointment of the Liangjiang governor-general in Nanjing to the position of director-general of the Lianghuai salt administration. The first incumbent of this position, a youthful Yinjishan (1696–1771), found salt surveillance in Jiangbei unsatisfactory. He directed local officials to enlist salt watchers from well-off families, on the questionable assumption that the wealthy were less likely to yield to the temptations of smuggling than were the poor. These watchers were to be stationed at critical junctions of transport routes. Watchers previously employed under the patrol merchants could also be retained if their performance had been satisfactory. In all, 464 watchers and 186 boatmen to man 89 patrol boats were enlisted from the combined dependencies of Huai'an and Yangzhou to patrol transport routes north and south of the river.[102] Well over half of these were stationed in Yangzhou prefecture, particularly in Jiangdu, Ganquan, and Yizheng coun-

Table 6.1
Salt Patrols Under the Yangzhou Field Administration, 1731–98

Jurisdiction	Watchers 1731	Watchers 1798	Boatmen 1731	Boatmen 1798
Jiang-Gan	24*	20 (1761)		
Jiangdu	39	32 (1761)	18	18
Ganquan	48	40 (1761)	20	12 (1761)
Yizheng	20	20	6	6
Gaoyou	40	40	22	22
Xinghua	13	23 (1768)	6	10 (1768)
Baoying	40	32 (1761)	12	12
Taizhou	75	40 (1769)†	30	20 (1769)†
Dongtai		32 (1769)†		14 (1769)†
TOTAL	299	279	114	114

NOTE: Dates in brackets indicate year of adjustment in the quota.
*"Land route" (*lulu*) or road watchers, probably located around Yangzhou city, which was divided between Jiangdu and Ganquan counties.
†The establishment of Dongtai county in 1769 by the subdivision of Taizhou resulted in a shift of a proportion of watchers and boatmen to the new jurisdiction.
SOURCE: *JQYZFZ*, 21.6b–7b.

ties (see Table 6.1). This is quite an interesting development from the point of view of bureaucratic organization in northern Jiangsu: the field administration was being drawn into the affairs of the salt monopoly, much as in the domain of water control the reverse was occurring.

Military surveillance was upgraded soon thereafter. At Qingshan, in the western part of Yizheng county, a special corps of 100 soldiers under three officers was established to catch salt smugglers moving from an untaxed into a taxed area. Qingshan was already the site of a substantial garrison of over 600 troops and was, perhaps not by coincidence, a standard rendezvous for salt smugglers. Further bodies of troops were stationed at market towns—Majiaqiao, Gongdaoqiao, and Shaobo—north of Yangzhou and on or near the Grand Canal. The Zhenjiang naval detail at Sanjiangying, southeast of Yangzhou, was replaced by a permanent body of 155 troops, and the Jiangning subprefect for Yangzi defense was reclassified as the Sanjiangying subprefect. In the meantime, merchants had tightened up controls on routes leading directly to Yangzhou. In 1732 embankments were built at ten different points along the main water routes running east and north of the city, apparently to force boats to halt in mid-passage. Day and night surveillance was to be maintained at these places, and an experienced head merchant aided by two patrol merchants was deputed to head the investigation of offenses.[103]

By the 1730s, then, Yangzhou was a rather heavily guarded bastion of the salt monopoly, and the major transport routes leading to its many gates were peopled by detectives, watchers, and soldiers whose express duty it was to prevent illegal salt entering the city. In 1735, the neighboring county of Taixing was converted from peddled to taxed salt, and in 1743 Gaoyou and Baoying, further up the canal, followed suit.[104] These changes should have had beneficial effects on official salt sales in Yangzhou, but in Gaoyou and Baoying the conversion to taxed salt was singularly unsuccessful, and the salt quotas were reduced within a couple of years because of the accumulation of unsold salt. Quantities of unsold official salt in Jiangdu, Ganquan, and other local ports were also mounting at this time.[105]

These failures to meet the Yangzhou salt sales quota are not a sign of a general malaise in the monopoly: as the Lu Jianzeng case was to show, merchant sales flourished in the middle years of the Qianlong reign. Rather, the failure of official sales in the local ports, particularly in Yangzhou city, suggests corruption on a minor scale on the part of salt watchers and petty officials. The symbolic importance was greater than the economic effects at this point. In the official mind, Jiangdu and Ganquan, Yangzhou's home counties, were the logistical key to successful distribution of official salt. In 1759, when Gao Heng requested to be allowed to divert unsold official salt from sixteen local ports, the Board of Revenue recommended that this not be allowed for Yangzhou's home counties. "The counties of Jiangdu and Ganquan," it submitted, "are close to the salt yards and are very much places where smuggled salt abounds. If [the unsold local salt] is allotted to the shipment salt supply area, not only will smuggled salt be flowing through at an accelerated rate, but we will see shipment salt ports overrun with local salt, and the salt monopoly will suffer great harm."[106]

In the second half of the Qianlong reign, the downward trend of official sales in Jiangdu and Ganquan became palpable. In 1765, security at the city gates was further tightened,[107] but despite heightened vigilance, unsold salt remaining from the 1768 quota was so great that the 1769 delivery was not even attempted. In 1770 and again in 1774, requests were made for the Jiang-Gan quota to be halved, a proposal apparently adopted as an interim measure in 1774 and then confirmed in 1793 after restoration of the full quota proved unsuccessful.[108] Taking into account the real increases in the value of the quota due to the increase in the weight of the *yin*, we may safely conclude that during the Qianlong reign the sale of official salt in Yangzhou declined in relation to the increase in the population.

If the experience of these two counties is any guide, the private salt trade was expanding rapidly during the Qianlong reign. Thomas Metzger, in his well-known apologia for the Qing state, has stressed the flexibility of Qing procedures for coping with the salt quotas, arguing that "salt quotas did to a considerable extent increase with population" and that the official trade was steadily expanding in the eighteenth century.[109] This could be put in a less positive fashion: that the increase in salt quotas failed to match the increase in population and that the growth of the official salt trade was proportionally modest. Whether the salt cellar was half full or half empty, the growth of the private trade to the point that in Yangzhou itself the salt quota had to be halved in the late eighteenth century must have been a response, as well as a factor contributing to, inefficiencies in the salt monopoly. Further, population growth clearly had implications for the salt monopoly other than the growth of demand. The increased numbers of salters in the Lianghuai salt production zone in the second half of the eighteenth century made possible increased participation in the illicit salt trade.

The peddler system undoubtedly complicated the control of the salt trade in Jiangbei. How was it possible to monitor the activities of the large number of hawkers who daily trekked to and from the salterns? After the introduction of the *laoshao* system, designed to effect the adequate supervision of peddlers, it was reported that the number of carts and pack animals being used to transport "catties of salt" (i.e., the legitimate small amount) reached "hundreds and thousands," evidence of the quantitative problem faced by the authorities.[110] Apart from the licensed peddlers, who must themselves have engaged in smuggling in greater and lesser amounts, there were those "crafty people from neighboring areas" who "secretly brought along carts and boats and made clandestine purchases [of salt] that they proceeded to sell elsewhere."[111] There is no way of ascertaining the numbers of such entrepreneurs, and the success of their activities is partly obscured by the continued growth of merchant sales. By the fifth decade of the Qianlong reign, however, they had made severe inroads into the Shandong salt market, which lay outside the Lianghuai salt production zone.[112]

The competitiveness of the private salt trade was not built on the piecemeal activities of individual smugglers. A case brought to the Qianlong emperor's attention in 1778 revealed the worrying phenomenon of large-scale criminal organization. On the nineteenth day of the fourth month of that year, 20 to 30 salt boats were observed approaching the west gate of Yancheng, a city located right in the salt production zone. Their illegality must have been manifest, because this was a peddler area and peddler salt could not be transported by boat. Carriage by cart or pack animal had also been prohibited.[113]

According to the report, the bold district magistrate emerged to investigate the situation, supported by a number of military personnel from the local garrison. He found himself confronted by a belligerent body of smugglers, perhaps 100 in number, who mounted the canal brandishing weapons and then set fire to the bridge outside the gate. In the ensuing conflict, one soldier was killed, and three garrison runners were wounded. The defenders finally resorted to light artillery, and the smugglers fled. They left fourteen of their number, male and female, together with sixteen vessels, in the hands of the authorities. Another three smugglers had been killed.[114]

The size of this band of smugglers and the fact that it included women (suggesting an internal familial structure) implied the existence of an underworld organization highly threatening to the state. An extensive inquiry revealed that one of the leaders of the band was a salt watcher. Since watchers were drawn from the local populace, it seems probable that the man was a native of the area with extensive local contacts and that the band itself was of local origin. The investigation also revealed that the smuggled salt had been purchased in amounts of between 200 and 500 catties directly from salt masters and salt peddler stocks.[115] By this time, the *laoshao* registration system was held to be at the root of the proliferation of smuggled salt, and authorities in Guangdong had already called for its abolition.[116] The problem with the system was probably the relative immunity it gave to those who were able to gain peddler registration fraudulently. The Yancheng incident spelled the end of *laoshao* registration in the Lianghuai zone but not of the peddler system nor, as we have seen, of smuggling.[117]

Competition between the legal and illegal sectors of the salt trade did not pit wealthy, "law-abiding" merchants backed by a stern bureaucracy against poor coastal dwellers struggling to keep body and soul together by selling a peck of salt here and there. Licensed merchants and officials were as likely to engage in smuggling or its equivalent—underreporting of sales—as were salters and salt watchers, and they were in a better position to do so. Their possession of a franchise, together with their access to credit, banking facilities, and influence within the bureaucracy undoubtedly strengthened their position in the private salt trade vis-à-vis the "hundreds and thousands" of petty traders, peddlers, and the aged and infirm, who lacked these advantages. Li Cheng, chronicling the problems of the salt monopoly in the 1820s, listed merchant smuggling as the most serious, involving the connivance of merchants with officials, boatmen, and salt porters.[118] "Alone, the ports are not capable of damaging the merchants," he wrote; "the merchants have damaged themselves."[119]

A competition of sorts between Yangzhou and the coast, between Huizhou merchants and local smugglers, appears more obvious. The sudden proliferation around the salt yards of carts and pack animals bearing away salt for sale after the introduction of the *laoshao* system shows the potential of the salt trade to generate local commercial activity, which was otherwise extremely subdued throughout much of Jiangbei. The activities both of "peddlers" and of smugglers such as those involved in the Yancheng incident constituted a counterclaim on the control of salt and the rights to the profits from its sale, a claim that was to some degree successfully made as the monopoly began to weaken. By the 1820s, a large volume of Huaibei salt had long been controlled by smugglers organized in groups of up to several hundred members, each under its own leader and with its own salt-marketing area.[120]

Smuggling in Huainan never reached the scale apparent in Huaibei, but the problems were substantial enough and more serious because the Huainan sector was so much larger. From reports of smuggling cases during the 1820s, it appears that Huainan merchants were competing with intricately structured local alliances of minor gentry colluding with salt workers, bandits, and soldiers to profit from a trade that was being conducted in boats and carts by sometimes sizable bands of armed men dealing in tens to hundreds of thousands of catties of salt at a time.[121] At this time even candidates traveling to Beijing to sit the metropolitan examinations were identified as likely carriers of smuggled salt: they brazenly assumed that their goods would not be checked at customs.[122]

Throughout the eighteenth century, the salt administration and the salt merchants between them were successful in maintaining their monopoly over salt sales. The steady extension of salt surveillance around Yangzhou from the second quarter of the century on shows an intensifying struggle for control over resources, but merchants still held the upper hand. The effect of this in Jiangbei was to promote the untoward growth of Yangzhou as a center of wealth and power: salt came from the coastal zone, but the money went largely to Yangzhou. "How," Rowe has asked, "can we reconcile the view that sojourning was basically extractive with the widely reported physical splendors of such sojourner-dominated localities as Yangchow and Hankow?"[123] The answer appears to be that sojourning was extractive only to the extent that urbanization itself is extractive. Sojourner families in Yangzhou certainly did not send all or even most of their money back to the ancestral place, but neither did they send much to other places in Jiangbei. One obvious sign of the highly uneven distribution of resources is the disparity in levels of merchant support for schools and benevolent institutions.[124]

City, Salt, Infrastructure

The official salt trade, as noted in Chapter 2, was of fundamental importance to the structure of communication routes and the urban hierarchy in Jiangbei. Yangzhou's pre-eminence in the region was ensured by its administrative importance within the trade and the residence in the city of the monopoly merchants. Further, the fixed route for the salt transport, running from the Huai'an county town of Funing in the northeast down the Salt Yards Canal and then westward to Yangzhou, influenced the shape of infrastructure in Jiangbei, determining, among other things, the location of a number of important market towns.

Yangzhou's place in the salt transport, however, was limited to administration, taxation, and rites. At the start of every shipment, the salt controller reported the selection of an auspicious day to the salt censor. On that day, the salt censor, after conducting a divination, gave orders for the shipment to proceed past North Bridge on the Baita Canal, east of the city. At Wantou, on the canal, sacrifices were made to the river god, the Baita Canal salt inspector presented his report, a sample package of salt was selected and weighed, and the transport was launched. This was Yangzhou's only immediate involvement in the salt transport, with the very minor exception of an inspection station outside the Quekou Gate, which, during the autumn and early winter, checked that returning grain boats were not smuggling salt (see Fig. 9). During the Ming, Huainan salt was checked and repackaged at a depot just outside the city, but the practice was abandoned early in the Qing.[125]

Unlike Hankou, the great wholesale center for a large proportion of Huainan salt, Yangzhou was not a salt-marketing center except for that consumed by the city's population. Its role in the salt monopoly, although not inappropriate to its geographical position, could easily have been played by some other town, as had been the case during the Northern Song when Yizheng was the more important center.[126] In the Ming-Qing period, Yizheng was a busy site during the salt transport. It was the collection point for Huainan salt, the bulk of which was shipped along the Yangzi. Indeed, the repackaging of salt for the various ports was a major industry at Yizheng.[127]

Fig. 9 The salt transport bypasses Yangzhou: (*above*) officials stationed at the Baita Canal inspect salt being transported from the coast to Yizheng; (*below*) random checks for salt are conducted on boats traveling along the Grand Canal to the east of Yangzhou (SOURCES: prefatory illustrations in *JQLHYFZ*).

Yangzhou's allotted position in the administration, then, was the reason for its wealth. Salt merchants lived in the city because the salt administration was headquartered there. Commerce flourished within its walls and traders flocked to it because the salt merchants had a lot of money. At certain nodes of the salt trade within Jiangbei were to be found pale reflections of its prosperity. Huai'an, independently important as prefectural capital and benefiting from the presence of river officials as well as salt officials and merchants, boasted a garden suburb dominated by properties of the Cheng clan of Huizhou.[128] It was dubbed "Little Yangzhou."[129] Yizheng, the exit port for the Huainan salt transport, was similarly described.[130] On the coast, yet another Little Yangzhou was to be found in Dongtai, for "at the gateways and in the main thoroughfares of the towns were teahouses, taverns, and bathhouses, and in [Dongtai] city this was even more the case."[131]

All these Little Yangzhous confirm the centrality and formative influence of Jiangbei's premier city on life in the virtual province of Lianghuai. After the decline of the salt monopoly reduced Yangzhou itself to a Little Yangzhou, however, there was nothing much left for these other places to be.

SEVEN

Controlling the Waters

In the second half of the nineteenth century, Dominic Gandar, S.J., embarked on the useful project of documenting the history and present circumstances of the Grand Canal. Gandar was engaged in missionary work in Yangzhou and its environs during the 1870s, a task that took him up and down the canal from Guazhou on the Yangzi to Qingjiang, at the confluence of the canal and the Huai River in the north.[1] *Le Canal impérial*, the scholarly book that emerged from his labors, is based on, among other works, the Chinese classics, local gazetteers, and merchant route books. It also shows a familiarity with the terrain of Jiangbei born of his years of residence and extensive travel experience.

Of all the districts of Jiangsu province, Yangzhou prefecture was, in Gandar's view, "the best favored in terms of the distribution of the waters." To the east, the sea washed the shores of Dongtai county; to the south, the Yangzi provided for Jiangdu and Yizheng counties; further north Baoying, Gaoyou, and Xinghua were endowed simultaneously with waters from Heaven and from the Huai, collecting in 33 lakes, ponds, marshes, canals, and innumerable streams. East of Yangzhou, the department of Taizhou benefited both from the sea and from the rivers.[2] Within the prefecture, only Ganquan county, with its higher elevations, had often to endure the sight of parched fields, "its inhabitants obliged to seek more fortunate sites."[3] Terrifying tidal waves occurred occasionally on the coast, but these had been rare in recent years, and the incidence of floods in general had declined since the Yellow River had shifted course in 1850.

Such were the circumstances surrounding the "canal at present." Historically the situation had been quite different. A century earlier, as his study also indicates, the battle to control the Yellow River and prevent floods in Jiangbei was an annual struggle that occurred just before the summer harvest was due. The sequence of developments traced in Gandar's book faithfully follows the outline in the 1810 prefectural gazetteer for Yangzhou. The four chapters of the gazetteer devoted to waterways signaled to newly arriving officials and posterity the great significance of waterway management within the prefecture. Although the hydraulic system in Jiangbei operated at peak efficiency for the greater part of the eighteenth century, Yangzhou prefecture was, from the bureaucracy's point of view, a jurisdiction bedeviled by problems of water control.

Nearly two centuries before Gandar arrived in Yangzhou, in 1684, an imperial flotilla traveled down the Grand Canal transporting the Kangxi emperor on his first southern tour. According to the Yangzhou gazetteer, he was appalled to see the effects of the previous year's floods in Baoying and Gaoyou: "People and dwellings, fields and beasts, were all lost under water; his heart was filled with commiseration."[4] He had a good view of the disaster because the royal barge was floating some distance above the level of the surrounding land. William Alexander, draftsman with the Macartney embassy, made the same trip in 1793 and produced an engraving of the scene at Baoying Lake. A narrow bank divides the canal from the lakes to the west, and the waters of both are perilously high. Residences are apparent on the broad eastern dike, which was meant to protect the lowlands to the east from floods (see Fig. 10).

In 1684 the eastern dike was sturdy enough. Over the preceding decade, an immense effort had been put into restoring waterworks along the Grand Canal.[5] The problem was that when the waters

Fig. 10   The Grand Canal at Baoying Lake as depicted by William Alexander in 1793 (SOURCE: Alexander, *Costumes et vues de la Chine*, 2: facing p. 32).

in the canal rose too high, they had to be released through sluice gates or temporary dams (*ba*). Nor was it rare for the dike to be opened manually or ruptured by floods. Gaoyou was struck by floods in seventeen of the years between 1645 and 1684, approximately one every 2.4 years, and the greater number of these were caused by overflows from the Grand Canal (see Appendix C).

No floods are recorded in Gaoyou for 1684. What the emperor beheld were waters that had failed to disperse after the floods of 1683. On seeing the devastation, he reportedly turned to the director-general of river conservancy, Jin Fu (1633–92), and interrogated him on the causes of the flooding, possible solutions, and the likely costs. In Jin Fu's view, "the cost [of dredging Xiahe's waterways] in silver and grain would be more than one million taels," no mean sum. On the other hand, if the dredging were to be undertaken by the local people, summoned for labor by the magistrates, it "would certainly be over ten years before it was finished." As the emperor noted, by the end of that period, "who knows what state the canals would be in?"[6]

This exchange marked the beginning of a period of intensive imperial intervention in the Huai-Yang water control system,

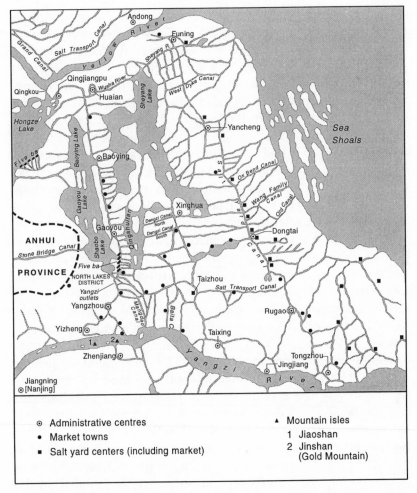

⊙ Administrative centres	▲ Mountain isles
• Market towns	1 Jiaoshan
■ Salt yard centers (including market)	2 Jinshan
	(Gold Mountain)

Map 4 Waterways and important market towns in Qing Jiangbei. Note that the salt yard centers incorporated markets. Major features of the hydraulic control system developed under the Qing included the development of Qingkou, designed to force the waters of Hongze Lake into the Yellow River with sufficient force to flush out the accumulating silt; and also the five release dams (*ba*) on the southeastern tail of Hongze Lake, along with five matching dams on the eastern dike of the Grand Canal, south of Gaoyou, as indicated. This map is based on an original of 1863. By that year the Yellow River no longer followed the route indicated here, and many of the key hydraulic installations had long since ceased to function (SOURCE: *Huangchao Zhong wai yitong yutu*).

which is generally accredited with restoring Jiangbei to conditions of stability and even prosperity in the eighteenth century. Surveying conditions there in 1984, Fei Xiaotong reflected on his own prejudice against Subei as "a sorry place, poor, backward and without prospects" but noted that "in reality, this prejudice only reflects the history of the last hundred years before liberation, and does not correspond with the longer period of history."[7] For Fei, the most evident sign of the region's well-being in the eighteenth century was that it had "a web of canals resembling those seen south of the Yangzi today"[8] (see Map 4). He did not know how little was transported along them, how often they silted up, and how necessary they were to the drainage of the Grand Canal, even though they were often useless for this purpose.

After the salt trade, water control was the most significant aspect of administration in Jiangbei. The problem of the water control system in this region is a familiar aspect of late imperial history because of the importance of the Grand Canal and the difficulties posed by the junction of the canal with the Yellow River and Hongze Lake. Less well known are the effects of the system on agriculture in Yangzhou's hinterland, particularly in the low-lying region known as Xiahe, which incorporated the greater number of Yangzhou's subordinate counties. Flood statistics show the fragility of the agricultural sector throughout the Qing and confirm Yangzhou's dependence on the salt trade for its prosperity. Further, water control became a significant domain of activity for the salt administration and salt merchants and contributed to the city's prominence as the organizing, integrating center of economic, social, and bureaucratic activities within Jiangbei. In the eighteenth century, when the intervention of the salt sector in hydraulic maintenance was at its most extensive and the hydraulic system at its most efficient, Yangzhou's dominant position among the administrative centers of Jiangbei was clear.

The Hydraulic Infrastructure

All the key elements of the hydraulic infrastructure in Qing Jiangbei were legacies of earlier times. The Grand Canal had been constructed over a period of many centuries; the Yellow River, which had formerly spilled into the Bo Sea north of the Shandong massif, was divided into a northern and a southern stream in the twelfth century and then diverted in its entirety to a southern course in 1495;[9] the Huai River, denied its natural outlet when the Yellow River turned south, amassed near Huai'an to form Hongze Lake. This lake, which received waters from both the Huai and the Yellow rivers, was the proximate source of most serious floods in Jiangbei. It was contained within a dike known as the Gaojiayan or Gaoyan, but the incursion of silt-laden waters from the Yellow River meant that its bed gradually rose over the centuries.[10] The same was true of the bed of the Grand Canal, which received waters from the lake. The dikes on both lake and canal had in turn to be elevated to prevent floods and, in the case of the canal, to preserve navigation. The Ming imperial commissioner for river control Pan Jixun (1521–95) summarized the danger of the system to Xiahe:

The Gaoyan stands one *zhang* and eight *chi* [approx 6.3 meters] or more above Baoying, two *zhang* and two *chi* [7.9 meters] or more above Gaoyou. The Gao-Bao dike [on the Grand Canal] stands one *zhang* [3.6 meters] or so above the fields of Xinghua and Taizhou or [at least] eight or nine *chi*. If the Huai descends south, with the land [dropping away by] three or more *zhang* (10.7 meters), one thousand *li* of lowlands will be inundated. How could the many districts of Huainan [continue to] exist?[11]

Since breaches in the dike system invariably resulted in massive flooding, upkeep of the dikes was of paramount importance to flood prevention. To safeguard the dikes in times of high water, it proved necessary to relieve the pressure on them by draining water out of Hongze Lake down the Grand Canal and through Xiahe to the sea. For this to be accomplished, the seaward routes had to be kept clear through dredging, something achieved only on rare occasions through large-scale maintenance projects. Further, between Xiahe and the sea stood the Fangong dike, through which the seaward routes had to pass; yet the sluice-gates in the dikes could be opened only at the risk of admitting seawater during high tide, which would adversely affect the cultivated lands to the west.[12]

The foundations of the hydraulic system in this problematic basin were laid by Pan Jixun in the second half of the sixteenth century.[13] Pan's theory in essence was that the waters of Hongze Lake had to be so strongly contained that the Huai River would rush through the exit to the Yellow River, at once preserving the lowlands of Jiangbei from flooding and helping scour the bed of the Yellow River of its silt deposits. Full containment of the lake was not attempted, even if it had been possible, because the Grand Canal needed a source of water. Pan's aim was to release 70 percent of the Huai's waters into the Yellow River and 30 percent into the Grand Canal. This formula was maintained during the Qing, even though, as Gao Bin (1683–1755) memorialized in 1736, "the waters that enter the canal are . . . great in volume and difficult for the canal to receive."[14]

Given adequate control of the confluence of the Grand Canal, Hongze Lake, and the Yellow River, the other major difficulty presented by the Jiangbei hydraulic system was draining the Grand Canal. From the late seventeenth century till the middle of the nineteenth, officials in charge of hydraulic management in Jiangbei pursued one of two conflicting policies of flood defense. The first was to drain excess water east to the sea through Xiahe, thus protecting the dikes of the Grand Canal. The second, aimed at preventing the accumulation of flood waters in Xiahe, was to ensure that the Grand Canal should contain the full volume of waters received from Hongze Lake and disburse them into the Yangzi via the various small tributary canals running through Jiangdu and Yizheng.[15] Drainage eastward was advocated by Jin Fu; drainage into the Yangzi by Zhang Pengge (1649–1725), who set in place the dike and dam system governing the balance of waters in the early eighteenth century.[16] In the course of the century, however, any major crisis in the system tended to lead to a reversal of the reigning policy, and sometimes one principle was adhered to, sometimes the other.[17]

The hydraulic history of Jiangbei under the Qing basically conforms to the pattern of development-(crisis)-recession, or "phase A, phase B" described by Pierre-Etienne Will, although echoes in a minor key (a-b) could arguably be inserted into the formula.[18] In the early decades of the seventeenth century, the late Ming period, the existing difficulties of drainage in Jiangbei were compounded by the neglect of waterways, and during the early Qing flooding was endemic. Under the Kangxi emperor, the waterway system underwent a major overhaul in the period 1684–1705. The emperor inspected the results during his southern tour of 1705 and found them good. "My river works," he declared, "have been brought to a great completion,"[19] a statement that proved premature from the point of view of flood prevention in Xiahe. Shortly after the emperor returned to Beijing, he received reports of massive flooding throughout Jiangbei.[20]

Although Zhang Pengge's system was not fully successful in preventing floods, it made possible a period of relative stability and reduced the incidence of disasters. In the 1720s, upstream breaches in the Yellow River dikes led to renewed crises in the system and a second, minor phase of reconstruction under the Yongzheng emperor. Between the 1730s and the 1760s, management of the Yellow River was on the whole effective, but serious problems of drainage and flooding were experienced in Xiahe. Another period of stability in the system can be observed between the 1760s and 1790s, when the incidence of floods in Xiahe lessened, but from around the turn of the nineteenth century the entire inland waterway system was visibly under threat from a combination of environmental and managerial factors.

Governing the Waters

One consequence of the imperial government's attempts at flood control in the lower Huai–Yellow River basin was the enlargement of the bureaucracy within Yangzhou prefecture. In other large-scale hydraulic schemes such as those at the Han-Yangzi confluence in Hubei and around Dongting Lake in Hunan, intensive involvement by the Qing government was limited to initiatives and investment during the phase of reconstruction early in the dynasty. After the reconstruction phase, management and maintenance of these schemes partially devolved onto local bodies—typically a combination of gentry (or at least landlords) and local officials, often locked in conflict with each other.[21] Although there were some parallels to this pattern in Jiangbei, the size, complexity, and fragility of the hydraulic system and the competing interests it was designed to serve meant that water-control administration in this basin evolved in highly distinctive ways. More rather than fewer officials became involved as time went on, although this was partly through duties being added to the briefs of existing official positions.

A comparison of water-control arrangements in different basins reveals two further characteristics of the hydraulic society in Jiangbei. The first is the prominence of salt officials and salt merchants as players in water politics. The second is the relative weakness of the local gentry in Jiangbei. In rural China, water control was one of the most important domains of activity for local gentry, but the literature on the hydraulic system in Jiangbei rarely mentions this significant social group. References to the undertakings, petitions, and responses of local people are not to scholar-officials (*shenshi*) or local gentry (*yishen*) or even powerful people (*haoyou*) and large clans (*daxing*),[22] but to commoners (*min*), scholars and commoners (*shi min*), or soldiers and commoners (*bing min*).[23] The use of the word *shi* (scholar) is ambiguous, but in the absence of an accompanying *shen* (official) it cannot signify much more than holders of the first degree, who enjoyed some social prestige but not gentry status. (Examination results show higher-level graduates to have been spread thinly in Jiangbei outside Yangzhou.)[24] The concentration of gentry was highest in Yangzhou's home counties, where flood defense and irrigation were less problematic than elsewhere in the prefecture, and where graduates were likely to come from salt merchant families with little or no interest in the land.[25]

In contrast to those basins where the management of water control became localized, the lower Huai River valley, containing the Yellow River–Grand Canal–Hongze Lake complex, remained a concern of the imperial government. The middle and lower reaches of the Yellow River, heavily fortified by dikes, commanded attention because of the river's capacity for inflicting damage on neighboring waterways and agricultural lands. The Grand Canal was important to the central government because it conveyed the annual grain tribute from the south to the capital. These concerns received bureaucratic expression in the form of the river administration, a tripartite body headed, as of 1730, by three directors-general of river administration, one each for the Jiangnan, Henan-Shandong, and Zhili jurisdictions.[26] Under each director-general was a large body of civil and military officials. During the Ming dynasty, labor for dike works and dredging along both waterways was conscripted from among able-bodied peasants during the slack season. These were replaced in the Qing by standing forces of soldiers and laborers directly under the authority of the river administration and charged with the care of dikes and other installations.[27] The size of the river administration grew steadily in the eighteenth century; the total number of civil officials, ranked and unranked, in the three sectors increased from 142 in 1689 to 304 in 1785.[28] The number of soldiers and laborers employed by the river administration in the middle of the eighteenth century has been estimated at around forty thousand.[29]

Xiahe, poor and flood prone, paradoxically rivaled the Yellow River and the Grand Canal in significance because its various departments and counties included the Huainan salt yards, one of the most significant sources of government revenues. The state of the waterways in Xiahe directly affected the cost of Huainan salt. On Jin Fu's calculations, the silting of transport routes in Jiangbei in the late seventeenth century added between a tenth and a fifth of a tael per year to the cost of a *yin* of salt, or around one-quarter of a million taels for the Huainan quota.[30]

Since each of these three concerns—control of the Yellow River, maintenance of the Grand Canal, and protection of the salt industry—was represented in the bureaucracy, Jiangbei was home simultaneously to officials of the Jiangnan branch of the river administration, of the grain tribute administration, and of the Lianghuai salt monopoly, as well as to officials in the field administration. Hydraulic management, the success or failure of which was of major importance to Xiahe, was a common concern to all three sectors. It constituted the raison d'être of the river administration, was directly relevant to the transport of salt and grain, and was a prerequisite for the raising and safe harvesting of crops. These interests were far from being mutually compatible. That river conservancy works should serve both the court, because of their importance for the Grand Canal and the grain transport, and the common people, because of their significance for irrigation and flood prevention, was an accepted principle. But there was some truth to the view in the local annals of Xinghua that contradictions in the hydraulic scheme amounted to contradictions between river officials, who wanted to drain the waters, and local officials, who were intent on protecting fields and dwellings.[31] In the drawn-out endeavor to maintain and improve the system designed by Zhang Pengge in the early eighteenth century, tensions between the interests of the court, local governments, merchants, and the common people were frequently evident.

On the whole, the political importance of the Grand Canal and the natural difficulties posed by control of the Yellow River meant that maintenance of the inland waterway system took precedence over the interests, particularly the agricultural interests, of Xiahe. This is illustrated in the limitations placed on access to the waters of the Grand Canal. These waters were needed both for transport routes in Jiangbei, including the Salt Transport Canal, and for irrigation of farmlands in Xiahe, but no water was available for irrigation during the fourth and fifth months when the tribute vessels were passing through. Moreover, at all times of the year, the water in the canal had to be maintained at a depth of five *chi* (approx. 1.8 meters).[32]

An excess of water was more consistently a problem than was a lack. As noted above, the Jiangbei water control system, as it operated for the greater part of the eighteenth century, was designed to protect Xiahe from the waters of Hongze Lake as long as these waters could safely be contained by the Grand Canal. When the waters were high, the outlets in the eastern dike of the canal had to opened to alleviate pressure on the dike; this happened at least two or three years out of every ten.[33] Since maintenance of the waterways in Xiahe was a prerequisite for the safe drainage of floodwaters into the sea, effective administration of water control at the local level was an essential element in the operation of the Jiangbei system.

In any analysis of the administration of hydraulic operations in Xiahe, the distinction between water control and river conservancy must be taken into account. Water control (*shuili*, lit. "water benefits") was an important area of concern for officials — prefects and magistrates — in the field administration. Within the Qing bureaucracy, it was distinguished clearly from river conservancy (*hewu*), which was the specialist concern of the administrators of the Grand Canal and the lower reaches of the Yellow River. Xiahe hovered uneasily between the two domains of authority. Its western border, the Grand Canal, fell clearly into the province of the director-general of river conservancy, but its eastern border, the Salt Yards Canal, was defined by the Yongzheng emperor as belonging to "the territorial affairs of the governor-general and governor."[34] This distinction had been blurred by the fact that in 1692 the Kangxi emperor, in response to a memorial from Jin Fu, had taken the unusual step of appointing river officials to the administrative seats of Tongzhou, Taizhou, Rugao, Xinghua, and Yancheng, all of which were situated to the east of the Grand Canal and well outside the normal domain of the river administration.[35] The rationale behind these appointments was the need for closer supervision of waterways in Xiahe, and the appointments had the effect of pulling most of the county-level jurisdictions east of the Grand Canal into the ambit of the river administration.

The authority of the river administration within Xiahe was, however, weak, partly because of certain characteristics of the river administration as a whole and partly because of limitations on the administration's presence in Xiahe. The river administration was less sharply demarcated from the general bureaucracy than, for example, was the salt administration. It was structurally integrated into the field administration, its rank and file consisting of assistant prefects, assistant magistrates, registrars, and lesser mortals, all with specific duties in river conservancy.[36] Regular officials in the field administration who held incumbencies along the Yellow River and the Grand Canal also had extensive responsibilities for river conservancy. Thus in 1728, a year of great administrative reform, the Yongzheng emperor felt it essential to point out the necessity for maintaining a balance between the interests of river conservancy and those of the locality:

River conservancy works are certainly very important, but all department and district officials have responsibility for the local populace and its welfare, in solving matters of great weight. When a vacancy occurs in the riverside departments and districts, the governor-general and governor put forward men who have exerted themselves in construction work. Although they may well be able to carry out river conservancy, matters concerning the populace and its welfare are most important, and such officials are not necessarily well versed [in these concerns].[37]

In Xiahe itself the position of the river officials was not one of strength. At one appointment per county-level jurisdiction, they were few in number, and they were, moreover, mostly *zuoza*—the lower and unranked officials. Only one official ranked higher than assistant magistrate. Further, in contrast to the situation in the main part of the conservancy system, they had no standing force of river troops or river laborers to support their efforts.[38] This meant that these river officials were forced to rely on local magistrates to summon the manpower for dredging and dike maintenance.

Not surprisingly, although records confirm the continuation of river conservancy posts in Xiahe throughout the eighteenth century, documents of this period make few references to their responsibilities or activities. An illustration of their low profile can be found in a memorial of 1727 concerning the overhaul of the waterways in Xiahe. The suggestions of the governor-general, on the whole approved by the emperor, were that the administration of this project was to be divided between the Huai-Yang intendant and three high-ranking extra-territorial officials, with the prefect of Yangzhou and the salt controller playing a role in the organization of materials and finances. Lower-level administration was to be entrusted to twelve deputies of the director-general of river conservancy and 31 officials from the Board of Works. Local river officials barely escaped being ignored completely: if the project proved too great for the above-mentioned officials to administer, then from among the *zuoza* of Yangzhou prefecture "those well versed in river conservancy should be selected to assist in the management of the various repairs."[39]

Since this was a special project, these arrangements say nothing about the routine operation of the water control system in Xiahe, but it is worth noting that the prefect of Yangzhou was a more obvious choice than any of the river officials for the tasks of collecting materials and auditing finances. Both these tasks were essential aspects of river conservancy work but probably tested the authoritative reach of the local river officials, who were both lower in rank than the prefect and lacked his powers of territorial jurisdiction. It is notable also that the appointment of the Huai-Yang intendant as director of the project was made not in respect of his role within the river administration but on grounds that he was the "highest-ranking territorial official" (*difang dayuan*).[40]

During the 1750s, when drainage through Xiahe was revived as a central element in hydraulic management in Jiangbei, it again became apparent that river officials in Xiahe lacked authority. The plan emphasized the importance of dredging, but implementation was to be entrusted not to the river officials but to the local magistrates. The magistrates were to "be given responsibility for inspecting the canals between the fields for places that are blocked up or restricted in passage" and then for "gathering together the local people" to rectify the problem areas. Annual reports on the waterways that had been cleared were to be presented to the governor-general and governor.[41]

Apparently, the river officials of Xiahe were by and large assistants to local magistrates, who, by virtue of their powers of territorial jurisdiction, were better equipped than officials of the river administration to handle water control projects. Early in the Qianlong reign, when the failure to resolve Xiahe's drainage problems led to pressure for strengthening the administration of the hydraulic system, no effort was made to upgrade the status or role of the river officials. The one change in the administration, made in the first year of the reign, was the appointment of a special official in Dongtai, on the Salt Yards Canal, to take charge of the canals in the eastern part of Xiahe and the sluice gates in the Fangong dike, both essential elements in the seaward drainage of the Grand Canal.[42] The nature of this appointment illustrates the blurred distinction between the river and field administrations. Although the incumbent's duties fell largely in the domain of river conservancy and he was subject to the authority of the Huai-Yang river intendant,[43] he was entitled *shuili tongzhi* (first-class subprefect for water control). This was in contrast to the title of, for example, the Huainan *haifang hewu tongzhi* (first-class subprefect for maritime defense and river conservancy) stationed at Miaowan to the north of Dongtai[44] and suggests that the appointment was in fact to the field administration, since responsibility for *shuili* (water benefits) fell to the field rather than the river administration. The *shuili tongzhi*, in addition to his responsibility for managing the waterways, was authorized to "investigate bandits and apprehend smugglers."[45] This role belonged to the domain of the field rather than the river administration and was typically performed by the assistant magistrate.[46]

Later in the reign there was a tendency to extend the role of the field administration and combine territorial and river conservancy duties. In 1765, the Huai-Yang and Huai-Xu river intendancies were reclassified to incorporate territorial duties.[47] In the same year, the governor-general of Liangjiang—the senior official in the field administration—was placed in supreme command of Jiang-nan river conservancy,[48] and in 1783 the rank of director-general of river conservancy, hitherto equal to that of a governor-general, was reduced by two grades from 1a to 2b.[49] This ran the risk both of administrative overload and of lack of expertise in river conservancy, factors identified by a later governor-general, Fei Chun (ca. 1737–1811), in a memorial in which he requested to be relieved of responsibilities for river conservancy.[50] The tensions between territorial and river conservancy concerns, never far below the surface, were subsequently manifest in the reluctance of statecraft officials to become involved in river administration.[51]

A desire to remove the occasion for conflicts between the different administrations and to reconcile the aims of river conservancy and water control was one factor behind the restructuring of the bureaucracy during the Qianlong reign. Another was undoubtedly the desire to control the spiraling costs of hydraulic management, which were regularly attributed to corrupt river officials.[52] This was a matter of concern to the Qianlong emperor. In 1738 acting Jiangsu governor Xu Rong (1686–1751) estimated that the prevention of floods in Xiahe would require a yearly outlay of two to three million taels. In the emperor's opinion, this figure could only have been based on local officials' fabricating projects as a means of self-enrichment,[53] but Xu Rong's projection may not have been too wild. In 1739 it was estimated that dredging the major waterways in Xiahe and repairing the Fangong dike on the coast would cost one million taels.[54] The rarity of suggestions for projects of this scale may have owed more to fiscal restrictions than to necessity. A similar project on a smaller scale in 1727–28 had cost almost 300,000 taels,[55] and the project, as Xu Rong pointed out, had done nothing to prevent floods in Xiahe.[56] Part of this project involved improvements to the Fangong dike, but further extensive repairs were carried out in 1733, and the dike was again in dire need of repairs by 1738.[57] In the same period dredging in Xiahe had had no noticeable influence on the efficiency of the system as a whole. In 1740, the recommended over-

haul of Xiahe's waterways was carried out on a relatively eco-
nomical scale at a cost of 670,000 taels, but by 1745 another large-
scale dredging project was required, at an estimated cost of
480,000 taels.[58]

These various projects were ad hoc and designed to keep the
existing system working. Since in themselves they did little to re-
dress the problem of flooding in Xiahe, it seems reasonable to
conclude that only a substantial increase in annual expenditure
would have maintained the existing water control system in
Xiahe at peak efficiency and that the elusive goal of full flood
prevention would have demanded an extraordinary financial investment. Although repeated floods in Xiahe led to the remission
of taxes and provision of grain at levels that clearly caused the
emperor some concern,[59] the river administration's contribution
to hydraulic projects in Xiahe during the Qianlong reign was limited. In 1740 the river conservancy treasury devoted a mere
70,000 taels of the 670,000 spent that year to the restoration of
Xiahe's waterways.[60] In 1744 cautious assent was given to a modest proposal that earthen banks be constructed around agricultural fields in Xiahe: "If this is successful," stated the emperor, "it
may be said to be beneficial and not a waste [of money]."[61]

This was consistent with approaches to financing control of the
Huai–Yellow–Grand Canal confluence in the same period. At the
end of 1742, Bai Zhongshan (d. 1761), an official with a solid
reputation for frugality, was moved from the office of director-
general of river conservancy in Shandong to the equivalent position in the Jiangnan administration.[62] The emperor was careful to
warn Bai of the dangers of overeconomizing, but the appointment was plainly motivated by a desire to limit the costs of river
conservancy. In 1745, the emperor frankly admitted that Bai's talents were limited to the area of finances and placed authority
over river conservancy in the hands of the Liangjiang governor-
general.[63] Bai, however, remained in office until early the following year and later returned as director-general first in Shandong
and then again in Jiangnan.[64] Meanwhile, in 1748, the emperor
endeavored to control the rising costs of river conservancy by de-
cree, demanding a reduction of annual spending in the Jiangnan
sector to 400,000 taels.[65]

The Salt Administration and
Hydraulic Management

During the emerging crisis in both fiscal and bureaucratic management of river conservancy and water control in Jiangbei during the Qianlong reign, Yangzhou—the largest and wealthiest city of the region—was seen as an obvious source of both money and administrative expertise. This was less because it was the prefectural capital than because it was the administrative center of the Lianghuai salt monopoly. Hydraulic maintenance in Xiahe was of immediate interest to the salt sector because of its bearing on the transport of salt to Yangzhou and Yizheng, and also on the welfare of the salters. Both the Salt Transport Canal, which ran eastward from Yangzhou through Taizhou, and the Salt Yards Canal, which ran parallel to the Fangong dike, were connected to the drainage system of the Grand Canal. The Salt Transport Canal took its waters from the Grand Canal, and the Salt Yards Canal received waters from a number of canals running eastward from the Grand Canal to the sea. These two waterways fell clearly into the sphere of interest of the salt administration and salt merchants, as did the Fangong dike. A third important waterway of interest to the salt sector was the Sancha Canal, which led from Yangzhou to Yizheng and served as the exit route for salt destined for transport along the Yangzi. All these parts of the infrastructure were clearly related to the salt industry, and salt merchants—unlike the peasantry—could afford to pay for their upkeep.

Early in the Qing dynasty there were few pressures on the salt merchants to assume responsibility for the upkeep of these waterways, but they did have some established obligations. At least from the latter part of the Ming dynasty, merchants had financed the upkeep of the Fangong dike.[66] An outstanding instance of merchant enterprise during the early Qing was the restoration of the dike in 1665, a project financed by three Shexian (Huizhou) merchants.[67] In 1689 the salt merchants were faced with the task of dredging part of the Salt Yards Canal, and although in the end they paid for only a small part of this operation, this appears to have been because they were specifically exempted from the greater burden, not because they were not expected to assume it.[68] By 1700, however, the salt censor "did not dare" suggest the diversion of central revenue for dredging the Salt Transport Canal. The cost of this operation was to be borne in full by the salt merchants.[69]

In the eighteenth century, as the official salt trade expanded and merchant wealth reached new heights, profits from the Lianghuai salt monopoly became an increasingly important source of revenue for hydraulic maintenance. There were four ways in which these profits were tapped. First, revenues from the salt tax were routed to the river conservancy budget, a measure initiated by the Yongzheng emperor.[70] Second, in addition to the regular tax, merchants had to pay an annual levy into the salt transport treasury (*yunku*), and a portion of these funds were diverted to the river conservancy treasury. In the middle of the Yongzheng reign, the river administration received around 50,000 taels from this source.[71] Its total budget at this time was 670,000 taels, of which 300,000 taels apparently came from the salt tax.[72] Third, the merchant body either made voluntary contributions or was solicited to contribute to the financing of a range of large-scale projects, particularly in Xiahe. Such contributions paid for dredging projects in Xiahe on at least three occasions in the second quarter of the eighteenth century, with the funds being processed through the salt transport treasury (see Appendix D). Finally, the salt merchants were directly responsible in full or in part for maintaining specific waterways and installations, such as the Sancha Canal and the Fangong dike.[73]

From the 1720s on, as shown by the record of expenditures, salt merchants were contributing substantially to the upkeep of the hydraulic system in central Jiangsu, particularly in Xiahe, paying between thousands and hundreds of thousands of taels for projects ranging from the repair of a sluice gate to the dredging of Xiahe's canals (see Appendix D). The timing of this development is not that much different from that of the "people's dikes" in Hunan, the first of which was constructed in 1734.[74] These dikes, "built with private funds and official permission,"[75] reflect much the same sort of interaction between officialdom and local elites as that evident in merchant funding of waterworks in Xiahe. Although the interests of the salt merchants in hydraulic maintenance were qualitatively different from those of landowners and peasants, and although there was a degree of formalistic altruism involved in merchant support of the hydraulic system, the principle of "user pays" (invoked in the case of dike repairs in Hunan)[76] was theoretically in force. Thus we find Ji Huang concluding his memorial of 1757 with the statement: "[As for] the silver needed for each [item of] work, what pertains to water control should be obtained from the Jiangsu provincial treasury; what pertains to the salt administration should be financed from the Lianghuai [salt] transport treasury; what pertains to river works should be financed from the river conservancy treasury."[77]

In practice, the salt merchants and the salt treasury appear to have borne much of the burden of large-scale hydraulic maintenance through the middle decades of the eighteenth century. The principle of "user pays" was difficult to apply with any consistency since in many and perhaps most cases, particular fixtures, waterways, and modes of hydraulic control were likely to benefit many categories of users. In these circumstances, it was often convenient for the state to rely on the combined strengths of the salt administration and the salt merchant body to manage and maintain hydraulic installations. The Fangong sluice gates offer a case in point.

The regulated opening and closing of sluice gates in the Fangong dike were essential both to the drainage of waters from Xiahe to the sea and to protecting Xiahe from the entry of seawater at high tide. Maintenance and control of the gates were thus related to the interests of the field administration, insofar as they bore on the harvest, and to those of the river administration, because of their place in the overall scheme of draining Hongze Lake and the Grand Canal. The obligations of the river and field administrations in the area were clearly stated in 1736 when 40 sluice gate hands (*zhafu*) under the control of the river administration but paid out of the land tax were stationed along the dike. Boats and machines used for dredging were stationed at two important outlets, with dredging operations to be carried out twice a year by a force of 132 men. The cost was to be covered by the river conservancy treasury.[78] Yet all repairs to the sluice gates were covered from the salt transport treasury; the operation of certain of the sluice gates had by the beginning of the Qianlong reign been added to the duties of the nearest salt receiver; and in

1761 all the sluice gate hands were transferred from the river to the salt administration.[79]

Revenues from the salt monopoly were not normally used to finance small-scale irrigation works. The construction of earthen banks to protect agricultural fields and the excavation of irrigation ditches remained the responsibility of the local people, with local officials assuming a role as directors and coordinators.[80] Nonetheless, the waterways that served the salt transport and drained the Grand Canal constituted the basis for the irrigation of the greater part of Jiangbei, and it was inevitable that the salt administration should occasionally be drawn into irrigation concerns. In 1753, a year of great floods in Xiahe, "villainous merchants" opened a dam near the large town of Hai'an, in Taizhou. This dam served both to sustain water levels in the upper reaches of the canal it spanned and to prevent the use of the canal as a thoroughfare to the Yangzi for the transport of smuggled salt. The breaching of the dam resulted in the rapid drainage of the canal, and farms in its vicinity were deprived of irrigation waters. In consequence, local officials from Tongzhou and Rugao pressured the salt controller to make good the damage, and a sum of 30 taels was issued from the salt transport treasury to cover the cost of sealing the breach.[81]

In that same year, the burden of hydraulic maintenance in Xiahe was shifted to the salt administration. The floods of 1753 had resulted from a dramatic collapse of the Grand Canal dike north of Shaobo. The catastrophe this unleashed on the lands east of the canal led a number of provincial and river officials to request the appointment of a special official to administer the management of waterways in Xiahe. On the grounds partly that it would "save the expense of appointing another official," it was successfully recommended that this unenviable appointment be added to the lot of the Lianghuai salt controller, to be held in conjunction with his other onerous duties and in association with the Huai-Yang river intendant.[82] The intendant was stationed at Huai'an, some distance from the geographical heart of the problem in Xiahe. Moreover, since he already bore responsibility for "defenses on the Yellow River, Grand Canal, and Yangzi, the matters involved were many and complex; he could not also care [for Xiahe]."[83]

In the second half of the eighteenth century, then, the administrative reach of the salt controller extended into a domain that in other basins was dominated by local gentry. The expanding role of the salt administration in water control in Jiangbei might be explained in terms of Will's thesis that the "the late-imperial Chinese State strived to define policies which would attain maximum effect with minimum direct involvement."[84] Although the salt administration was part of the Qing bureaucracy, its role within the machinery of the state was compromised by the necessity for intimate dealings between it and the salt merchants, who were not part of the bureaucracy. Just as local officials relied on or were persuaded by the advice of local gentry, salt officials relied on salt merchants. And salt officials were quite capable of acting in collusion with merchants at the expense of the state, as famously illustrated in the Lu Jianzeng case.[85]

Contradictions in water control systems managed by local gentry in other basins had their parallel in problems created by the salt merchants in Jiangbei, as noted by censor Xia Zhifang (*js* 1723):

Originally entry [of waters] to the Yangzi was provided by the Yangzi Canal, the Sancha Canal, and the Mangdao Canal. The Mangdao Canal is an important route with locks at the eastern and western ends. These locks are managed not by river officials but by salt merchants, who control the water as they need it for purposes of transport. When the locks are opened or shut, this is done with an eye to transport needs only; loss of homes through [consequent] flooding is common, and the people are placed under great strain.[86]

The peculiar position of the salt administration and the merchants in the bureaucratic organization of the Qing state and the social context of the greater Yangzhou region signified that the growing role of the salt sector in waterway maintenance was an ambiguous one from the point of view of state involvement. Nonetheless, the nexus between local water control and river conservancy also meant more or less continuous intervention in water control by higher-level bureaucracies, in contrast to the situation in the middle Yangzi, for example, or around Dongting Lake.[87] At the same time, the relatively formal organization of hydraulic administration in Xiahe was not markedly more successful in preventing floods than the less structured, more internalized system that operated in Hubei and Hunan. The striking factor that emerges from case studies of hydraulic management under the Qing is not the difference made by greater or lesser degrees of state commitment to and intervention in different basins but rather the synchrony of processes of development and recession in the efficiency of hydraulic systems. The case of Xiahe confirms Will's conclusion, based on a study of water control in Hubei, that more state involvement would have made little difference to the outcome.[88]

There were fewer natural disasters in Jiangbei during the eighteenth century than in either the seventeenth or the nineteenth century, but life and livelihood remained under constant threat. Records from Gaoyou show few periods of respite. In 1755, in the middle of what Will has termed the "golden age of famine relief," the price of rice in Gaoyou ballooned to 4,000 cash per catty after three successive years of floods. Relief provided over a period of three months failed to prevent a famine. "Countless numbers" died in the first half of 1756 before a good autumn harvest brought the price of grain down by 75 percent. The following year crops were again damaged by floods. The record continues in like vein. In six of the ten years between 1759 and 1768, Gaoyou was afflicted by flood, drought, or locusts, sometimes in combination.[89] This seems an adequate explanation for why the land tax in Jiangbei was so much lower than in Jiangnan. In Xiahe, it was always a question whether the crops could be harvested before the floods came. The counties here, Ho Ping-ti has noted, were "virtually the experimental farm for extremely early-ripening varieties [of rice]."[90] In terms of the economic productivity of Jiangbei, salt was always more important than crops.

City and Hinterland

Yangzhou, situated in the elevated southern part of Jiangbei, was not especially prone to floods, although the northeastern districts of Jiangdu county were vulnerable. The city's place in the hydraulic system of Jiangbei, like that in the salt trade, was as an administrative center, a place for the organization of money and manpower. Its problematic hinterland, however, helped shape the city.

The fact that so much of the farmland in its natural catchment was subject to natural disasters rendered the local economy uncompetitive with the wealthy cities of the south except while the salt monopoly flourished. The difference between Yangzhou and, for example, Suzhou in relationship to their respective hinterlands was marked. Suzhou sat in the middle of a prosperous, productive, highly urbanized region and drew upon its immediate hinterland for much of its prosperity. Yangzhou, probably wealthier than Suzhou in the eighteenth century, was situated in a poorer, much less urbanized region. It was far from the source of its wealth, the salt pans on the coast; and even further from its major markets in far-flung provinces. Although the salt monopoly made some returns to the city's hinterland, notably in the domain of water control, these were insufficient to produce the economic diversity necessary for Jiangbei's long-term betterment. To the extent that it used up much of the available space for economic activity—in terms of territory, infrastructure, and human resources—the salt monopoly arguably stymied Jiangbei's development.

Water control was another factor differentiating Suzhou and Yangzhou. There were some physical similarities in the hinterlands of the two cities. The "water districts" (*shuixiang*) in the Lake Tai area of Suzhou were not unlike those of the Xiahe area of Yangzhou prefecture, the canals being so many in both places as to make boats the most common form of local transport. The entire Lake Tai drainage area was low-lying, and floods were, as in Xiahe, the major source of natural disaster. Both places faced the problem of unstable waterways, prone to silting and shifts in course.[91] But the scale of such problems in Yangzhou prefecture well exceeded that in Suzhou. The administration of the Grand Canal south of the Yangzi needed no massive bureaucratic presence.

The contrast between conditions north and south of the Yangzi has survived into the twenty-first century, despite the relative decline of both Suzhou and Yangzhou in favor of Shanghai. In the 1990s, when Suzhou's rural hinterland was dotted with the mansions of the rural rich, Yangzhou's still featured humble gray-brick cottages with earthen floors that might have been constructed two or three centuries earlier. Flooding from the Huai River–Hongze Lake–Grand Canal complex has continued to wreak disaster at regular intervals, most recently in 2003. Far from being a departure from historical precedent, these circumstances are a long-term product of patterns of land use and methods of river conservancy that shaped the terrain in Jiangbei during the centuries for which the Yellow River followed a southern course.

In this environmental history, Qing Yangzhou was deeply implicated through a three-cornered relationship involving the salt trade, the hydraulic infrastructure, and the city's administrative roles. Forged in the early eighteenth century on foundations laid in the seventeenth, this relationship was to become if anything more obvious in the first quarter of the nineteenth century as the urban economy, the salt trade, and the regional water control system simultaneously entered into a period of crisis.[92]

EIGHT

Shaping the City

I n the course of the eighteenth century, Yangzhou's administrative functions multiplied, and the city grew richer and undoubtedly more populous.[1] It grew physically as well, expanding beyond its walls in ways that gave expression to its social composition, cultural ethos, and relationships with other places. The most notable feature of the built environment in this century was the great garden suburb that developed to the city's northwest, but internally, too, the city changed. Sites of consumption and leisure proliferated through the New City, keeping pace with the creation of the extramural gardens and in many cases exhibiting a greater longevity. A time line of these various developments would show that the city reached its zenith in the period of the Qianlong emperor's southern tours (1751–84).

The spatial organization of Yangzhou conforms broadly to the binuclear pattern observed in other Chinese cities of the period.[2] Yinong Xu's study of Suzhou, for example, affirms the applicability of the binuclear model while noting locally specific patterns of urban social differentiation.[3] As his discussion intimates, local details of urban space tell us much about the socioeconomic features of any particular city. In Suzhou, families engaged in the household production of textiles were concentrated in the northeast quarter of the city. That no comparable district can be identified in Yangzhou illustrates a fundamental difference between the two cities. The spatial organization of Yangzhou reveals what the administrative record suggests: that Huizhou salt merchants were both the dominant economic force and the social elite in the city. But changes during the eighteenth century also reveal differences in the visibility of the merchants, as well as in patterns of urban life.

Historical maps of Yangzhou suggest a certain continuity in urban form from at least the Southern Song (see Fig. 11). From the twelfth to the twentieth centuries, the western end of the city remained the domain of officialdom. The Great City (*da cheng*) of Southern Song times was rectangular in form and was divided along the north-south axis by a canal running parallel to the avenue linking the northern and southern gates of the city. All the offices of the civil administration, together with related institutions such as the Temple of the City God and the prefectural school, were situated to the west of the canal.

The early Ming city repeated the plan of the Great City.[4] The walls formed a neat rectangle, duplicating the shape of the Southern Song city.[5] The moat may have helped ensure a consistency of form through the intervening Yuan period. As in the Southern Song, so too in Ming times, the city was divided by an internal canal parallel to the north-south avenue, and the offices of the field administration were clustered on the western side. The office of the Salt Inspectorate (*chayuan*) was situated on the eastern side,[6] but this appears to have duplicated the Song arrangement. The continuing spatial differentiation between the field and the salt administrations reflected the ambivalent position of the salt administration in the overall structure of the imperial administration.

The east-west division of the city was duplicated on a larger scale in the sixteenth century with the construction of the new wall around the suburb to its east. This double city was henceforth to be internally differentiated as the Old City and the New City. The eastern stretch of the Old City moat now constituted a second internal canal that could be crossed by bridge, but passage

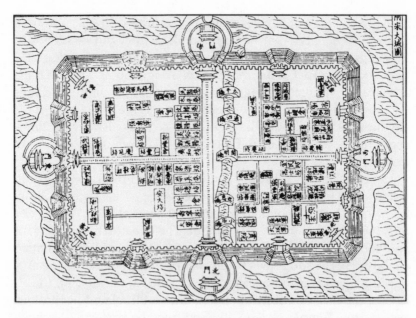

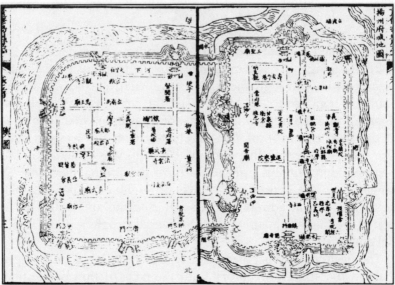

Fig. 11   The "Great City" of Song Yangzhou (*above*), compared with its Qing successor (*below*) (SOURCES: prefatory illustrations in *JQYZFZ*).

between the two sections of the city was restricted to the two gates in the wall that divided them. The Old City and the New City presented to the eye the contrast to be expected between a planned and an unplanned settlement. The streets in the Old City were laid out in a regular, almost geometrical pattern, whereas those of the New City were arranged more haphazardly. The inner city wall divided two societies: in the early seventeenth century, the population of the New City was composed of "wealthy merchants and great traders who esteemed wasteful extravagance," whereas in the Old City lived "many official families without much to do."[7]

After the sack of Yangzhou in 1645, the city was restored quite rapidly, with few apparent changes. The physical and the functional distinctions between the Old and the New cities were retained, and the original east-west division in the Old City was preserved. The only office of the field administration located east of the Old City's central canal was the Ganquan county yamen, constructed after the creation of the county in 1732. The offices of the salt censor continued to be situated on the east side of the Old City, but the rambling establishment of the salt controller was located in the New City.[8] The effect of the creation of Ganquan county was to divide the city bureaucratically between two magistracies, south and west falling to Jiangdu, north and east to Ganquan, but this made little difference to the internal logic of the city.

The imposition of alien rule by a dynasty preoccupied with martial prowess and otherwise ever watchful for signs of sedition should lead us to expect signs of Manchu military power in Yangzhou, and according to J. P. Du Halde's early eighteenth-century geography, there was a "Tartar garrison" in Yangzhou.[9] But Du Halde appears to have been incorrect. Security in Yangzhou was maintained by a regular battalion garrison (*ying*) of the Army of the Green Standard stationed in the northwest sector of the New City, with military posts (*xun*) distributed at strategic points along the city's perimeter, within and beyond the city gates. The garrison was located where its Ming predecessor had been.[10]

One change in military organization in Yangzhou was reflected in urban land use. After the Manchu conquest, the "small parade ground" or barracks (*xiao jiaochang*) in the southwest corner of the Old City, west of South Gate, was no longer used for military activities, and the land was converted to market gardens.[11] During the Qing dynasty, this quarter was probably Yangzhou's "urban sociological periphery," to use Yinong Xu's term, visibly distinct from the major sites of commercial activity and the city gates.[12] By contrast, the Parade Ground in the New City was centrally located amid the hustle and bustle of the western part of the New City.[13]

Spatial Differentiation in the New City

A much richer historical record exists for the New City than the Old, a fact in itself suggestive of the contrasts between the two parts of the city. Both sectors show some stability in spatial organization from the seventeenth to the nineteenth century, but the impact of a growing urban economy was more obvious in the New City. The courtesan quarter provides an example. Zhang Dai (1597–1689), connoisseur of *fin de dynastie* decadence, knew how to get to the city's brothels in the late Ming:

Pass the customs station, and after a circuit of 200 yards or so, you will find nine lanes. Originally they numbered nine, but now almost one hundred encircle and zigzag left and right, before and behind them. The alleys are narrow at the mouth, and winding. At every inch and turn there is an exquisite house with a secret doorway: these are the various dwellings of the courtesans and whores.[14]

The customs station stood at the southern gate of the New City, overlooking the Grand Canal. The inland customs was inaugurated in 1429, and Yangzhou, standing close to the junction of the Yangzi and the canal on a major north-south route, was among the first of the customs ports.[15] The gateway was one of three in the southern city wall, but it was the usual point of entry into the New City, and commercial activity in the vicinity was brisk.

The earliest available map of the city dates from 1883, but it was closely based on the account in Li Dou's guidebook.[16] With the exception of certain institutions created or relocated after the Taiping Rebellion, it must hold true for the eighteenth century, if not earlier. The written street directions accompanying the map, largely lifted from *Painted Barques*, mention the nine lanes (shown as ten on the map), which follow one another in a regular pattern along the inner side of the Old City wall.[17] Although Zhang Dai's account would seem to point to a New City location for the brothel quarter, it is plain that the nine lanes mentioned by Li Dou and the 1883 gazetteer as being in the Old City are the same as those referred to by Zhang. To get to them, the visitor to the city had to pass through Little East Gate, which led into the Old City, or alternatively further north through an underpass (*daocheng*) or Great East Gate (see Map 7, p. 184).[18] In the nineteenth-century *Dreams of Wind and Moon*, the author showed the way: out the backdoor of a teahouse in the Parade Ground in the New City, along Worthy Street to North Willow Lane as far as Celestial Longevity Monastery, down the incline to the banks of the canal, and through the underpass to "someone's place in the nine lanes."[19]

Although Li Dou listed the nine lanes and adjoining streets, he did not otherwise venture much into the Old City. His detailed account of courtesans shows that during the eighteenth century, if not before, a parallel row of pleasure houses developed on the other side of the wall, along the canal in the New City. Some change over time in this part of the city can be traced through his record of the Kang Family Garden, built by a wealthy salt merchant early in the century.[20] The garden stretched for a third of a mile or so along the western bank of the canal; its size suggests that land in the city was relatively easy to acquire at the time of its construction. This garden disappeared in the course of the eighteenth century, part of it being replaced by a teahouse called the Garden of Pleasures (Hexinyuan) run by Old Woman Lin, who did a thriving business in flaky pastries (*shaobing*). The teahouse's great draw, however, was the woman's daughter, who attracted so many visitors that the pair finally became quite rich. The garden-cum-teahouse thus served also as a courtesan's quarters.

After the old woman's death, the teahouse became a lodging house. Among the lodgers was one Wu Lunyuan, a fine flautist who made money by giving singing lessons to high-class courtesans, earning himself the nickname of "Professor Crow" (*wu shi*, i.e., an instructor of singsong girls).[21] The courtesans did not have far to travel for their singing lessons. According to Li Dou, many lived in Waichengjiao, in the shadow of the city wall on New City side.[22] South was the Garden of Pleasures, and north was the house of Wang Tianfu, a pimp with many consorts who ran a "most flourishing" establishment, his property half on the bank, half on a terrace over the water.[23] The courtesans must have done a lot of shopping in the streets just to the east of the canal, north of the customs station. Fabrics dominated trade in Satins Street (Duanzijie), a flourishing center of commerce in the city;[24] women's accessories were sold in Kingfisher Blossom Street (Cuihuajie); and the best-known outlet for cosmetics was in Ridge Street (Gengzijie)[25] (see Map 5).

Traveling eastward from the Customs Gate along a route parallel to the city wall, the visitor to Yangzhou would enter a quite different social space. Here were situated the salt merchants' houses, spread along an arc stretching around the southeastern corner of the New City wall. The wall, built in the sixteenth century, divided the houses from the waterway, but salt merchant properties must originally have flanked the canal itself. In the late seventeenth century, Wu Jiaji commented on residential patterns in this quarter:

> The crows of winter are not everywhere:
> Down by the canal live people
> In households rich: how elegant and refined!
> Suffer not they poor scholars for neighbors.[26]

The street flanking the southern city wall was called Nanhexia Street, which can with some latitude be rendered into English as "South Canalside Street." It led into "Central" and "North Canalside" streets (Zhong, Beihexia), flanking the eastern city wall. This was Yangzhou's golden mile (see Map 6). As noted in Chapter 3, Zheng Xiaru's Garden of Retirement, constructed in the early years of the Qing and covering an extensive area inside the eastern wall, was located in this neighborhood. In the late eighteenth century, a number of the richest merchants had gardens in the same quarter.[27] In the nineteenth century, as the number of wealthy merchant households declined, native-place associations sprang up here, some on premises that had been the mansions of salt merchants.[28] The extended families encouraged by Huizhou clan culture may have contributed to the scale of the houses. At the Zheng family mansion in the early seventeenth century, over a thousand people are said to have eaten at the same hearth; and the household of Wang Jiaoru, one of the more prominent of the eighteenth-century merchants, achieved the ideal of five generations under one roof.[29]

In contrast to the southeastern quarter of the New City, the northwestern section appears to have been the bohemian neighborhood during the eighteenth century (see Map 7). Many well-known artists lived for some period in this quarter, within easy reach of the city gates leading westward into the Old City and northward out to Tianning Temple and the northern stretch of the city moat that led to Rainbow (originally Red) Bridge and the lake beyond.[30] Artists also found the area north of the Tianning gate a convenient place to live, within sight of gardens and scenery on the one hand and close to the city on the other. Huang Shen (1687–1770) lived in Tianning Temple,[31] as did Zheng Banqiao during an early stay. Later Zheng lived in the Temple of the Bamboo Forest, just outside the north gate, and then in an abode next to the Temporary Retreat (Xing'an) garden of the Ma brothers, which was adjacent to Tianning Temple.[32]

Li Dou portrayed the northwestern quarter as lively. At night the air was filled till after midnight with the strains of the *qin*, played by an opera performer who lived in Master Dong's Temple—the abode of artist Gao Fenghan (1683–ca. 1748) at an earlier date. Next to the temple were a pair of two-story buildings occupied by Daoist monks, a medical dispensary, and the salt controller's yamen. Close by was a mounting shop, owned by a Suzhou man, and down the road was the book market, with "seats all around and a platform for books in the middle." The doorway to the market was hung with a sign advertising the name of the storyteller who would perform that day. Further south, in South Willow Lane, was the residence of the Hangzhou poet Chen Shouyi. The Little Qinhuai Canal (as the New City canal was called) ran parallel to South and North Willow Lane, bearing boatloads of tourists northward to the city moat past Tianning Temple and thence to Slender West Lake.[33]

The social profile of the northwestern quarter was mixed, with monks, courtesans, scholars, and officials of the salt administration living and working close to the Parade Ground, the noisy center of this part of town. The Parade Ground lay south of the

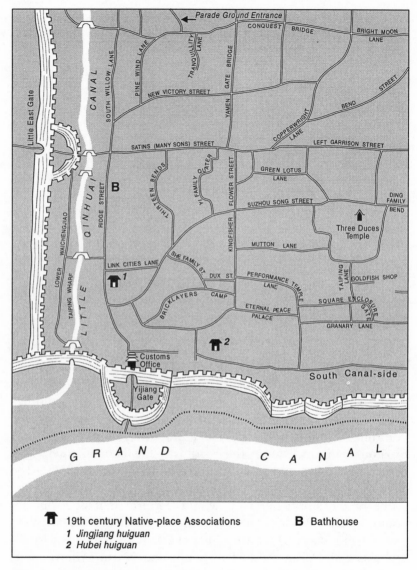

Map 5 (*above*) Southwest quarter of the New City (SOURCE: *GXJDXZ*, slightly adapted by Zhou Cun, *Taipingjun zai Yangzhou*, frontispiece). (*facing page*) Southwest quarter of the New City, showing the Customs Stations and major commercial streets (based on Chinese original).

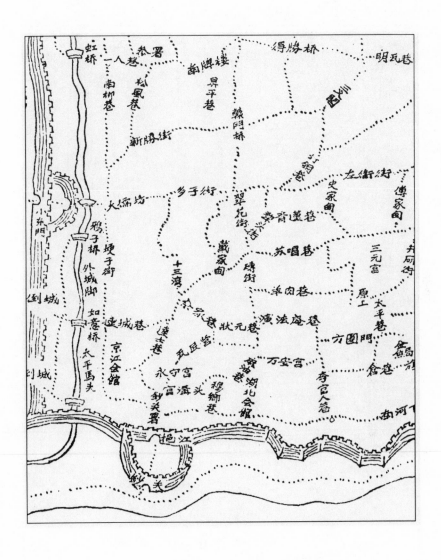

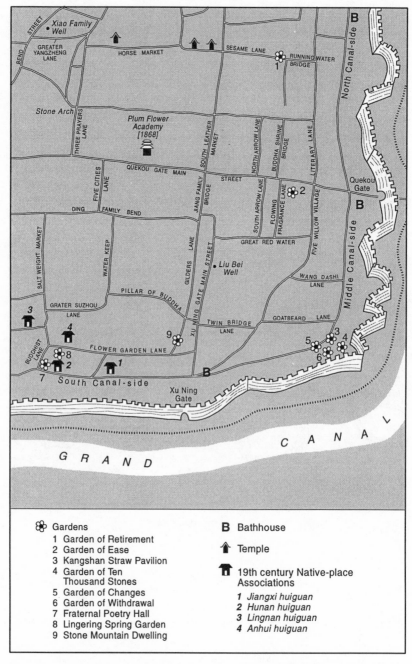

Gardens
1 Garden of Retirement
2 Garden of Ease
3 Kangshan Straw Pavilion
4 Garden of Ten
   Thousand Stones
5 Garden of Changes
6 Garden of Withdrawal
7 Fraternal Poetry Hall
8 Lingering Spring Garden
9 Stone Mountain Dwelling

B Bathhouse

Temple

19th century Native-place Associations
1 Jiangxi huiguan
2 Hunan huiguan
3 Lingnan huiguan
4 Anhui huiguan

Map 6 (*above*) Southeast quarter of the New City (SOURCE: *GXJDXZ*, slightly adapted by Zhou Cun, *Taipingjun zai Yangzhou*, frontispiece). (*facing page*) Southeast quarter of the New City, showing the major residential area of salt merchants in the eighteenth century (based on Chinese original of 1883).

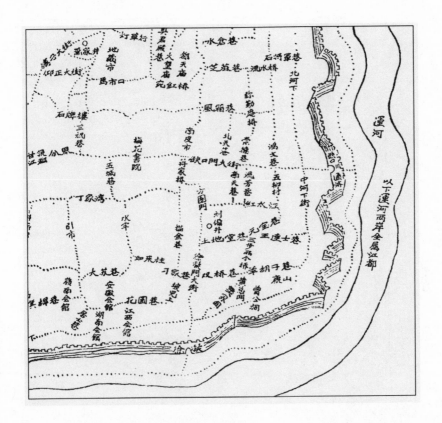

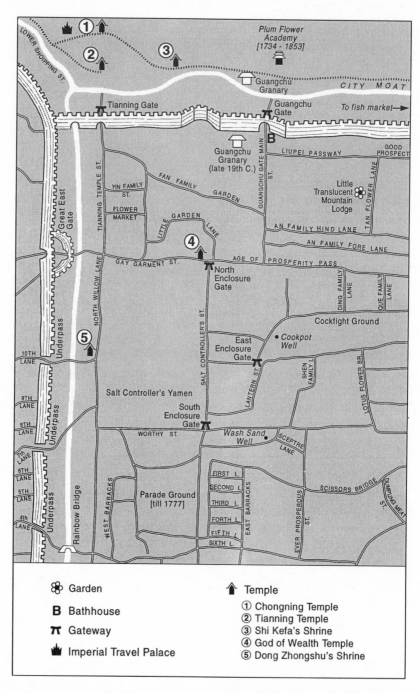

Map 7 (*above*) Northwest quarter of the New City (SOURCE: *GXJDXZ*, slightly adapted by Zhou Cun, *Taipingjun zai Yangzhou*, frontispiece). (*facing page*) Northwest quarter of the New City, showing the salt controller's yamen and the Parade Ground. The "nine lanes" are indicated to the left of the internal city wall (based on Chinese original of 1883).

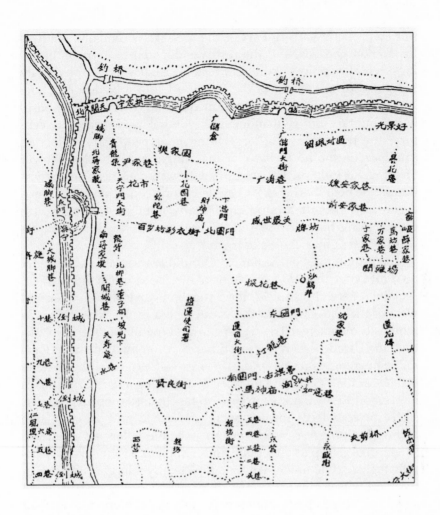

salt controller's yamen. Part of the area was originally reserved for market gardens, and the rest was devoted to garrison buildings, temples, a horse pen, and residences. It was not physically separated from the rest of the city and had a mixed population. Around 1739, it reportedly contained a total of over 1,200 households, military and civilian together.[34] In the late seventeenth century, the rehousing of a large number of people from flooded sandbars on the north bank of the Yangzi in Kingfisher Flower Street just a little further south had created considerable congestion in the neighborhood.[35] Crowded living conditions made fires a common hazard around the Parade Ground. Drainage was poor; during the rainy season the area would flood, and the air would be rent by a chorus of frogs.[36]

The propinquity of the Parade Ground to the courtesan quarter no doubt contributed to problems of discipline while the garrison remained here. But in 1777, the Yangzhou garrison was relocated west of the Old City, and the vacated area in the New City was rented to merchants. The name clung to the place, but from this time the Parade Ground was primarily a commercial site.[37] Thus in the opening pages of *Dreams of Wind and Moon*, Lu Shu, freshly arrived in the city for the purpose of obtaining a concubine, takes himself off to the Parade Ground on his first afternoon in the city. There he sees local theater being performed and western-style painting in progress, listens briefly to a storyteller and to a male singer dressed up as a woman, and has a look at a peep show (*xiyangjing*), a scene much like that depicted by a local artist later in the century (see Fig. 12).[38]

The salt controller's yamen was the other dominant feature of the quarter, and salt merchants must have been regular visitors. They might stop before the artist's modest residence, pay their respects, and request a painting. Not far northeast of the yamen lived the Ma brothers, and Shaanxi merchant Yuan Guotang must have lived somewhere in the vicinity, together with his artist-in-residence, Hua Yan (1682–1756).[39] It is quite possible that other salt merchant residences were found in this neighborhood. Overall, however, the mixed features of this part of the city leave an impression rather different from that of the southeastern quarter, with its succession of mansions and gardens.

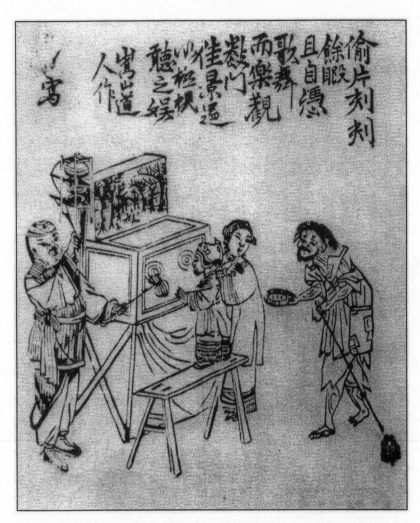

Fig. 12 A peep show in the late nineteenth century, printed from a wood-block held in the Yangzhou Museum (published in Fang Chen, "Yangzhou Qingdai muke nianhua," facing p. 85).

In the late eighteenth century, the city became increasingly crowded. This was one factor behind the relocation of the Parade Ground in 1777.[40] Boats proliferated on the waterways, so congesting the wharves that in the 1750s regulations were introduced to limit the use of wharves to registered vessels.[41] Markets spilled

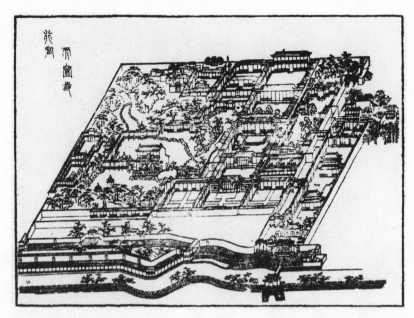

Fig. 13   Tianning Temple and the imperial travel palace, showing the row of covered markets north of the city, along the banks of the moat (SOURCE: *GLMST*, 8b).

out of the city gates to line the northern city moat opposite Tianning Temple (see Fig. 13).[42] Above all, the expansion of the city was evident to the eye in the spread of gardens along the banks of Slender West Lake.[43] The relatively detailed documentation available for gardens permits a close chronological account of the expansion of the city through the eighteenth century and its sudden contraction thereafter.

Yangzhou's Gardens and Extramural Expansion

The Yangzhou garden emerged as a culturally distinctive phenomenon in China in the same century as *le jardin anglais-chinois* was establishing its hegemony in Europe. Its distinguishing features have been described by Chen Congzhou as rockeries and the architecture of the garden buildings.[44] Great fortunes were invested in the development of these features. Yangzhou was poor in the resources needed for the construction of gardens. There were no forests: timber had to be imported, probably from Huizhou. There was no stone. Yangzhou rockeries were constructed of three types of stone: yellow rocks, available from southern Jiangsu, Anhui, and Jiangxi; streaky gray rocks from Anhui; and lake rocks from Suzhou and nearby areas, distinguished by their interesting, eroded shapes.[45] All were transported to the city by boat from considerable distances away and at great cost. One nineteenth-century visitor to the city heard that the Yu family of Shanxi had spent 200,000 silver taels on Taihu rocks for its lakeside garden, Rosy Evening Clouds by the Water (Linshui hongxia).[46]

The Yangzhou gardens were also distinguished by a particular pattern of ownership. Local people ("old Yangzhou" [*lao Yangzhou*] or "natives" [*tuzhu*]) feature not at all among eighteenth-century garden owners. Qin Enfu (1760–1843), Hanlin compiler and bibliophile, owned a small, tasteful garden in the Old City and looks at first to have been an exception to this rule, but, like most of the city's wealthy men, Qin was of merchant descent. His forbears had moved to Yangzhou from Shaanxi.[47] For Yangzhou people, the wealthier of whom probably lived off rents, land undoubtedly remained something to till and sow.

The owners of private gardens in Yangzhou, with hardly an exception, were salt merchants, and although an occasional West merchant appears among them, most garden owners were Huizhou people.[48] The Zheng brothers, each of whom owned a garden, were from the most eminent Huizhou family in Yangzhou in the seventeenth century. Their successors in the eighteenth century were the Huang brothers: Sheng, Lümao, Lüxian, and Lühao. The most famous of the Huang gardens, located on the lake, was the Mist and Rain by the Four Bridges (Siqiao yinyu), on which the Qianlong emperor in 1762 bestowed a title. Each of the four brothers also owned a garden along the golden mile in the southeastern corner of the New City.[49]

Pingshan Hall, on Shugang, provided the point of orientation for Yangzhou's garden suburb. In a study of the "eminent sites of Pingshan Hall," leading salt merchant Wang Yinggeng offered a carefully constructed relationship between the walled city and temple complex on Shugang. The work begins with poems on the Little Qinhuai Canal, in the New City of Yangzhou, follows the painted barques through another set of poems to the Red Bridge and the Han Garden of yore, and culminates with poems and memoirs on the various holy sites of Shugang.[50] Of these, the most notable was Pingshan Hall itself, constructed in 1048, renovated in 1673, and in 1705 adorned by the imperial brush with the three characters for Pingshan Hall and the four characters for "worthy [men] protect pure customs" (*xian shou qing feng*).[51] Further east, on Gongdeshan (the highest point of Shugang) stood the Guanyin Temple, a Ming reconstruction of a Song temple. Below Shugang, spanning the narrow width of Baozhang Lake, was the famous Four-and-Twenty Bridge, and by it the Yuan dynasty Lianxing (later Fahai) Temple. The aesthetic advantages provided by the lake, in combination with the distant hills, old bridges, and ancient temples, made this the more or less obvious zone for the establishment of the gardens for which Yangzhou became known in the eighteenth century. The gardens eventually formed a continuous stretch between the city and Shugang. The distance was only three or four *li* in a straight line, but the winding water route covered eight or nine *li* (four to five miles).

The historic buildings on this stretch of land provide a hint of its ritual significance in the world of Yangzhou. Zhang Dai has left a charming sketch of the scene during late Ming celebrations of Qingming. The Qingming Festival falls around the fourth month of the year in the solar calendar. One hundred and six days after the winter solstice, when the willow leaves were green and the peach flowers red, people in Yangzhou packed their baskets and took up their brooms to visit the graves of their forbears. Rich and poor, young and old, male and female, they headed out of the city, bound for the site or sites of the family tombs. The scene described by Zhang Dai resembles that in Zhang Zeduan's famous scroll painting *The Festival of Qingming Along the River*, and he explicitly compared it to that painting, the vista unfolding before his eyes just like a scroll and from certain vantage points certainly visible in close to its entirety. A remarkable range of social types and activities are portrayed:

On this day, people from the four corners gather here: merchants from Huizhou and traders from the Western provinces, Yangzhou courtesans, and a few rowdies. In the lush grass by the long embankment, people ride their horses and fly their hawks. High up on the plateau, they organize cockfights and kickball. In the dense woods and cool shade, the zither is played. Energetic youths wrestle together, and children fly their kites. Old monks preach about the penalties for ill-doing. Blind musicians recount tales, with a musical accompaniment. People stand in crowds as thick as trees in the wood and squat together in great clusters. As the sun sets gloriously, the road is crowded with horses and sedan chairs returning home, and even women from good families have the curtains open to gaze on the last sights of the festive day.[52]

In the early Qing dynasty, much of the land traversed by these late Ming revelers appears to have been reduced to a wasteland. It was "covered with rubble and abandoned graves," according to Kong Shangren, and the area where a plantation of flowers once flourished by Red Bridge was now vacant land.[53] Despite Kong's statement, there is some evidence of commercial horticulture in this area in the second half of the seventeenth century. Late in life, Shitao painted a vista of the scenic stretch beyond Yangzhou out toward Pingshan Hall and made reference in an accompanying verse to "women from the flower nurseries drawing water from a stream."[54] His older contemporary, the poet Zong Yuanding, left a short account of the life of a flower seller (himself) who lived in a three-roomed thatched hut behind Hortensia Temple, west of the city, and made a living from flowers he grew on two *mu* of land. He had more than ten different varieties under cultivation and in the morning sold them by the Red Bridge to passing literati, who would also exchange poems with him, or to the common herd at a greatly inflated price, which enabled him to go and get drunk at an inn.[55]

During the eighteenth century, small gardens just outside the city were still used commercially. Wang Xiwen, a Suzhou tea merchant, used his Shao Garden just north of the city for horticulture and fish-farming.[56] But a shift away from commercial uses, and also away from ownership by locals, is unmistakable. The inlet north of the Lotus Flower Bridge (not constructed till 1757) was the property during the Kangxi period of "a local man named Huo" but was later incorporated into the garden of the salt merchant Huang Sheng.[57] Further north, by the Four-and-Twenty Bridge, lay a large garden that was owned during the Kangxi period by a local man who cultivated peonies and plum trees and sold tea to sightseers who came to gaze upon the flowers in bloom. This entire property was sold in 1716 to Cheng

Mengxing, who had just retired from office as a Hanlin compiler to attend to the family business in the salt trade. The land was now redeveloped in quite a different way—planted extensively with flowers and trees, to be sure, but with a close eye to the clustering of plants for aesthetic affect. A plantation of cress was replaced by one of water lotus, and a pavilion was constructed over the water to allow full view of the flowers; 100 plum trees were planted, and another pavilion constructed in their midst; a grove of bamboo was established, and within it a gazebo; earth was piled up to form a hill, and rocks used on it for decorative effect. Each site and scene was given a name.[58]

The Bamboo Garden (Xiaoyuan) was one of the "eight great gardens of the Kangxi reign," most of which were located close to the city. Chen Weisong (1626–82), active in Yangzhou in the Shunzhi and early Kangxi reigns, commented on the large number of gardens in Yangzhou, but most of these must have been gardens in the city, around the temples, or perhaps in the immediate vicinity of Red Bridge.[59] The record of the Bamboo Garden shows that the spread of gardens out along the lake to the northwest must have been mainly a phenomenon of the eighteenth century. In his choice of a site, Cheng Mengxing may have been inspired by his brother, Cheng Zhiquan, who in 1697 embarked on the construction of a garden on an island in the middle of the lake. This garden, which took more than three years and 200,000 silver taels to build, centered on a temple of Lord Guan and was thus in keeping with the pre-existing character of the lake area as developed through the construction of temples over preceding centuries. It was also located rather nearer to the city than the Bamboo Garden.[60] With the creation of the Bamboo Garden, at a point well on the way to Pingshan Hall, we begin to see the development of the lake area and the effects of garden building on communication routes between Yangzhou and the plateau of Shugang, where Pingshan Hall stood.

The Bamboo Garden was one of a number of ambitious gardens created before the great garden-building frenzy of the Qianlong reign. Another was He Junzhao's Eastern Garden (Dongyuan), also known simply as the He garden. Commenced in the Yongzheng reign and completed in 1746, this was located on the eastern side of the Lianxing Temple, again out on the lake.[61] This garden was depicted fantastically by local artist Yuan Yao, in a hand-scroll painting that begins with a wilderness area of sky, lake, and mists, giving way in the middle of the scroll to a river, a hamlet, and a mountainous isle before the eye is finally led to the garden itself, depicted in fine architectural detail.[62] It was, albeit anachronistically, included among the "eight great gardens of the Kangxi reign," and the impressive dimensions portrayed in Yuan Yao's painting are no doubt true to its scale. Certainly it had an impressive roll of visitors.[63]

A third garden developed during this period was Yu Yuanjia's Garden of Ten Thousand Stones (Wanshiyuan), located in the southeast corner of the New City amid the residences of the salt merchants. In the second month of 1731, Yu held a literary gathering in his garden, attended by the Ma brothers and a number of mutual acquaintances. Immediately on entering the gateway to the garden, visitors were confronted by one of the best examples of rock architecture in Yangzhou: an artificial mountain large enough to incorporate "several hundred large and small stone caves," by which we should understand miniature caves, giving the impression of a fairy grotto.[64]

Finally, there was the Ma brothers' Temporary Retreat, which flourished in the same period and of which more will be said in Chapter 10. This garden was located near the city, on the north bank of the canal. Of the famous gardens of these decades, then, the Bamboo Garden was unusual in being located at such a distance from the city. At the time of its construction, the garden was reached only with difficulty because the lake was silted up. Cheng himself must have found this inconvenient, for in 1732, sixteen years after he had purchased the property, he organized finances—probably through the salt-merchant body—to dredge the lake as far as the city moat and plant peach and willow trees along both banks. Until this time, the painted barques had taken passengers only as far as Lianxing Temple; now they were able to proceed around the next bend of the lake to Cheng's own garden.[65]

To these eighteenth-century gardens must be added Zheng Xiaru's Garden of Retirement, constructed in the early Qing but descending through the Zheng family for several generations and mentioned by Li Dou as one of three gardens noted for literary gatherings.[66] The other two in this grouping were Cheng Meng-xing's Bamboo Garden and the Ma brothers' Little Translucent Mountain Lodge (Xiao linglong shanguan), but the Garden of Retirement and the Bamboo Garden were "the most flourishing" in this respect.[67] Li Dou described the Garden of Retirement but, in contrast to his descriptions of other great gardens, did not provide details on the people who visited it. It is possible that his access to information about it was slight. In 1773, one of Zheng Xiaru's descendants published a book about the garden, featuring various memoirs and poems by men who had visited it, but the work appears to have been proscribed during the literary inquisition that commenced that same year.[68]

In retrospect, it seems that if the three gardens noted for literary gatherings in the eighteenth century were to be considered as representative of Yangzhou's gardens, we would be left with the impression that the great age of the Yangzhou garden was over by the middle of the eighteenth century, or soon afterward. This proved not to be the case. Yuan Mei (1716–97), who like most other southern literati periodically crossed the Yangzi to visit Yangzhou, wrote in 1794 that 40 years earlier, "I started the journey by boat from Tianning Gate along a narrow river barely two yards wide, with hardly any pavilions on the banks. I saw mostly cascades and wild bushes." He goes on to describe the changes since then: "From the year *xinwei* [1751], when the emperor conducted a tour to the south, tremendous human and material resources were directed to large-scale constructions in the city. Now the rivers are broad and winding, and the hills are grand and magnificent."[69]

Mathematics may not have been Yuan Mei's strong point, since 1751 is more, not less, than forty years before the time of writing. Perhaps he meant the year *dingchou*, 1757, the occasion of the Qianlong emperor's second southern tour. In that year there was certainly an extraordinary flurry of activity around the lake, with the construction near the Lianxing Temple of the Lotus Flower Bridge (now commonly known as the Five Pavilion Bridge), and the carving out of a waterway to make Slender West Lake navigable as far as the foot of Shugang. In 1765, the year of the fourth southern tour, salt controller Zhao Zhibi published *The*

Illustrated Gazetteer of Pingshan Hall and was able to mark every point on the waterway between the city and Shugang as the site of a garden (see Fig. 14). In a memoir of the Eastern Garden published in this gazetteer, Qu Fu described the panorama thus created: "Fifty years ago I climbed up to Pingshan Hall. The gardens north of the city were like a tapestry. Only beyond the Guangzhuangmu Shrine was there a stretch of wilderness. Last year in spring I went there again to find that what had been neglected was now beautiful, what had been in pieces was prospering . . . [and] sightseers were everywhere."[70]

The Qianlong emperor's visits to the south, as Yuan Mei's preface suggests, stimulated garden building in Yangzhou. A tale is told that when the emperor visited Little Gold Mountain, he thought it a pity that the outlook included no building close at hand:

The merchant Huang Shifu then began building on the wasteland of the [nearby] mulberry garden, collecting workmen and preparing materials. In one evening the Shrine of the Three Worthies was built. Next day the emperor mounted his throne and, glancing out, observed it. In surprise he asked an official in attendance, "How was it built with such speed?" The official replied that Huang had built it. The emperor said with a sigh: "I am no match for these wealthy merchants!"[71]

An almost identical if even less probable story is told of the overnight construction of the famous white stupa by Lotus Flower Bridge in 1784, the year of the sixth southern tour[72] (see Fig. 15).

These myths appear to have been simply elaborations of the truth. Recorded dates for the renovation or redevelopment of the more prominent gardens often coincide with the imperial visits (see Appendix E), and some expensive projects were undertaken at these times. The Gaoyong Mansion in the garden Dawn over Shugang (Shugang chaoxu) belonging to the Zhang family from Lintong, Shaanxi, was built in 1762, the year of the third southern tour.[73] It stood more than ten *zhang* high (approximately 117 feet). The emperor composed a couplet in praise of the mansion when he visited the garden that year[74] (see Fig. 16). It was probably in preparation for the same visit that Huizhou merchant Wang Changxin transported nine great rocks across the Yangzi from

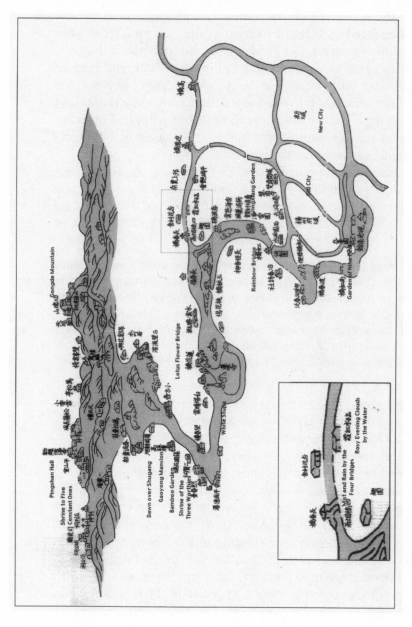

Fig. 14 Sites along Slender West Lake, as depicted in a woodblock print in Zhao Zhibi's *Gazetteer of Pingshan Hall* (1765), adapted from the reproduction in a 1980 Yangzhou tourist guide. Places identified in the text and shown in the diagram have been translated into English. Not shown are later gardens and older gardens and sites that had already disappeared.

Fig. 15 The White Stupa, 1980
(photograph by the author).

Taihu in 1761.[75] These were the most notable feature of his famous Garden of Nine Peaks (Jiufengyuan), which was distinguished among the great gardens by being located on the south side of the city, a consequence of the early date of its construction by Changxin's father. It was previously known simply as the Southern Garden.[76] The emperor visited this garden on all subsequent tours and ended up taking some of the rocks back to Beijing with him.

With the imperial tours of the Qianlong reign, the style and social function of gardens in Yangzhou appear to have changed, the

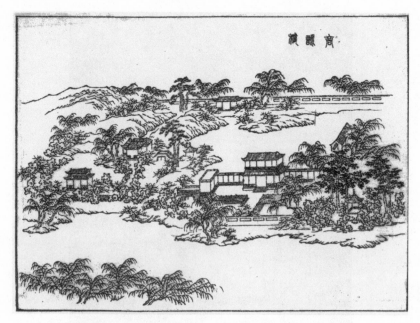

Fig. 16 The Gaoyong Mansion, a palatial garden building
constructed in 1762 (SOURCE: *GLMST*, 38b).

essentially private scholar-garden of the seventeenth and early
eighteenth centuries giving way to the imposing, palace-style
garden constructed for public display. In a reflection of northern
influence, multistory mansions as well as the standard pavilions,
gazebos, belvederes, and covered walkways came to be important features of Yangzhou gardens. The four Huang brothers of
Shexian, who had extensive property holdings in Yangzhou
through the middle decades of the eighteenth century, reportedly
invested a thousand taels in a secret book of garden design that
instructed them in "the method of creating palaces" (*zaozhi gongshi zhi fa*).[77] Their ambitions may be deduced: known as the "four
chief treasures" (*si yuan bao*), these were the princes of Yangzhou
in their time, and like the Zheng brothers in an earlier century,
they competed with one another in garden building.

Some examples of deliberate emulation of Beijing architecture
in Yangzhou can be dated to this period. The most obvious is the
white stupa next to Lianxing Temple, modeled after the stupa
constructed by the Shunzhi emperor in Beijing's Western Park a
century earlier (see Fig. 15).[78] Another example was to be found

in the Almond Garden (Xingyuan), constructed as part of the imperial travel palace (*xinggong*) next to Tianning Temple. The
Hanlin Academy and the Eight Banner offices in the capital provided the inspiration for an artificial hill just within the main gate
and also for a covered walkway in the garden.[79]

The Almond Garden incorporated two earlier private gardens,
the Ma brothers' Temporary Retreat and the Humble Plot (Rangpu), owned by two close associates of the brothers.[80] The change
in ownership of these gardens and their transformation into a
northern-style retreat for the emperor marked a new era in Yangzhou. While the imperial tours were in progress, pomp and cir-
cumstance prevailed in the city. In the years between the tours,
visitors flocked to the city to see the marvels that had been created in the emperor's honor. The grand gardens of the second
half of the eighteenth century set a new tone for urban life.

The Gardens and Urban Culture

One way of looking at a garden is from within its walls, and from
this perspective Yangzhou's gardens can be viewed as having provided highly specific social spaces within which significant rituals
of literati life were played out.[81] Li Dou has left a detailed
description of such literary gatherings: writing desks set out in the
garden with calligraphic tools ready to hand; tea and fruit available for refreshment; brushes flying till the poems were written;
the poems put aside for corrections before being sent off to the
printer; wine and delicacies rewarding each poet for his labors;
and the day ending with musical performances. Sometimes the
poetry session would take the form of a word game: small tiles inscribed with characters were distributed to the guests, who competed to compose poems with the words provided.[82] This was
probably all hearsay on Li Dou's part, since he would have been
too young and probably of too humble a status to attend such
gatherings himself,[83] but other gatherings are documented by
paintings, poems, and colophons produced within the gardens.[84]

The physical environment of such activities is depicted in a
portrait of the Little Translucent Mountain Lodge by the Zhejiang
painter Zhang Geng (1685–1760), now known principally for his
biographical record of Qing painters.[85] Li Dou was apparently

unfamiliar with this painting, but it nicely complements his record of this garden's cultural and social functions, which were performed in an environment designed expressly for literary activities. Ma Yuelu's (1697–after 1766) colophon both describes the garden's features and reveals the garden's purpose. Of the two main buildings, one was designed to provide a view of the distant Yangzi and its island mountains, Gold Mountain and Jiao Mountain, and the other was meant for "browsing through books." Two balconied pavilions were intended for, respectively, "enjoying the cool air" and "gazing upon the shades of evening." There were a plot of peonies, a plantation of plum trees, including both red and green varieties, a grove of bamboo, and seven peaked rocks by a straw-roofed pavilion (*cao ting*). A long corridor wound its way around the garden, "through which one can stroll when at leisure, delving into the very bowels of poetry and chanting verses in accordance with the matter at hand. This has hence been inscribed as the 'The Corridor of Searching out the Phrase.'"[86]

Plants, rocks, water, and buildings were thus carefully composed to form a miniature landscape topographically defined in terms of the enjoyment of literature and nature. The name by which the garden became known derived from its most famous feature, a large stone from Taihu, which Ma Yuelu apparently obtained while the garden was being built or soon after its completion. It could not be displayed upright because the neighbors complained. As Ma explains: "My brother and I had the neighborhood divined and certainly did not wish to put out our good neighbors by strange peaks for visitors to gaze upon. So although the hall takes its name from the stone and the painting has the stone as standing upright, in fact the stone is reclining and will remain thus for the future."[87]

The primacy of the literary function of this garden is indicated by the first name inscribed on the garden gate, the Bookroom on the South Side of the Street (Jienan shuwu), which continued in use beside the more allusive poetic name. One of the two main buildings in the garden was used to house the brothers' impressive book collection, held in 100 great bookcases and including so many rare and fine items that Ma Yuelu's son was able to offer

776 titles to the Four Treasuries literary project (Siqu quanshu) launched by the Qianlong emperor in 1773.[88]

The Ma brothers' gardens, as Ginger Hsü has well described, provided a context for their patronage of painting.[89] The same was true of the Bamboo Garden and the Eastern Garden, both of which had artists-in-residence[90] and were otherwise visited by noted painters. Even the irascible Zheng Banqiao occasionally graced salt merchant gardens. *Jinshi* graduate, sometime district magistrate, poet, and calligrapher, as well as one of the most famous of Yangzhou's eighteenth-century painters, Zheng possessed a visceral hatred for salt merchants. As magistrate of Weixian in the 1740s, he was wont to "order his retainers to kick them, or have them seized by the head, branded like criminals, and thrown out."[91] But to the Ma brothers he presented a fan inscribed with a poem,[92] and for Cheng Mengxing he composed a couplet for inscription in the Bamboo Garden.[93]

Zhang Geng's painting of the Ma brothers' garden shows the most obvious and direct impact of gardens on painting: garden owners liked to have their gardens depicted and commissioned artists for the task. Wang Yun (1652–after 1735) painted the Garden of Retirement on a horizontal scroll in twelve parts, taking the four seasons as his theme. The painting shows in detail the garden's pavilions and belvederes, meticulously placed in their arboreal settings, rendered with such care that one might well suppose the artist to have spent days measuring the garden's dimensions (see Fig. 17).[94] Yuan Yao, as discussed above, provided a splendid "mountains and waters" setting for his fantastic depiction of He Junzhao's Eastern Garden. And at the other end of the spectrum in both scale and interpretation is a gentle, moody rendering of a scene in the Bamboo Garden, painted by Luo Ping (1733–99) while he was in Beijing in 1773 (see Fig. 18).

Luo Ping's late, rather nostalgic work draws attention to the fact that well-documented examples of literary and artistic activity in garden contexts do not date much earlier than the middle of the eighteenth century. Although Jiang Chun is regarded as the Ma brothers' successor as a convener of literary parties,[95] the different trajectories of art history and garden history in Yangzhou suggest that these were decades of social change. The peak period

Fig. 17 Wang Yun, *Xiuyuan* (Garden of Retirement), 1715–20, detail (Lüshun Museum collection; published *Zhongguo meishu quanji*, 10: 103, no. 102).

of painting, as judged by the critical acclaim of artists active in the city and by the fame of their extant works, preceded the peak period of garden construction.[96] Of the individualist artists associated with Yangzhou, most were born in the middle years of the Kangxi reign and died in the middle years of the Qianlong (see Appendix F). The second half of the Qianlong reign, from the 1760s through the 1790s, boasts no really celebrated Yangzhou artist apart from Luo Ping.

The competing demands on merchant fortunes at this time are not difficult to see: both gardens and opera companies flourished at the time that merchant patronage of artists was waning.[97] Investment in palatial constructions such as the Gaoyong Mansion in the Dawn over Shugang Garden and the white stupa by the Lotus Flower Bridge, together with the development of the Kun opera companies, suggests that Yangzhou high society was peculiarly responsive to signals from the capital in the period of the southern tours and intent on providing the emperor (and perhaps everyone else as well) with displays of magnificence.

The imposing buildings of lakeside gardens certainly seem designed for a sort of viewing practice different from that fostered within the Temporary Retreat, where on the Double Ninth in 1742 a few select friends were invited by the Ma brothers to view a painting by Qiu Ying (1494/95–1522). This garden was separated

Fig. 18 Luo Ping, *Drinking in the Bamboo Garden*, 1773.
© The Metropolitan Museum of Art collection, 2004,
13.220.34

from the outside world by a high wall. As Quan Zuwang noted,
"Were we to step outside the [garden] gate several paces, we
would cringe to see the dusty trails and polluted streams. Within,
the quiet atmosphere stirs one to otherworldly thoughts."[98] The
lakeside gardens, by contrast, seem fully open to the view of
passersby, whose numbers increased greatly over the period of the
southern tours.

The Great Age of Tourism

From the feast of the God of Wealth in the first month of the year through the Double Ninth Festival in the ninth month—the date of the famous gathering of literati in the Ma brothers' garden—Yangzhou was crowded with visitors. Apart from the standard holidays, there was a festival to mark the blooming of various flowers: in spring the plum and peach trees, followed in summer by the peonies, chrysanthemums, and lotuses, and in autumn by the cassia and hibiscus.[99] Flowers were everywhere in Yangzhou: "Yangzhou people rich and poor all wear flowers."[100]

The flower festivals are of special interest if, as Li Dou suggested, they were comparable to other festivals in terms of the number of visitors they attracted to the city. Their high point probably occurred at the time he was writing, in the later decades of the eighteenth century. Horticulture was well established around the city before the eighteenth century, but as a place for viewing flowers Yangzhou must have gained greatly from the development of the great merchant gardens along the lake. The flowers were not only lovely to gaze upon: the aesthetic pleasures they offered were grounded in a well-established iconography. "Weave a wattle fence and grow chrysanthemums, like Warden Tao in days of yore," advised Ji Cheng in his *Craft of Gardens*. "Hoe the hillside and plant plum trees, following in the ancient footsteps of Lord Yu."[101]

The visitors were carried around the city and out to the lake by the painted barques that gave Li Dou's famous guidebook its name. These boats illustrate the direct effects of garden building on the tourist trade, for they increased in number in tandem with the gardens. Li listed by name three craft from the Shunzhi period, five from the Kangxi period, six from the Yongzheng period, and twelve from the early years of the Qianlong reign. Then, "from this time on, with the dredging of the Lotus Flower Canal, the waterway through to Pingshan Hall became a great thoroughfare, and the painted barques increased in number daily."[102] Traffic on the canals grew so dense that regulations were introduced; boats had to register to operate from one of twelve wharves in and around the city. Li Dou found 235 registered boats at the time of his writing, and there were numbers of unregistered craft that, although prohibited from calling at the wharves, were free to seek passengers elsewhere.[103] The total number of boats available for hire increased during the busiest seasons, and the price of hire soared.[104]

The effects on the urban economy were incremental. From Waichengjiao in the southern part of the New City, where three separate wharves serviced passengers, north to Tianning Gate, residences on both sides of the Little Qinhuai Canal were interspersed with brothels, teahouses, and taverns.[105] Outside Tianning Gate, petty entrepreneurs on Lower Shopping Street (Xia maimai jie) made a living by hiring out lanterns at eight coppers apiece for use on the boats.[106] The name Shopping Street originally referred only to the line of markets on the north bank of the canal, which traded in a variety of rare and precious goods. In the later eighteenth century markets expanded to the south side of the canal, by the city wall, necessitating the differentiation of "Upper" and "Lower" Shopping Street.

Boats bearing sightseers turned west along the canal from Tianning Gate and were poled or rowed to Rainbow Bridge, the gateway to Slender West Lake. The grounds of a tavern here had been converted into an official garden at the time of the imperial tour of 1757, but the original proprietor was contracted to maintain the tavern. It operated around the clock: flags advertised it by day and lanterns by night, and wines from Shaoxing and various Yangzhou counties were served.[107] From Rainbow Bridge, the boats proceeded on to the lake itself, winding their way through inlets and around islets. Wherever they went, teahouses and taverns were to be found. A secluded grove of deciduous pines beyond Buddhist Sea Bridge (Fahai-qiao) became a popular retreat after the Lively Ladle Tavern (Yishuang jiusi) opened there.[108] As the custom of "eating in the wilderness" (*yeshi*) became popular, restaurants were also established outside the city walls: Han Garden, Stay-Your-Step, Guo Hanzhang's Place, Suzhou Refreshments, The Overflowing Goblet, among others.[109]

The gardens on the lakeside were probably accessible to casual visitors. According to Wu Woyao (1867–1910): "Originally Yangzhou had the most gardens of any place, all built by the salt merchants. In the morning people were allowed to wander around [in them] and enjoy themselves. In the afternoon, the garden owner would come to the garden and invite guests along, or put on a theatrical performance."[110] But some measure of control was exercised over who entered the gardens. Ruan Yuan long harbored a resentment that in his impecunious youth he had not been allowed access to the gardens in Yangzhou.[111] This reveals not only an expectation that gardens should be accessible but also the fact that there were obstacles to people entering them at will. During the southern tours of the Qianlong emperor, the Lianghuai salt controller organized guards to patrol points of access to the gardens.[112] A diminished form of surveillance may have been the order of the day at other times.

A feature distinguishing the lakeside gardens, however, was their relative openness to the view and apparently the entry of passersby. Gardens within the city were invariably surrounded by high walls. This did not mean that they were shut off from the passing parade: entry could be had by casual visitors, as shown by well-documented cases in the late Ming.[113] But the lakeside gardens visibly offered greater accessibility. They opened onto the lake itself and were easily entered. Among those who took a boat to view the gardens were women, some from the pleasure quarters of Yangzhou, others in boats equipped with screens to protect them from lewd glances.[114]

Extramural developments were replicated inside the city. A number of well-known intramural gardens developed in the later eighteenth and nineteenth centuries.[115] Buildings and landscapes within the walls, moreover, duplicated elements found at other sites. This might be on a small, domestic scale: Li Dou, who planted more than ten plum trees outside his house in Kingfisher Blossom Street;[116] the artist Gao Xiang (1688–1752), who lived near the Ma brothers, dwelt in "a crude abode in one of the back alleys," but nonetheless "gladly surrounded himself with trees and flowers."[117] In teahouses and brothels, the garden style could be fully elaborated in miniature. The general principles of teahouse design as impressionistically set down by Li Dou include many elements found in any garden, including a view:

The Double Rainbow Terrace is a teahouse at the North Gate Bridge. It is a two-story building with five pillars. From the eastern wall, a window opens onto the canal, and one can gaze into the distance. The teahouses of my native district are the finest in the empire. Many people make their living out of them, setting themselves up either by coming up with the money to create a garden or by selling the old family house and doing away with the garden. With storied buildings, terraces, pavilions, and huts; flowers, trees, bamboo, and stones; tumblers, plates, spoons, and chopsticks—there is nothing that is not elegant and beautiful.[118]

Customers crowded such places during the holiday seasons. In Tianning Street, visitors would sometimes spend a day at a time in the Green Lotus Vegetarian (Qinglianzhai) Teahouse. The proprietor was a monk from Anhui, who served his customers tea brewed from leaves plucked in his native hills.[119] And at the Hit-the-Crock Tavern (Pugangchun jiusi), again in Tianning Street, holidaymakers could "drink and talk their fill" over fish cooked with bamboo shoots.[120] Restaurants (*shisi*) were generally attached to noodle houses (*mianguan*), and the proprietors would accept advance orders for an evening meal to be enjoyed by sightseers after their return from a day out on the lake.[121] The New City was the major beneficiary of such economic activities. Of thirteen teahouses regarded by Li Dou as the most popular within the city, only one was in the Old City, a vegetarian teahouse located at West Gate.[122]

Bathhouses proliferated alongside the teahouses and structured a way of life encapsulated in the saying "In the morning, skin holds water; in the afternoon water holds skin."[123] According to Li Dou, the fashion for bathing originated with a bathhouse in the market town of Shaobo, north of the city. In Yangzhou, the first bathhouse was built outside Xu Ning Gate, southeast of the city. The date of founding is not known, but the location shows that it was strategically located to catch the lucrative salt merchant market. By the late eighteenth century, a large number of bathhouses had been established within the walls, of which at least four were again positioned to cater to the salt merchant families. Bathhouse life was directed at the provision of luxury and could be very costly. On the eve of going to meet his bride (*qinying*), a man might spend tens of taels of silver at a bathhouse, no doubt entertaining his friends as well as himself.[124]

The quickening of leisure in the later eighteenth century, spurred as it was by the seasonal influx of visitors, sheds some light on the altered profile of painting in the city. Although the famous painters of Yangzhou were active mainly up to the early 1760s, there seems to be no evidence that the number of painters in the city declined after that time. "The calligraphers and painters of Yangzhou are very numerous," wrote Li Dou, who proceeded to provide a list of 149 artists active in Yangzhou during the Qing up to his own time, including the "eight eccentrics," although they were not then known as such.[125] A cohort of artists emerged from a painting school at the Daoist Shrine of the Three Worthies around the middle of the eighteenth century and must have supplied the local art market through the later decades.[126] Yet in 1768 a plum blossom painting by the acclaimed Jin Nong (1687–1764) was sold by a poor old man at a market stall for a mere 100 cash. Paintings at least of this genre would appear to have been losing their value within a few years of the passing of the artist.[127]

At precisely this time, however, other genres of art were very popular. They included portraits of "lovely ladies" (*meiren*) by Shi Panzi, who was able to charge 30 taels of silver for paintings rendered on a scale anywhere between miniatures and close to life size; pictures of donkeys by Shi Yuan, dubbed "Donkey Shi" to demarcate him from "Lovely Lady Shi"; and portraits by a popular father-and-son team, Ding Gao and Ding Yicheng, natives of Danyang in Zhenjiang prefecture, just south of the Yangzi.[128] These were not the works that connoisseurs collected and literati praised, but they were certainly those that might have appealed to holidaymakers out for a stroll in the town. In an oft-quoted commentary on a little-known Yangzhou painter, Wang Yun noted: "Of old, there was a saying about painting in Yangzhou: 'For gold, paint portraits; for silver, flowers. If you want to beg for your food, paint landscapes.'"[129] When this saying originated is not clear, but artists such as Zheng Banqiao and Jin Nong had done very well on works that did not feature portraits. Lovely ladies, donkeys, and portraits, on the other hand, seem precisely the sorts of subjects that would sell well to tourists. If they supplanted the bamboo paintings of Zheng Banqiao and the plum blossoms of Jin Nong, it may have been because they appealed to a different market.

The City in Retreat

Over the eighteenth century, a pattern of first gradual and then rapid expansion of property development outside the city walls is apparent. The chronologies of three major gardens developed during the Yongzheng period provide the beginnings of an explanation for why the lakeside gardens did not last very long, for they show that it was only the continual rise of new families and new money that ensured the ongoing development of a garden culture in Yangzhou through the eighteenth century. Cheng Mengxing's Bamboo Garden had fallen into disrepair by the 1760s. It was restored and rented by salt controller Lu Jianzeng and then later purchased by another salt merchant, Wang Ting-zhang.[130] He Junzhao's Eastern Garden was reduced to half its size in the later eighteenth century and was incorporated into the grounds of the Lianxing Temple.[131] The Garden of Ten Thousand Stones lasted no more than Yu Yuanjia's lifetime.[132] Afterward, the stones were procured by head merchant Jiang Chun for his city garden at Kangshan, located in the southeastern corner of the New City, just within the walls.[133]

Sometimes it is not known what happened to a garden, or even where it was located. This forgetfulness is remarkably apparent in the case of the Ma brothers' Little Translucent Mountain Lodge, the precise location of which was long unknown to historians of Chinese gardens. The discovery of Zhang Geng's painting of the garden in the early 1980s led to a most surprising solution to this puzzle, for colophons on the painting showed that the Ma brothers' garden ended up in the hands of Principal Merchant Huang Zhiyun. In the 1820s Huang used the site to construct the Ge Garden, the largest and best-known of Yangzhou's extant gardens and one frequently studied by garden historians.[134]

The limited life-spans of these famous gardens illustrate the dynamism and also the fragility of the Yangzhou garden. It existed in a particular social context, which appears to have been in part generational. Some Yangzhou gardens passed from father to son, but most changed hands, were broken up, and were redeveloped in other contexts by other men. The period of the Qianlong emperor's southern tours, between 1751 and 1784, accounted indeed for not much more than a generation. In these years, judging by the places visited by the emperor, the gardens of Jiang Chun, Wang Chengxin, Hong Zhengzhi, and other salt merchants of Huizhou ancestry flourished (see Appendix E). Soon after the passing of this generation, the number of salt merchants in the city began to decline.

The history of Yangzhou gardens suggests a more nuanced periodization of the city under the Qing than is often apparent in references to it. Saeki Tomi, in a discussion of the lavish lifestyles of the salt merchants of Yangzhou, drew on claims in two nineteenth-century editions of the Jiangdu county gazetteer to make a general point about the city's prosperity. In the 1811 edition, he noted, Yangzhou's gardens were described as "the greatest in the empire," and in the 1883 edition, the Qianlong-Jiaqing period was referred to as an era in which the salt trade flourished and "many of the merchants created gardens." In elaborating on these references, he pointed to particular gardens, including Cheng Mengxing's Bamboo Garden and Zheng Xiaru's Garden of Retirement, neither of which lasted even into the last decades of the Qianlong reign.[135]

This blurring of historical periods is facilitated by the lack of clarity in the exact dating of gardens and the lives of garden owners, by the length of the Qianlong reign and its overlap with the Jiaqing reign, and by the dramatic developments in international relations during the Daoguang reign (1821–50), which caused Chinese scholars to differentiate rather sharply between the Qian-Jia (1736–1820) period and what followed. A closer examination of the chronology of garden construction and concomitant developments, including tourism, points to a few decades in the late eighteenth century as the critical period of urban expansion and to a period of contraction beginning soon thereafter. In the 1840s the Parade Ground was still humming and the nine lanes were still busy, but when visitors to the city took a boat out onto the lake, it was to behold the gardens in ruins.[136] When we consider that all the gardens built in the nineteenth century were constructed within the city walls, it does not seem too far-fetched to describe the city as retreating into itself at this time.

CR

PART IV

Hui City, Yang City

F rom economic and administrative foundations laid in the late
Ming and redeveloped in the early Qing, Yangzhou burgeoned
during the eighteenth century to become one of the most beautiful, culturally productive, and socially vigorous cities of the era.
Throughout the century, although waning in importance by the end, the
Huizhou merchant-gentry were a dominant social force in the city.
Their prominence points to a fundamental characteristic of Yangzhou
society as it evolved between the late sixteenth and the early nineteenth
centuries: a social division between the very rich and everyone else, between salt merchants and those who were not salt merchants, and between Huizhou people and Yangzhou people.

There is an easy slippage between the place-name "Huizhou" and the
occupational term "salt merchant" because they were so often aligned.
Taken together, these terms signify a social grouping that differentiated
one section of the urban population of Yangzhou from the others. There
were Huizhou people who were not salt merchants, and there were salt
merchants who were not from Huizhou, but there were no salt merchants who were natives of Yangzhou. In other words, Yangzhou society featured a cultural division of labor.[1] Huizhou people made up the
merchant and gentry elites; local people lived off land rents or supplied
services, or both.

Of the chapters in this part of the book, Chapter 9 identifies the critical place of Huizhou in the discursive structure of Yangzhou as a city that produced women. The roles of women in the local economy reveal the relative weakness of local society in face of the economic and organizational strengths of the Huizhou diaspora. Chapter 10 shows that Huizhou men, demonstrably significant as urban property owners, were equally important in the development of urban institutions. Their sustained prominence in Yangzhou society through the seventeenth and eighteenth centuries paradoxically provides grounds for a reassessment of the thesis of "blurring of social boundaries" so often advanced in studies of eighteenth-century China. And, as we shall see in Chapter 11, by the early nineteenth century, writings by local scholars were giving expression to a strong consciousness of native place at a time when the Huizhou presence in the city was weakening. Rather than leading to a long-term weakening of native-place particularism in Yangzhou, interregional trade and migration appear to have produced an awareness of native place.

C03

NINE

City of Women

Speak of 'Yangzhou,'" wrote Zhu Ziqing, "and people immediately think of a place that produces women."[1] Zhu grew up in Yangzhou, and in a well-known essay written during the anti-Japanese war, when he was living far away in Kunming, he would claim Yangzhou as his hometown.[2] He was not, however, a native of the city. The family was registered in Shaoxing, and he was born in Haizhou, on the northern coast of Jiangsu. They moved to Yangzhou in 1902, when Zhu was four years old, and he lived there long enough to observe and absorb much of the culture of the city while it was yet under Qing rule.

The city of Zhu's childhood was not quite the flamboyant place of a century before. By early twentieth-century standards it was a conservative place, hardly touched by the new technologies and the social changes that had young women cycling along the streets of the port cities of Shanghai and Tianjin. Zhu was frankly bemused by Yangzhou's reputation for beautiful women. "I grew up in Yangzhou," he commented, "and I never noticed any remarkable women around. Perhaps at that time there were as yet very few women who went out onto the street."[3]

Zhu's retrospective impressions, based on his memories of what he had witnessed as a boy, are not necessarily reliable. This comment draws attention, however, to the relative visibility of women as a measure of time and a marker of place. In the late sixteenth century, women in China were "secluded and virtuous," according to one European visitor. They tended to "keep themselves close," and were generally speaking a rare sight "unless it was an old crone" or perhaps "some light huswives [sic] and base women."[4] By the late nineteenth century, it was still praiseworthy for people in a north Chinese town to be able to say of a neighbor's daughter that they knew nothing of her; they had never seen her.[5] The saga of Zhu Ziqing's own betrothal bears this out: four matches were negotiated for him with girls unseen by him or his parents. The fourth girl, who became his first wife, was doubly unseen: she had hidden when the matchmaker came to inspect her, and another girl had been shown in her place.[6]

As Zhu Ziqing realized, the women for whom Yangzhou was known were not its secluded and virtuous wives and daughters. The phrase "producing women," he explained, "actually refers to concubines and prostitutes; the word 'produce' is used in the sense of 'producing wool' or 'producing apples.'" He had in mind here the raising of young girls for sale into concubinage or prostitution, a practice known as "raising thin horses" (*yang shouma*): "A passage on 'the thin horses of Yangzhou' in *The Dream Memories of Tao'an* records this phenomenon, although it is not something with which I am at all familiar. But the customs of taking concubines and consorting with prostitutes are gradually disappearing. I think the phrase 'producing women' will sooner or later lose all meaning."[7]

Despite his reference to changes in China's sexual mores, Zhu confirms in these lines the sustained power of the association between Yangzhou and women. Zhang Dai's *Dream Memories of Tao'an* is a well-known collection of reminiscences on late Ming society as recalled by the author after the fall of the dynasty. Zhu's citation of Zhang's work suggests how readily such writings leaped to the minds of educated Chinese. In the 1950s, Lin Yutang (1895–1976) translated Zhang Dai's passage on the "thin horses" for inclusion in one of his English-language anthologies designed to introduce Chinese culture to the West.[8]

Concubines and prostitutes are among the most common social phenomena mentioned in association with Yangzhou in the late Ming and Qing periods. The frequency with which they were described was a function in part of the social positions they occupied. They were relatively visible and accessible to the men who wrote of them and sometimes, by virtue of their training, were able themselves to write, although not necessarily of the same matters as men. They were associated with certain social spaces in the city and particular social activities. Considered alongside less visible women, the virtuous wives and daughters of Yangzhou, they provide some insights into the hierarchy of native places in the city, that hierarchy's relationship to local economic activities, and the ambivalences attendant on historical memories of one of the great cities of Ming-Qing times.

Women on View

What does it mean for a place to be described in terms of "beautiful women"? This cannot be a self-description: the observer, the outsider, the lascivious procurer of someone else's daughter, is immediately apparent. Also apparent is the flow of money — used for the acquisition of sexual services among other things — and of the men who made it. The buyers of thin horses were probably merchants or traveling officials, members of that small but important sector of Chinese society known as sojourners. In 1604 it was said that non-natives of Yangzhou outnumbered locals by twenty to one.[9]

Yangzhou was the quintessential sojourner city, and the local sexual economy was shaped by the exogenous character of its male population. It is probable that there were far fewer women than men in any late imperial city. Officials, merchants, carters, boatmen, rascals — in brief the sort of men who made their living from the cities — were by nature mobile. Their wives and children, if they had them, were as likely as not to be living in their home village. The situation was favorable for the establishment of secondary households, and also of brothels, which in the early seventeenth century appear to have proliferated in Yangzhou.

The late Ming was reportedly a time when prostitutes were everywhere to be seen. "In the big cities they run to the tens of thousands, but they can be found in every poor district and remote place as well, leaning by doorways all day long, bestowing their smiles, selling sex for a living."[10] Thus wrote the well-traveled Xie Zhaozhe. Commenting on the situation in Yangzhou, Zhang Dai made mention both of courtesans (*mingji*), who "hide themselves away from the public view," and of common whores (*waiji*), whom in Yangzhou he estimated to number between 500 and 600. The *waiji* were true women of the street; they emerged from their quarters as evening approached and hung around the doorways of teahouses and taverns in search of business. Late at night they were to be seen tripping after their clients to where they would entertain them, the way lit by lantern bearers in accordance, as it happened, with the stipulations in the *Book of Rites* for the nighttime excursions of women.[11]

The thin horses were neither *mingji* nor *waiji* and belonged to a different category of the sexual economy. The geographer Wang Shixing (1547–98), writing late in the sixteenth century, recorded:

In Guangling [i.e., Yangzhou] there are families who bring up slave girls, a practice commonly termed "raising thin horses." . . . Their foster mothers train them in female propriety and the forms of politeness. The superior ones can perform on the zither, play chess, and sing. The very best are accomplished in calligraphy and painting, the less able in embroidery and other female arts. In their readiness to serve a man's wife and show deference to their peers, they are most seemly, neither flouting convention nor giving way to stupid quarrels. They captivate men's hearts and minds. So all the men who want concubines go to Guangling.[12]

To judge from this early commentary, the trade in thin horses emerged as a notable phenomenon of Yangzhou no later than the last quarter of the sixteenth century.[13] It is widely noted in writings of the late Ming and early Qing periods.[14] In consequence, thin horses became a metonym for the city. The characterization was made quite pointedly in Kong Shangren's famous *Peach Blossom Fan* (completed 1699), in which icons are invoked to represent various places:

> Now in the ancient palace of Wu,
> The women's quarters open anew,
> And in Yangzhou the thin horses
> Are one by one put through their paces.
> Drums of Huaiyang, Kunshan strings,
> Suzhou girls and Wuxi songs.[15]

A defining characteristic of the thin horses, apart from the fact that they were bought and sold in an unusually frank way, was their education. Wang Shixing described this as consisting of instruction in instrumental music, singing, painting, calligraphy, and chess—a set of accomplishments similar, as Dorothy Ko has shown, to those regarded as desirable in the gentrywoman.[16] The emphasis on acquisition of such skills marked the thin horses as products of their time. The spread of literacy was a feature of the late Ming and the Qing dynasties, and the idea of the accomplished woman, although not uncontroversial, was gaining currency in this period.[17] Yangzhou was not among the great producers of women's writing in the late imperial period. That honor went primarily to the Jiangnan cities of Hangzhou and Changzhou, followed by the eastern Lake Tai districts,[18] places that, unlike Yangzhou, boasted strong, stable gentry societies. The definitive "accomplished woman" of Yangzhou in the late sixteenth and seventeenth centuries was the thin horse.

Shen Defu mocked the accomplishments of the thin horses as limited to mastery of one or two tunes, the opening moves in a game of chess, and a few well-chosen characters in calligraphy, but the level of education acquired by a particular girl was immaterial to the significance attached to the accomplishments themselves.[19] They added to a girl's value. Hence a ditty popular in early Qing Yangzhou:

> At thirteen she's painting and studying chess,
> At fourteen it's music and verse.
> Who cares about talk of "raising thin horses"?
> The girl's an adornment to the family house![20]

All this training suggests much social activity centered on the thin horses and a commercial enterprise of some scale. According to Zhang Dai, "hundreds of people in Yangzhou made their living from 'thin horses.'"[21] His scathing and hyperbolic account of the practice might suggest that the raising of thin horses was occurring on the social margins, in the shadowy world frequented by roués such as himself. Shen Defu's more sober account places the thin horses deep within the heart of Yangzhou society. According to Shen, "even officials and rich families keep girls for profit, and some raise dozens." He also observed the complex kinship ties created by the removal of a girl first from her natal home and then from her foster home: "There were sometimes wealthy visitors [in Yangzhou] searching for their mothers' relatives. Grief and joy were frequently to be seen at the one time."[22]

The urban scene in late Ming Yangzhou was enlivened by the spectacle of thin horse weddings in progress. Shen Defu "day and night saw the bridal chairs coming and going ceaselessly to the sound of drums and music."[23] Zhang Dai evoked a picture of color and noise and movement, with "bridal chair and lanterns, lights and torches, sedan carriers and bridal attendants, paper candles, sacrificial fruits, meats and sweets" all being rushed through the streets within minutes of the conclusion of the transaction.[24]

Also on view in late Ming Yangzhou were the "official prostitutes" (guanji), the descendants either of Mongols who had remained in China after the defeat of the Yuan in 1368 or of condemned Chinese officials. These unfortunate women were registered either as musicians (yuehu) or as beggars (gaihu), déclassé social categories that allowed no upward mobility.[25] During the spring festival, the guanji were on public show. According to local custom, on the eve of the festival the prefect would travel to the Fanli Shrine east of the city to welcome the spring. The guanji accompanied him in solemn procession, dressed up as officials and servants.[26]

Wu Qi, recording this social practice from the other side of the dynastic transition, may have felt he was writing of a lost world. The Manchus were not impressed with the courtesan culture of the Ming and early endeavored to sunder the link between prostitutes and officials, which is to say between the demimonde and the dynasty.[27] In 1673, the use of prostitutes in ceremonies to welcome the Spring Ox was prohibited throughout "the provinces, prefectures, departments, and counties."[28] This evidently signaled the complete disestablishment of the official prostitutes in Yangzhou. Li Dou, writing in the late eighteenth century, referred to his tales of the guanji as "accounts of things in bygone years." Even private houses of prostitution were prosecuted at this time, and the women fled elsewhere either to avoid punishment or to find a more congenial environment for their trade.[29]

References to the "thin horses" also decline over time. It is

unlikely that the practice disappeared completely, since it is mentioned without ambiguity in other places in the Lower Yangzi region, but it ceased to be the most obvious point of reference for the
city. The novelist Wu Jingzi (1701–54), who lived for a while in
Yangzhou, may have been thinking of the practice when he drew
attention to a concubinal marriage arranged in Yangzhou between
an educated girl and a salt merchant.[30] His contemporary Zheng
Banqiao, who spent many years in the city, appears to allude to the
thin horses in a verse,[31] and Xu Ke, writing in the Republican era,
averred that the practice was still flourishing in the 1860s.[32] But Li
Dou, writing in the later eighteenth century, and Zhou Sheng, in
the early nineteenth, made no mention of thin horses. A reduced
profile for both the *guanji* and the thin horses in Yangzhou is consistent with the changes in courtesan culture described by Dorothy
Ko, who identifies the early Qing as the end of a period when courtesans had relatively high status, were intimately a part of the literati world, and could attain recognition as writers.[33]

The impact of such changes on urban life should not be over-
stated. The demimonde in Yangzhou shows some strong continuities from the early seventeenth through the early nineteenth
centuries, with common prostitutes still evident on the city streets
and high-class courtesans continuing to entertain officials and
scholars. In Li Dou's time, a century after the disestablishment of
the *guanji*, street girls regularly came and went in the neighborhood of an artist who lived by the inner city canal. He liked to
take their portraits and produced "no less than a hundred and
several ten" in all.[34] Huang Shen's portrait of a woman carrying a
qin, her sash trailing almost to the ground, may have been produced in similar circumstances (see Fig. 19). On summer days,
women from the pleasure quarters could be seen out on the lake,
enjoying the sights from the distinctive boats used by courtesans.
And in the gardens high-class courtesans provided company for
well-born men. The courtesan Su Gaosan once attended an archery contest in the Garden of Pure Fragrance (Jingxiang yuan),
built by prominent salt merchant Jiang Chun in the middle of the
eighteenth century.[35] She asked to participate and scored three
bull's-eyes in a row. The scholar Lin Daoyuan, a relative of
Jiang's and the host of the contest, organized a poetry-writing

Fig. 19 Huang Shen, *Xieqin shinü* (Lady carrying a *qin*), 1724, painted during the artist's first year in Yangzhou (Taizhou Museum collection, published in Lu Hao, ed., *Yangzhou baguai huaji*, p. 39).

session in celebration of the event, and eminent scholar-official Ruan Yuan later wrote some matching verses.[36]

Su Gaosan's circumstances suggest that in eighteenth-century Yangzhou courtesans had a prominent, recognized, and normalized place. She lived in a commodious house by the canal at Waichengjiao:

Within the gate was the dwelling proper, of three [main] rooms. Left and right were the side buildings, and in the middle a piece of open ground about fifty paces [wide]. A water gate opened out from the middle of the enclosure facing the canal. There were seven rooms upstairs [in the main building], and each of the side dwellings had two rooms, providing alternative lodgings. One was a sitting room for guests, the other for sleeping. Both side dwellings were connected to the central building. In the latter, two rooms were for entertaining guests. The other room was partitioned by a wind screen of green glass and was Gaosan's place of rest and relaxation.[37]

The repeated references in this account to the reception and entertainment of guests allow the reader to visualize Su Gaosan as very much in charge of her own fortune and destiny, within the confines of what was possible for a woman of her class. With such a large house at her disposal, she must have had a large number of servants and perhaps girls in training for when her own time as a courtesan was past.

Su Gaosan's circumstances seem generally consistent with those of the courtesans in both the preceding and the following century. Zhang Dai would no doubt have found familiar the elaborate hierarchy of prostitution in Yangzhou described by Zhou Sheng in the 1840s. At the summit of the demimonde were those who "kept the light on," sang and danced for their clients, or puffed on opium or tobacco pipes as they conversed with them through the wee hours. Their sexual favors could not be bought and were bestowed only on those whom they liked. "Although they were not registered," wrote Zhou, "they behaved like official prostitutes." Their guests were high officials and men of well-placed families. Traveling merchants, lesser officials, and yamen underlings found their pleasures among the second rank of courtesans, who might openly invite a client to spend the night. Courtesans of the third rank, distinguished by their lewd manners, catered to the city riff-raff (shijing wulai), but they were evidently riff-raff with means, for they could aim higher than street-walkers.[38]

Zhou Sheng also attested to the sight of common prostitutes waiting for custom in the streets. In a vivid pen portrait, he described the sensations they stirred in his breast: "Returning [from our outing] at twilight, we would find the ladies just emerged from bathing, sitting each before her door. Their short jackets, silky trousers and sashes trailing to the ground, the flowers of their headdresses dancing like butterflies, and their scent detectable from ten paces away were enough to intoxicate one."[39] The scene cannot have been too different from that beheld by Zhang Dai two centuries earlier.

Overall, the documentation of courtesan life in Yangzhou during these two centuries suggests that, like the city's gardens, this was a steady feature of urban culture during the period of the Huizhou ascendancy, although outlasting it. If the character of the Yangzhou demimonde changed in some respects during this period (see below), its sustained visibility from the late sixteenth to the mid-nineteenth century was testament to a strong consumer ethos born of the salt trade, as well as to a pattern of gender relations in which major disruptions finally amounted to no more than minor fluctuations over time.

Consumers and Producers

Women played distinctive roles in the consumer economy of Yangzhou. Concubines and prostitutes, arguably even wives, were themselves major consumer items; but women from all walks of life were active patrons of the Yangzhou marketplace. Most obviously, their attention to personal adornment stimulated a demand for cloth, jewelry, and hairpieces, significant items in the local retail trade.

A decline in fashion has been posited for the Ming-Qing transition. Specifically, Kenneth Pomeranz has suggested that in the "long eighteenth century" Qing women may have competed with one another by writing poetry rather than by dressing up.[40] The evidence from Yangzhou suggests, to the contrary, a steady consumption of fashion from the seventeenth to the nineteenth centuries and beyond. "In all the empire," complained Li Gan in the 1690s, "it is only in Yangzhou that clothes have to follow the times."[41] To the extent that this comment concerned women, fashion would seem to be related to the relatively high level of visibility of women in the city. Li Gan was privy to the fact that "concubines leaving their wealthy homes to go out wrap themselves in *yu* silk [and] protect their calves with brocaded and embroidered silk."[42]

The eighteenth century showed no diminution in attention to fashion. Yangzhou scholar Wei Minghua has exhaustively documented sartorial practices in the city in late Ming and Qing times, and although he fails to discriminate consistently between one century and the next, commentaries on trends in the long eighteenth century show a society preoccupied with personal appearances. Most tellingly, Li Dou, writing in the late eighteenth century, had precisely the same observation to make about Yangzhou dress that Li Gan had in the seventeenth: "Clothes worn in Yangzhou are always in the newest style (*xinyang*)." He provided no details on the actual cut of clothes but did itemize the different sorts of cloth that went in and out of fashion, silks and satins in different colors succeeding each other in popularity with dizzying rapidity.[43] Dyers flourished in this market, "the Yi household at Little East Gate being the best [in the business]."[44]

Apart from clothes, extraordinary attention was paid to hairstyles, coiffures with names such as "butterfly," "pair of flying swallows" and "flower basket" rivaling one another for attention. Depicting women earlier in the century, Yangzhou painter Yu Zhiding paid meticulous attention to their different hairstyles, right down to the closely observed hair ornaments (see Fig. 20). According to Li Dou, Yangzhou women dressed their hair differently from women in other places.[45] The market for false hairpieces and hair adornments was thriving, and one shop specialized in them.

In the greater Jiangnan region, "Yangzhou style" competed with "Suzhou style," and in the early years of the nineteenth century, both were in evidence in Shanghai.[46] By this time, new sorts of garments and new cuts in existing styles were proliferating in Yangzhou, as documented by Ruan Yuan's kinsman Lin Sumen.[47] Especially notable was the growing width of the sleeves on women's gowns;[48] the subsequent spread of the fashion to the court and women of the banner families invited imperial denunciations.[49]

Fig. 20 Fashions from the early decades of the long eighteenth century. Different styles of dress and contrasting hairstyles were depicted by Yangzhou painter Yu Zhiding. (*Left*) *Qiao Yuan zhi sanhao* (The three good things of Qiao Yuan), n.d., detail (Nanjing Museum collection, published in He Gong-shang, ed., *Lidai meiren huaxuan* [Taibei: Yishu tushu, 1984], p. 106). (*Right*) *Shuangying* (Pair of heroines), 1710, detail (Qinghua Fine Arts Academy collection, published Liu Rendao, *Zhongguo chuanshi renwu minghua quanji*, 2: 351, fig. 341).

Cosmetics, shoes, yarn for embroidery, flowers for their hair, fans, and tobacco and associated paraphernalia such as pouches and pipes were among the other items that women purchased, along with bridal wear when they married and lingerie such as the stomacher or vest and the cicada jacket, worn when they were at their toilet.[50] Tailors, shoemakers, and the practitioners of a variety of other crafts in Yangzhou must have done very well out of women's wear.

As producers of goods, women in Yangzhou were much less active, a fact with profound implications for any analysis of the urban economy. In the late Ming, the practice among poor families of selling their daughters to be raised as concubines or courtesans in Yangzhou was attributed to the fact that women had no gainful employment.[51] Gainful employment for women in the

Ming and Qing dynasties basically meant spinning and weaving. Late imperial China, despite its impressive cities, had a largely agrarian economy. Women throughout China at this time contributed in varying degrees to farmwork, in some places only in minor ways as the season required, but elsewhere regularly, particularly in the double-cropping wet-rice areas of southern China. They were otherwise responsible for the manufacture of cloth and clothing for the household. Hence the saying *nan geng, nü zhi*: "the man plows, and the woman weaves."

In the sixteenth and seventeenth centuries, these skills were increasingly being turned to the supply of a growing commodity market in the Lower Yangzi Delta, of which Yangzhou could be considered part. Female labor contributed significantly to economic growth and urbanization at this time. As spinning wheels hummed, market towns proliferated, and the waterways of Jiangnan were crowded with boats carrying goods to buy and sell.[52] Tanaka Masatoshi sums up the regional specialization in textile production in Jiangnan thus: "Cotton textiles were produced in Songjiang and Suzhou prefectures, flax in Changzhou and Zhenjiang prefectures, silk textiles and raw silk in Huzhou and Jiaxing prefectures."[53]

What were women spinning and weaving in nearby Yangzhou? Whatever it was, it was in such small quantities and of such little commercial importance that Yangzhou may be deemed a nonparticipant in the great handicraft revolution of the sixteenth century. Scattered references to household industry in Yangzhou prefecture in the Ming and Qing periods suggest a rural economy operating fairly close to a subsistence level and boasting little diversification. Zheng Banqiao once complained that women in his native Xinghua spent too much time playing mahjong. His recommendation for their moral improvement draws attention to the low level of textile technology in the average household in Xinghua in the middle of the eighteenth century: "The women of our town cannot weave coarse silk or cotton, but they can still cook and sew and do their part nobly."[54] In the neighboring prefecture of Huai'an around this time, the prefect grumbled that no one in his jurisdiction knew how to grow cotton, let alone process it.[55] One of his successors blamed women for the lamentable state of silk production in the locality. "Few people [in Huai'an] know anything about [sericulture]," he observed, "and those who do are unwilling to undertake it, and so very few mulberry trees are planted." Part of the problem, in his opinion, was the women, who "are lazy and just like having a good time."[56]

Elsewhere in Jiangbei, women were obviously industrious but at no great profit to themselves. In Baoying, women embroidered and sewed in their spare time.[57] In the coastal district of Dongtai, women bred silkworms as a leisure activity, but the market for the silk produced was very small.[58] Cloth was imported from Jiangnan. In the late eighteenth century, merchants from Zhenjiang handled the retail trade in silks and satins, and Wuxi merchants sold "no less than several tens of millions" of cloth pieces to "Huai'an, Yangzhou, Gaoyou, Baoying, and other places."[59] Spinning and weaving as commercial enterprises were virtually unknown along the entire length of the Grand Canal area from Yangzhou in the south to Huai'an in the north, throughout Xiahe, and along the greater part of the coast. Such was the case throughout the period under study. Only the districts east of Yangzhou along the Yangzi departed from the norm. As a new prefect discovered on his arrival in 1844: "All the nearby districts have women working. To the east, in the dependencies of Tongzhou, they weave all sorts of cloth. In the neighboring southern region of Wu [Jiangnan], they embroider dragon robes. Only in Yangzhou [prefecture] do they go about gaily with nothing at all to do."[60]

The cultivation of cotton in Jiangsu province north of the Yangzi was almost entirely limited to the southeastern jurisdictions of Tongzhou and Haimen. Here a busy cotton industry flourished during the Qing. In the second half of the eighteenth century, "outside merchants" (keshang) flocked to Tongzhou to buy raw cotton, and in Haimen in the same period "three hands engaged [in spinning] could support eight people."[61] In the nineteenth century, Jingjiang also produced good-quality cloth, although "not as good as in Jiangnan." There was a marginal northward extension of cotton growing into the elevated areas of Dongtai, but during the late imperial period secondary production apparently did not extend beyond the districts along the Yangzi east of Yangzhou. Even here, notes Kathy Le Mons Walker, "raw cotton rather than cloth remained the dominant item of trade [and] Tongzhou-Haimen's annual cloth output never surpassed the production levels reached in Songjiang in the sixteenth century."[62]

Biographical data on "faithful widows" from the Yangzhou area show that training in handicraft skills was part of a girl's education. Some virtuous women reportedly shouldered the burden of running the family farm,[63] but spinning and weaving as well as sewing and washing offered more common means of self-support. Yet weaving, which was the great staple of "women's work" (nü gong) in the late imperial period, accounts for only a small proportion of cases—around one in six of the sample surveyed.[64] The greater number of spinners suggests that the yarn produced by these women was bought by others, who wove their own cloth. Household enterprise in Yangzhou, to the degree that it flourished at all, was undoubtedly directed primarily at supplying household needs, and the labors of spinning and weaving widows were probably undertaken as services to great households.

As a nonproducer of cloth, Yangzhou enjoyed no profits from the great interregional trade in cotton, estimated to have accounted for nearly a quarter of all long-distance trade on the eve of the Opium War. Nor, except as tribute to the court, was it an exporter of grain, the largest item of interregional trade. It owed its profile as a great commercial city to the third most significant item in long distance trade, salt, which constituted perhaps 15 percent of all such trade.[65] Patterns of women's employment within Yangzhou, including as matchmakers and courtesans, were not a direct outcome of the salt trade but were powerfully affected by it. Although the Huizhou merchants had a reputation for miserliness, it was also said that "with regard to concubines, prostitutes and lawsuits, they squandered gold like dust."[66]

Women reveal the extent to which the prosperity of the city was based on salt. According to Zhou Sheng, when the transport merchants began to lose their monopoly on profits from the salt trade, the consequences were visible above all in the demimonde:

Since the change in the salt trade to the ticket system, the number of prostitutes inside and outside the city has increased by 3,000 or more. These people benefited from the privileged position of their forbears, routinely receiving set emoluments, enjoying drunken songs and lascivious dances. The women, too, reveled in fun and flirted with smiles, knowing only how to adorn themselves. Lazy and extravagant as they long were, when one morning they were deprived of their resources, were without food and clothing, had nothing to which they could turn a hand, and moreover were unused to hardship, these people found themselves without recourse. With their wives they devised plans, and only this means was relatively easy. They hardened their hearts to undertake it. Alas! Such are the consequences of eating the bread of idleness.[67]

Zhou's comment nicely illustrates the parallel between the salt trade and the position of women in the urban economy that was early apparent in the simultaneous rise of the thin horses and the Huizhou merchants as defining features of Yangzhou at the end of the sixteenth century.

Yangzhou and Its "Others"

The contrasting positions of Huizhou men and Yangzhou women shed some light on the different ways in which the city has historically been described and characterized. It would seem from the reputation of Yangzhou women as thin horses and prostitutes that for Huizhou men, Yangzhou was nothing to write home about. It is possible that they felt slightly appalled at beholding the patterns of life and death in their adopted home. Huizhou society was in principle very pious. Temple culture centered on the Neo-Confucian figure of Zhu Xi; close attention was paid to ritual; extended family and clan obligations were maintained through clan temples, fields, and schools; the piety of women was promoted and rewarded. The same could not be said of Yangzhou, where Buddhist and Daoist temples abounded, funerary practices were accompanied by superstitious and unorthodox practices — including the viewing of a woman's body by her natal family before burial — and kidnapping for the purposes of prostitution was endemic.[68]

Female virtue was a matter of intense concern to Huizhou men in Yangzhou. From Zheng Yuanxun, who established a register of virtuous widows and filial sons in the second quarter of the seventeenth century, to Wu Shihuang, who set up the Hall of Chastity two centuries later, this issue was firmly on the agenda.[69] The Shrine to the Five Constant Ones (Wuliesi), a famous site in the eighteenth century, testified to its significance. Construction of the shrine was underwritten by a Huizhou man of Jiangdu registration, the salt merchant Wang Yinggeng.[70] The Qianlong emperor honored his family by visiting the garden of his grandson, which, like the shrine, was located on Shugang.[71]

The shrine was essentially a monument to widow fidelity. It commemorated the virtuous suicides of five ordinary women, whose ghosts were no doubt surprised to find themselves in rather elevated company: two neighboring shrines honored the energetic Song scholar-official Fan Zhongyan (998–1052) and his contemporary the Taizhou philosopher Hu Yuan (993–1059) (commonly known as Hu Anding after his family's place of origin in Shaanxi).[72] The hagiographies of these five women are instructive for an understanding of the views of Huizhou people. One of the virtuous women honored at this shrine was surnamed Cheng and was obviously of Huizhou origin; her husband, a merchant, also seems to have been from Huizhou. Another woman was a native of Nanjing. The remainder appear to have been local women.[73] Of the five, four took their lives soon after the death of a fiancé or husband. None had any children, a fact that confirms Susan Mann's observation concerning the strict conditions for widow suicide at this time.[74] The young woman from Nanjing explained her suicide to her father on the grounds that she had no parents-in-law to care for, no heir to raise, and no desire to remarry, a statement that puts the conditions succinctly.

The biography of the fifth woman, who hung herself when both her husband and her mother-in-law were still alive, is unusual. The woman, named Yi, was a peasant from Huangjue Bridge in Beihu. Her father died when she was a child, and she lived with her mother till, at the late age of 24, she married a townsman called Sun Dacheng. Sun was a beancurd maker: his days and nights were spent soaking beans, straining them, grinding them, and then cooking up vast cauldrons of soybean milk. On this sort of work, as the local record observes, "the sun never sets." Sun's mother and sister led lives of abandon, and the woman Yi finally ran away to rejoin her mother. Forced to return to her husband's hearth, she hung herself rather than succumb to the pressure to compromise her honor by consorting with other men.[75]

This story provides a glimpse of how Huizhou observers might have seen the Yangzhou *hoi polloi*, their lives characterized by hard manual work in the service of the urban economy, by poverty, irregular habits, and the promiscuity of women. The biography of the woman with the Huizhou name offers a pointed contrast with the poor beancurd maker's wife. Her merchant husband met a premature death in Guangxi, but her parents-in-law dissuaded her from taking her life before the news—uncertain at such a distance—was confirmed. A year later, once the confirmation was obtained, the parents-in-law apparently had no further objections.[76] The woman paid her respects to them, shut her door, and died. In this case, unlike that concerning the woman Yi, familial organization was all that it should be, the virtuous widow's suicide completing the picture of harmonious and ethical family relations.

Given the association of both Huizhou and Yangzhou with salt merchants, and the corresponding association of salt merchants with courtesans and concubines, the initiatives of Huizhou men to promote female virtue in Yangzhou can be read as attempts to improve their self-image. The image of the Huizhou merchants, at best an ambivalent one, had two female counterparts. On the one hand, a reputation for extravagance and self-indulgence was paired with the sexual availability of women in the form of thin horses, concubines, or prostitutes. Wu Jingzi encapsulated this association in a tale of an educated woman who resists incorporation into the household of a salt merchant in Yangzhou. "We salt merchants employed by the state take six or seven concubines a year!" blusters the merchant, stupefied by the woman's resentment of the casual reception accorded her at his house.[77] On the other hand, the moral laxity of Huizhou merchants was translated across the gender divide in Huizhou society into faithfulness on the part of their wives—hence a corresponding story by Wu Jingzi of a young Huizhou widow who follows her husband to the grave.[78]

There was a fundamental contradiction between the stereotypes of Huizhou men and Huizhou women.[79] The men were above all merchants: simultaneously spendthrifts and misers, overweening in their ambitions and vulgar in their dispositions.

The women were to all intents and purposes identical to the home-bound, God-fearing wives of Flemish merchants. The story is told of one Huizhou wife that every year her husband was absent from home, she purchased a pearl with the proceeds from her embroidery. When three years after her death, he returned home, he found twenty pearls in a casket: a precious tear for every year of his absence.[80]

There is a ready explanation for this contradiction: the image of Huizhou merchants was crafted by men from other places, that of Huizhou women by Huizhou men. If merchants were to be absent from home for years at a time, it was necessary for them to cultivate a culture of constancy for their women, both for their own peace of mind and for the good order of Huizhou society. Huizhou publications in the female piety genre, unusually numerous, served both to promote this culture on the home front and to advertise it abroad.[81] Illustrated versions ensured that the genre would be attractive to a broad audience.[82] The aims of the publishers, insofar as they included development of a particular image of Huizhou society, were easier to achieve because Huizhou was not an immigrant city: Huizhou women were thus not subjected to the gaze of prurient sojourners in the way that Yangzhou women were.

Just as the contrasting images of Huizhou men and women were complementary, so, too, were those of Huizhou and Yangzhou women. Wives and mothers in Huizhou, concubines and courtesans in Yangzhou: these women, actual and figurative, were products of the same history, playing different but corresponding roles in a drama centered on the successes of the Huizhou diaspora. The profound importance of women as signs of community and social status is apparent in the conflicting images of Huizhou wives and Huizhou merchants, but it takes Yangzhou women to make sense of Huizhou women. Huizhou wives waited faithfully at home raising the children and serving the parents of men who dallied with Yangzhou girls.

Huizhou was one of several geocultural references used to construct particular understandings of Yangzhou in the minds both of outsiders and of Yangzhou people themselves. Another was Suzhou. In the Ming-Qing period, Yangzhou and Suzhou are commonly mentioned in the same breath, both places being associated in popular lore with beautiful women. Yang and Su, Huai and Wu, the one north and the other south of the Yangzi, close to yet far from each other, present as paired referents in the literary and oral traditions related to women in general and prostitution in particular. "Wu Sangui had two daughters," goes the story; "one lived in Yangzhou and the other in Suzhou."[83] "The clouds of Wu [Jiangnan] and the rains of Huai [Jiangbei] frequently mingled in the courtyard," wrote Zhou Sheng, in clear allusion to sexual play in the Yangzhou brothel he was describing.[84] The saying "Suzhou head, Yangzhou feet," a reference to the fickleness of prostitutes, juxtaposed Suzhou hairstyles, renowned for their elegance, with Yangzhou feet, famous for their smallness.[85]

Suzhou was firmly established as the organizing, productive center of Jiangnan in the early Ming. When Yangzhou was only just beginning to show its colors, Suzhou was already China's quintessentially cultured city.[86] In the late sixteenth century, it was performing something of the role played in Europe by Paris, setting the tone at least regionally for new and extravagant styles of dress.[87] It encapsulated the idea of Jiangnan. When in the late eighteenth century Shen Fu (1763–?) visited a Yangzhou "flowery boat" in Guangzhou, managed by a Yangzhou woman and her daughter-in-law, he found the women dressed in Suzhou-style clothing, a comment surely on the dominance of Suzhou style in the representation of Jiangnan elsewhere in China.[88]

Suzhou was a source of prostitutes in Yangzhou. In the seventeenth century, as recorded by Wu Qi, the "Suzhou prostitutes (*ji*) were called 'Su canals' and the local sing-song girls (*tu chang*) were called 'Yang canals,'" vulgar terms that had their origins in local slang for the fishing boats of Taizhou.[89] Even the "Yang canals" might include nonlocals, as we learn from Li Dou's example of Little Precious, who was born in Suzhou but "sold as a daughter" to someone in Yangzhou.[90] The demarcation of Suzhou and Yangzhou prostitutes in Wu Qi's account brings to mind the parallel situation in Shanghai at a later date, and the ranking—Suzhou higher than Yangzhou—may well have been the same.[91] Wang Hongxu (1645–1723), the Kangxi emperor's official confidant and private eye, reported in 1707 that salt merchants from Yangzhou went in swarms to buy girls in Suzhou.[92]

A third dichotomy involving Yangzhou differentiated it from other parts of Jiangbei. Zhou Sheng, writing in the 1840s, ranked the accents of Yangzhou people in a hierarchy that placed Yangzhou's home counties of Jiangdu and Ganquan at the top, followed by the neighboring county of Yizheng. Taizhou, probably signifying both Taizhou proper and Dongtai ("eastern Tai"), was placed at the bottom. Prostitutes from different parts of Jiangsu were also given an implicit ranking: those from Suzhou (or more broadly Jiangnan, *Wupai*) were gentle and cultured, and those from Nanjing were open and generous. Women from the salt zone on the coast (Dongtai, Yancheng) were immoral, spoke crudely, and behaved without restraint.[93]

In the contemporaneous novel *Dreams of Wind and Moon*, written by a native of Yangzhou, a Yancheng identity is attributed to many prostitutes featured in the story.[94] Yancheng, a vast county centered on a coastal city, stood at a polar extreme from Suzhou and Huizhou: poor, flood-prone, isolated from major trade routes except for the salt transport route, featuring a mixed population of farmers, fishers, and salt makers, producing few scholars. The fictional representation of the Yangzhou demimonde in this novel provides a new stereotype of prostitutes but does not necessarily reveal a new phenomenon. On the one hand, the term "Yang canals," cited by Li Dou in the preceding century, may well have referred to women from Jiangbei at large; on the other hand, little was written on the origins of street girls (plausibly from Yangzhou's hinterland) as opposed to the high-class courtesans trained in the cities of Suzhou and Yangzhou. The number of Yancheng women in the Yangzhou demimonde may, however, have increased in the first half of the eighteenth century. When *Dreams of Wind and Moon* was written, the salt industry was in decline, agriculture was in crisis, and the attractions of Yangzhou — an established place of refuge for people from northern Jiangsu — were undoubtedly enhanced.

At the same time, it seems unlikely that more women from Yancheng than from the neighboring areas on the coast (including the Yangzhou counties of Xinghua and Dongtai) were supplying the prostitution industry in Yangzhou. Yancheng might perhaps have stood for the impoverished northern hinterland in general, including counties of both Yangzhou and Huai'an prefectures. Xinghua is also identified in the novel as the native place of one prostitute.[95] But by naming Yancheng as the major source of prostitutes, the author of this novel, a native of Yangzhou, removed the issue of prostitution from his native place. Yancheng belonged to Huai'an prefecture, not to Yangzhou. And if Yangzhou was not quite Jiangnan, neither was it to be confused in the common mind with its neighbors to the north. This last point was perhaps to be particularly emphasized at a time of economic crisis and social malaise.

Like Suzhou and Huizhou, Yancheng served to define Yangzhou along borders created by a gendered essentialization of native-place characteristics. For local people as well as for outsiders, women in Yangzhou served as communal boundary markers, as signs of the different aspirations and identities of local and exogenous communities and also of the social circumstances that promoted women to a position of prominence at the intersection of these communities. They provided contemporary society with one of the most fully elaborated metonyms for place.

Silently invoked by the image of Yangzhou as a place that "produces women" is its converse: family, propriety, decorum, frugality, steady attachment to one's native place. In Yangzhou, no less than in other places in late imperial China, family norms were promoted. Although the early disposal of daughters was commonplace through practices such as infanticide, transfer in childhood to the in-laws' home, and sale into prostitution and slavery, the alienation of a daughter for purely monetary gain ran counter to ideals of family conduct.

This city had a place within the Confucian moral order, but the hardworking widow who took in washing to support her children and the gentry widow who instructed her sons in the classics, the sort of women honored in the local gazetteers, were fed by the same market that supported the thin horses. The money passed from hand to hand. It was not a city of great industry: weaving—the classic "womanly work" of late imperial times[96]—was not among the arts imparted to the thin horses nor was it widely practiced by other women of this city. Rather, pride of place in Yangzhou was given to literati skills and pastimes such as poetry and painting. These, the accomplishments of the cultured or literate person (*wenren*), were fetishized in the form of the thin horses and high-class courtesans, who embodied an entire cultural aesthetic.

In literary works and popular lore, Yangzhou women occupy a position subordinate to that of the travelers and sojourners who purchased them. The daughters of Yangzhou were bought by men from elsewhere. In the words of one balladeer: "Married off to men from distant parts, / They broke their hearts with grieving."[97] As the products of a city that had few other products to offer, brought into being at a time of economic and social change in the Lower Yangzi Delta, "beautiful women" were not inappropriate as symbols of Yangzhou in the late imperial period. Their place in the iconography of Yangzhou was rivaled only by that of the sojourning merchants, whose wealth brought into being the idiosyncratic urban culture of which the beautiful women were a part.

CR

TEN

The Huizhou Ascendancy

I n 1840 a hostel for fallen women was established in Yangzhou by Huizhou merchant Wu Shihuang, supported with
funds from the salt monopoly. Known as the Hall of Chastity
(Lizhentang), it was located in Left Garrison Street (Zuoweijie),
south of the Parade Ground and not far from the courtesan quarter.[1] It was the second of two institutions in Yangzhou founded
expressly for women; the Widow Support Society (Xuli gonghui)
had been established in 1805.[2]

These institutions were late products of a social welfare project
that had been under way in the city for some two centuries.
Among its many claims to fame, Yangzhou was the location of the
first society for the protection of orphans, established in the late
Ming, and of the first orphanage, established in the early Qing.[3]
Along with poor relief centers (*yangjiyuan, pujitang*), schools, and
academies of higher learning, such institutions helped transform
urban society in the late imperial period. The implications were a
sharper differentiation between urban and rural than had earlier
been apparent, a recognition of the established nature of the urban
population, and more complex forms of urban governance.

In Yangzhou, these institutions arose largely through the initiatives of salt merchants or were facilitated by them. Although one

account of the establishment of the city orphanage in 1655 credits a local man for the initiative, he was supported by Li Zongkong, a scholar-official from a West merchant lineage, and by the prominent merchant-philanthropist Min Shizhang.[4] Insufficient funds led to a deterioration of the premises, but in 1717, "local men" Min Tingzuo and Zhang Shimeng, together with gentry and merchants, bought land outside the North Gate and moved the orphanage there from its original site outside Little East Gate. The salt controller rallied the salt merchants to fund an annual budget of 1,200 taels. In 1734 Min Tingzuo mustered the gentry (*shen*) to support establishment of 79 nurseries. Further expansion was effected through the salt administration during the Qianlong reign.[5]

Although, as Liang Qizi has pointed out, officials and local people participated in initiating such projects, Huizhou surnames figure prominently in lists of founders and supporters. In most instances in which a "man of the locality" (*jun ren, yi ren*) or a "local gentryman" (*junshen*) is identified as a benefactor, he proves to be from a Huizhou merchant family.[6] In a variant account of the founding of the Yangzhou orphanage, five of the six supporters of the venture have Huizhou surnames.[7] Among the later benefactors, "local man" Min Tingzuo was very probably the son or grandson of the Huizhou merchant Min Shizhang. His surname points to Huizhou ancestry as clearly as O'Reilly does Irish ancestry.

The occasional offspring of a West merchant family makes an appearance among philanthropists, a prominent early example being Li Zongkong. "Local man" Zhang Shimeng bears a surname that was common among Shaanxi families. But the West merchants, apart from their smaller numbers, were more reluctant to change their household registration. These two factors explain their relatively low profile in Yangzhou. In Hankou, as William Rowe has observed, the West merchants were a highly distinctive and rather insular group.[8]

The Huizhou merchants, by contrast, entered enthusiastically into the remaking of Yangzhou. Their munificence supplied the city not only with benevolent institutions and schools but also with roads, bridges, lifeboats, and fire-fighting equipment. They effectively created the urban infrastructure, just as they created the city's gardens.

The undertakings of the Huizhou salt merchants in Yangzhou are documented in a variety of historical studies in various languages and will come as no surprise to historians of this period. In general, however, historical studies have failed to distinguish "merchant" from "salt merchant," Huizhou from West merchants, and the exogenous salt gentry from local landowning gentry. In the absence of such distinctions, the dynamics of social change in eighteenth-century Yangzhou remain unclear. As far as the establishment of philanthropic institutions is concerned, it is clear that Yangzhou was unique only for the early start it had over other places, but in Yangzhou, these institutions demonstrate above all the organizing, controlling capacities of the Huizhou merchant families in their adopted home. [9]

The Huizhou Colony

To establish the dominance of Huizhou families in Yangzhou poses no empirical difficulty. As summed up by Chen Qubing (1874–1933): "The prosperity of Yangzhou was created by Huizhou merchants. A Huizhou colony developed in Yangzhou. Hence of the great Huizhou clan names of Wang, Cheng, Jiang, Hong, Fan, Zheng, Huang, and Xu, none but was to be found there."[10] The distinctiveness of these surnames permits easy identification of Huizhou people in Yangzhou (see Table 10.1).[11] There is, of course, regional overlap in names: Zheng, Liu, ordinary Wang (king) are common names that are not highly place-specific. But the ancestry of someone in Yangzhou named Cheng, Wang (water radical), Fan, Fang, Jiang (water radical), Wu, Xu (language radical), or Huang was almost certain to be Huizhou; so, too, with the more unusual surnames Ba, Bao, and Min. These names are constantly to be found in association with one another. Charlotte Furth's account of a Huizhou physician working in Yangzhou in the early seventeenth century is thick with Huizhou names—Cheng, Fang, Wang, Hong, Bao, Wu, Luo.[12] Such common local surnames as Sun, Li, Liu, Chen, and Yang are, with the exception of Liu, rarely if ever to be found among Hui people.

Huizhou people in Yangzhou tended to cluster together. This was in part dictated by common economic activities; the salt mer-

Table 10.1
Huizhou and Yangzhou Surnames

Common and distinctive Huizhou surnames				
Wang 汪	Cheng 程	Jiang 江	Hong 洪	Fan 范
Zheng 鄭	Huang 黃	Xu 許	Ba 巴	Bao 鮑
Min 閔	Wu 吳	Luo 羅		
Strong Yangzhou surnames				
Sun 孫	Li 李	Liu 劉	Chen 陳	Yang 楊

chants, for example, were naturally linked by business. Thus, of the Jiang clan from the Shexian village of Jiyang, "many took up business under the salt statutes and gathered together in the city of Yangzhou."[13] But ties dictated by local groupings in Huizhou also remained strong in Yangzhou. The Wu clan from Shexian county was divided into four different lineages, based in four different locations in the county, "and those who sojourned in Yangzhou formed factions according to the villages where they lived."[14] If even members of the same surname group from the same county were divided among themselves, the lines between Huizhou and Yangzhou natives may be imagined.

In many cases, a Huizhou person who purchased land in Yangzhou changed his household registration; this did not, however, mean that he forgot his ancestral place. Wu Jiaji, in a poem bidding farewell to Wang Zuoyan as he departed for Huizhou, referred to him in the title of the poem as "going back" (*gui*) rather than simply "going" (*zhi*) to his destination, even though Wang's family, originally from Xiuning, had for three generations officially been natives of Jiangdu county. The verses themselves identify the place to which Wang was returning:

> The wanderer's thoughts turn to the mountains of home,
> Crystal-clear the tears that drench his gown.
> The family tombs with grasses overgrown:
> Long have they waited for grandson or son.[15]

Graves and family shrines tied immigrants to their place of origin. This was especially the case for Huizhou people, whose clan system was probably unrivaled in China. Those who changed their registration to Yangzhou imported their custom of

worshipping at family shrines and their preoccupation with genealogies to a place where such practices were only weakly established. The shrines of the great merchant families became landmarks in the city.[16] Other than land for their gardens, Huizhou
merchants purchased land around Yangzhou mainly as endow-
ments for the newly established clan shrines.[17]

In Huizhou itself, links with the emigrant families remained
strong through the early Qing. Lists of successful examination
candidates from Shexian included large numbers of students registered in Jiangdu and Yizheng, and since the lists were selective,
including some Yangzhou graduates but not others, it is probable
that those listed continued to have families based in Huizhou.[18]
The decline in the numbers of Yangzhou graduates in these lists
over the course of the eighteenth century suggests an attenuation
of the relationship as well as an easing of the outflow from Huizhou, but the biographies in Huizhou gazetteers recording the
achievements of Huizhou descendants in Yangzhou and other
places show how important the diaspora was to local identity in
Shexian, and more generally in Huizhou prefecture.[19]

Pride in place goes some way to explaining why Huizhou men
took such an interest in Yangzhou society, an interest manifest in
the establishment of philanthropic institutions. The roots they established in their new home counties—Jiangdu, Ganquan, Yizheng—were thoroughly urban. When they changed registration,
it was on the criterion of their engagement in the salt trade rather
than on the basis of landownership. Those who did change registration, if still conscious of the old hometown, were now also defined by reference to a new and quite different sort of native
place. Their initiatives in founding and supporting urban institutions show that they did not fail to take an interest in its workings.

Native-Place Associations

Before we turn to the institutions established in Qing Yangzhou,
it is worth noting one characteristic urban institution surprisingly
absent from the city—the native-place association, or *huiguan*.
The distribution of native-place associations is a good index to
the interregional trading activities of different merchant groups

across the empire because such associations were usually established when substantial numbers of merchants from one locality became active in another area. Native-place associations are commonly listed in local gazetteers, but for the period up to the middle of the nineteenth century, the Yangzhou gazetteers include no listing of *huiguan* at all.

Ho Pingti has concluded on the basis of Li Dou's passing mention of "all the departmental and county native-place associations" (*zhu zhou xian huiguan*) that Yangzhou probably had one of the greatest concentrations of native-place associations in the empire, because for a *huiguan* to represent a department or a county—as opposed to a province or prefecture—suggests a great concentration and diversity of sojourners.[20] But the *huiguan* premises mentioned by Ho, probably used as lodgings for people from other departments and counties in Yangzhou, occupied only a single site north of the city walls, and the reference is to the Ming period. In the eighteenth century, this site was occupied by the Meihua Academy.[21]

The absence of *huiguan* would seem to indicate a fundamental difference between Yangzhou and great centers of trade such as Suzhou and Hankou, where native-place associations flourished during the seventeenth and eighteenth centuries.[22] Not until the second half of the nineteenth century is there clear evidence of *huiguan* in Yangzhou. The 1883 city map indicates the existence of six provincial associations: Shan-Shaan, Hunan, Jiangxi, Hubei, Anhui, and Lingnan (Guangdong); along with three prefectural associations representing Zhenjiang (Jingjiang), Jiaxing, and Shaoxing. Of these, the Anhui association is known to have been established after the suppression of the Taiping and Nian rebellions in the 1860s,[23] and the Lingnan and Hunan associations occupied the premises of former salt merchant mansions, an indication that they were founded after the decline of the Huizhou merchants (see Fig. 21).[24]

It seems probable that native-place associations in Yangzhou began to proliferate during the third quarter of the nineteenth century. Yangzhou historian Zhu Jiang has traced the history of the premises of the Hunan *huiguan* from its origins as a garden site through a number of changes of ownership to the Taiping

Fig. 21 The façade of the Hunan *huiguan*,
formerly a merchant dwelling (from
Wang Hong, *Lao Yangzhou*, p. 141).

Rebellion, when it served as the residence of a Taiping general.[25] The Lingnan *huiguan* is well documented *in situ*. A stele at the site where it once stood, in the south of the New City, gives the date of establishment as the eighth year of the Tongzhi reign, that is, 1869.[26] This stele, incidentally, shows that Cantonese merchants had usurped Huizhou's position in the Lianghuai salt trade, although without ever achieving their predecessors' significance in Yangzhou society. The *huiguan* was supported and subsequently renovated through donations from individual merchants and by levies based on the amount of salt traded.[27]

Ho's data on Yangzhou in fact suggest that the city did not host large numbers of sojourning merchants other than salt merchants and that the salt merchants not so much mixed with as constituted the urban elite. Apart from the fact that salt merchants rarely formed native-place associations,[28] the absence of a Huizhou or Shexian native-place grouping suggests that such organizations were simply not needed, because in the eighteenth century Yangzhou was in many respects a Huizhou city. Their philanthropic activities declared the fact.

Philanthropic Activities

The institutional history of Huizhou philanthropy in Yangzhou begins with the founding of the Yangzhou orphanage, which is well documented in an essay by the seventeenth-century writer Wei Xi on its principal benefactor, Min Shizhang. In discussing Wei's essay, Joanna Handlin Smith has argued that the Ming-Qing transition witnessed a shift in the motivations for philanthropy from personal mission to social obligation, and that in any case the increasing corporatization of philanthropic activities under the Qing reduced the scope for individual initiative by philanthropists.[29] It is difficult to assess the validity of the second point. On the one hand, gazetteers are full of biographies of men who performed charitable deeds. On the other hand, as Liang Qizi has shown, the establishment and regulation of orphanages, poorhouses, famine-relief granaries, and academies of learning owed much to the direction of the Yongzheng emperor, whose controversial succession to the throne in 1723 ushered in a period of high-level bureaucratic involvement in domains of activity that had hitherto been relatively autonomous, private, and localized.[30]

The Yongzheng emperor's well-documented interest in philanthropic activities needs to be examined closely. In light of recent work on the ethnic nature of Manchu governance,[31] it is worth wondering whether philanthropic activities in early Qing Yangzhou owed something to a Huizhou consciousness of the needs of Chinese society in the context of recently imposed alien rule. The long list of Huizhou men associated with the founding of the Yangzhou orphanage—including Zheng Yuanhua, who had lost a brother and a nephew in the dynastic transition—would seem to suggest as much. The Yongzheng emperor, acceding to the throne in controversial circumstances, was unusually anxious to harness such energies and to forestall the possibility of dissent within the empire. His encouragement of the chaste widow cult, discussed by Susan Mann, suggests that his intervention in welfare in the 1720s may have been part of the same project of personal and dynastic legitimation.[32]

Whatever the merits of such an argument, the institutionalization of philanthropy over the course of the eighteenth century does complicate analysis of private motivation and initiatives in this domain, and in Yangzhou the salt officials anyway had a critical role to play in organizing merchant philanthropy from an early date in the Qing. As noted in Chapter 5, the salt censor rallied salt merchants to contribute 22,670 taels of silver to the relief of starving refugees in 1672.[33] The sum of money involved in this operation was nothing compared to the amounts merchants were later to pour into various projects in Yangzhou, but in the 1670s it may be counted as substantial. The scale of the project indicates the organizational capabilities of the monopoly, and the role of the salt censor points to the important part played by the bureaucracy in shaping the merchant body, coordinating its philanthropic actions, and generally ensuring its centrality to the workings of Yangzhou.

As was the case with water control, so, too, in the domain of philanthropy, salt officials and salt merchants played roles in Yangzhou that elsewhere were performed by local officials and gentry. This was a defining feature of charity in Yangzhou, and it signified that the prefectural city would be the great beneficiary of the philanthropic impulse in mid-Qing China. Famine relief provides the outstanding example. The Lianghuai salt charity granaries (Lianghuai yanyicang) were, writes Pierre-Etienne Will, "the best-known and most important" of charity granaries established with merchant capital.[34] The salt merchant body subscribed 240,000 taels of silver to the establishment of four granaries in the city in 1726.[35] The number of granaries was justified on grounds of Yangzhou's strategic location: grain could be shipped from there to other parts of Jiangsu and Anhui in case of famine.

More granaries were subsequently established in the coastal counties and salt yards.[36]

The implications of these developments for the increased powers of salt merchants in Yangzhou and the enhanced position of Yangzhou in Jiangbei are clear. First, salt merchants were heavily involved in the grain trade. Their profits from this charitable enterprise were, according to Will, "enormous," with the result that the granaries operated successfully for an unusually long period of time.[37] Second, the salt charity granaries in Yangzhou's hinterland were managed by Yangzhou merchants, usually of Huizhou ancestry. Such was the case in Tongzhou, a jurisdiction administratively divorced from Yangzhou in 1727, when it was made an independent department (*zhilizhou*). As of 1755, however, the salt charity granary in Tongzhou was run by "Yangzhou merchants Jiang Zhuzhou and Xu Shangzhi." The granary at the nearby Shigang salt yard was managed by "Yangzhou merchant Hong Chongshi."[38] (All three "Yangzhou merchants" have Huizhou surnames.) Finally, the accounts for the Yangzhou city granaries were originally kept by the salt censor, and those for other granaries within the Lianghuai zone were kept by the salt intendant (*yanwudao*). In 1741, the intendant's office was abolished, and all the accounts were turned over to the offices of the salt controller.[39]

The famine-relief granaries developed at the very time that the major philanthropic institutions in Yangzhou were being established or refounded. Apart from orphans and famine victims, the main objects of salt merchant charity were the very young or elderly without means of support, prisoners, the sick, and those without means to bury their relatives. The provision of coffins, food, money, and medicines was organized through the characteristic benevolent institutions described by Liang Qizi. One was the Welfare Center (Yangjiyuan), located outside the North Gate. This institution had a history dating to the early Ming and was restored by Min Shizhang in the early Qing, but it had long been in a state of collapse by the time of the Yongzheng reforms. In 1733 Ma Yueguan, together with two fellow Huizhou merchants, underwrote its renovation.[40] The other was the General Relief Hospice (Pujitang), established in 1700 and supported through the salt monopoly from 1744. The hospice was situated on the east side of the Grand Canal outside Quekou Gate.[41] The availability of land may have been a consideration in determining the venue, and there was a certain spatial logic to a southeastern location given the location of the Welfare Center on the northwestern side.

Liang Qizi notes that outside the prefectural city, *pujitang* in Yangzhou prefecture tended to focus on helping the sick rather than on other relief measures.[42] This difference in function might be explained by the broader base of social support in Yangzhou. There were many doctors in the city, as well as other sources of medical charity, notably the Hall of Mutual Benevolence (Tongrentang), which gave out both medicine and coffins; a free medical dispensary set up by salt merchant Huang Lüxian outside his garden in the southeast quarter of the city; another dispensary next to the salt controller's yamen; and no doubt others at different times in the city's history.[43]

Fire fighting was another area of merchant involvement. Fires were a hazard in the crowded residential areas of the city, most notably around the Parade Ground, and they periodically threatened major landmarks in and around the city. The drum towers at the prefectural offices were destroyed in one fire,[44] and the Literary Peak Pagoda, south of the city, was damaged in another.[45] A group of gentry and merchants supplied Yangzhou with fire-fighting equipment, which was stored in the Hall of Mutual Benevolence, an institution otherwise devoted to providing coffins and medicines to the poor. The fire-fighting equipment installed at the yamens of civil and military officials and at each of the city gates must also have been funded by the salt merchants, for the salt censor Sanbao is credited with their establishment.[46] Sanbao's tenure in Yangzhou, 1737–40, points to the early Qianlong years as the date of this initiative.

Mention of Sanbao's name is a reminder that officials of the field administration figure relatively rarely in the institutional history of Yangzhou city, *pace* the prominence accorded Chen Hongmou (1696–1771).[47] The salt administration had clear advantages over the field administration in advancing philanthropic activities in Yangzhou. These advantages lay not only in the superior resources of the salt monopoly but also in the ease with which these resources could be tapped. A levy based on the amount of salt traded by a merchant could readily be imposed at the time of tax payment. Monies thus collected were apparently itemized under pre-existing budgetary categories and issued to local institutions from the salt treasury. Thus in 1732–33, the grant for the Guazhou and Shaobo hospices was entered in the salt treasury ledger under the category "abandoning or retaining boats and carts used during the salt-boiling period" (*caicun huoyou zhouche*).[48]

The field administration was no match for the salt administration in organizational powers and financial resources. The history of Yangzhou's famous academies reveals the imbalance in resources. Of the three academies functioning in Yangzhou during the Qianlong reign, two became famous by virtue of the noted scholars who taught there and some almost equally famous graduates. These were the Meihua Academy and the Anding Academy, founded in close succession in 1734 and 1735, in both cases on the premises of existing, nonteaching academies. Their creation was a response to a call by the Yongzheng emperor for the establishment of academies throughout the empire[49] and can be viewed as one of the many acts by which the salt merchants confirmed their relationship with the throne.

The Anding Academy, located in the Old City next to the recently established Ganquan country yamen, was a typical salt monopoly institution. The salt merchant body funded its development through a joint subscription of 7,400 taels of silver, its ongoing costs were paid through the salt transport treasury, and the authority to invite teachers and appoint directors was vested in the salt administration.[50] The Meihua Academy, by contrast, had close links with officials of the field administration. This institution stood outside the New City walls, beyond Guangchu Gate, on a site occupied by a succession of academies since the sixteenth century.[51] Although initially financed by Ma Yueguan, inspiration for its revival is credited to subprefect Liu Chongxuan (*js* 1737). In the early years, Liu himself, and subsequently other officials, taught classes. Liu may initially have used some of his own resources to support the academy, for it was not until 1739, after he left Yangzhou, that the salt administration began to assist it with funds in the form of "candle allowances" (*gaohuo*), or student stipends.

Lacking the stable financial base of the Anding Academy, the Meihua Academy's operations faltered after Liu Chongxuan's departure, and in 1743 it was incorporated into the Anding Academy. In 1777, a petition by Ma Yueguan's son, Zhenbo, for its restoration as an independent institution managed by local officials prompted the salt administration to take action. Under the direction of Zhu Xiaochun (*juren* 1762), who held office as salt controller in 1776–78, the academy's premises were renovated, and its organization placed on the same footing as that of the Anding Academy. The authority for all appointments was invested in the salt administration. Eminent scholars were invited to teach and an academic director was appointed. Yao Nai (1732–1815), the first person to hold the position of director, was a personal friend of Zhu Xiaochun.[52]

By the end of the Qianlong reign in 1795, the two academies between them had 260 registered students, whose stipends alone required more than 6,000 taels per annum.[53] These academies stand in sharp contrast to institutions supported by the field administration. A charity school founded in 1712 by the prefect of Yangzhou was raised to the status of an academy during the Qianlong period, and given the name Guangling Academy. Devoted exclusively to training candidates for the first, or lowest, degree, this institution survived into the nineteenth century but apparently only with the support of the salt monopoly. In 1809, it was the salt censor, not the prefect of Yangzhou, who instigated increases in the student quota and the student stipend.[54] The Red Bridge Academy (Hongqiao shuyuan), founded by Governor-General Yu Chenglong in the Kangxi period, was no longer in existence by the beginning of the nineteenth century, nor was the Jiaoli School (Jiali xueshe), founded in 1735 by the Jiangdu county magistrate.[55] Although the Anding and Meihua academies were destroyed during the Taiping Rebellion, they were restored in different locations in the New City during the 1860s and continued to be financed through the salt monopoly.[56]

The history of the Meihua Academy illustrates a process of dialogue and negotiation between members of an elite Huizhou family in Yangzhou and officials of the field and salt administrations and points to the sustained importance of individual initiatives in the creation and maintenance of urban institutions. Evidence from eighteenth-century merchant biographies does reveal strong institutional contexts for philanthropic undertakings but tribute continued to be paid to the individual philanthropist. Wang Yinggeng, whose interest in the promotion of female piety has already been discussed, reportedly saved the lives of tens of thousands of people during a famine in 1740. Apart from this, he indulged in many good works, "such as decocting medicines for charity, repairing Confucian temples, supporting the poor, assisting orphans, encouraging widow chastity, building bridges and boats, helping travelers, and saving lives. [On such works] he used well over 100,000 taels, at a guess."[57] His younger contemporary Huang Lümao, one of the "four treasures," had at least three different financial strategies for supporting good causes. One was to donate grain: in 1736, he handed over the year's harvest from more than 300 *mu* of land to the Yangzhou Welfare Center.[58] Another was to pay cash on a regular basis. This he did in the case of the hospice in Shaobo, paying over 30 taels of silver per month for a period of five years between 1736 and 1740. The third suggests he had been talking to his accountant: in 1741, he invested 1,500 taels of silver with a pawnbroker, with the monthly interest going to the support of widows and orphans, again through the Shaobo hospice.[59]

Activities by the Ma brothers, the Huang brothers, and Wang Yinggeng, leading figures in Yangzhou society, show that the Yongzheng reign and the early Qianlong years—a period of around two decades from the 1720s to the 1740s—were an important era for urban development along lines somewhat different from those evident in the making of the lakeside gardens. In these years the major academies were founded, the Welfare Center was re-established on a new footing, the orphanage's finances were taken over by the salt administration, the salt granaries were established, and an efficient fire-fighting force was created. Leading merchants initiated or shouldered personal responsibility for a range of public works, providing a fine example of community leadership for their successors in later decades. Bao Zhidao, whom Liu Sen regards as the most astute salt merchant after Wang Yinggeng, was nearly as active as Wang had been, although he lacked the accumulated family fortune of the Wangs. In the late eighteenth century he instituted a mutual help system among the salt boatmen, had the streets repaved between Kangshan in the southeast corner of the New City west to the Customs

Station and then north to Little East Gate, established charity schools at each of the city's gates, had the premises of the Yangzhou native-place association in Beijing restored, and made various contributions to his native-place in Shexian as well. Principal Merchant for twenty years from around 1781, he was showered with official honors on his death.[60]

City and Prefecture

The high profile of Huizhou merchants in philanthropic endeavors in Yangzhou suggests that they had few rivals for local leadership. In economic terms, the locals were unable to compete with them. This brings us to a final point concerning the philanthropic activities of Huizhou merchants: they were very heavily concentrated in the prefectural city. In Yizheng, which, as the export port for Lianghuai salt, enjoyed a strong merchant presence, the orphanage was supported through the salt monopoly, but the orphanage down river in Guazhou was dependent on rents from the Yangzi shoals where reeds were produced for fuel. A single injection of merchant funds is recorded.[61] The hospice in Guazhou was developed though the efforts of a succession of officials and local gentry; Chen Hongmou played a particularly important role in expanding its premises after he took office as prefect in Yangzhou in 1729.[62] The hospice in Shaobo was founded by prefect Yin Huiyi (1691–1748). Arriving to take up office in 1731, Yin was struck by the contrast between the prosperous prefectural capital and nearby Shaobo, where people had been living in a state of unrelieved immiseration since floods earlier that year.[63] It was probably through his agency that the Shaobo hospice and the corresponding institution in Guazhou were eventually funded through the salt monopoly.

The examples of these two hospices show the highly contingent nature of salt merchant philanthropy: it was exercised in the first instance in their home location, which for most of them was the city itself if not some village in Huizhou. Officials of the field administration had to plead eloquently for a wider distribution of merchant wealth even within Yangzhou's home counties. More distant places were out of sight and often out of mind. The academy in Taizhou, founded by the departmental magistrate in 1718, was supported by land rents. Not until 1797, when it was badly dilapidated, was the salt controller asked for the several thousand taels needed for its repair. It may have been from this year that the salt transport treasury annually contributed a modest 240 taels for its support. In 1820, when further work was needed, the magistrate and local gentry raised the money.[64] The Wenzheng Academy in Tongzhou's Shigang yard was supported during the Kangxi era by the Tongzhou sub-assistant controller; subsequently it failed for lack of funds, but it was revitalized in 1793 by the Tongzhou yard merchants.[65] An orphanage in Dongtai—in the heart of the salt production zone—was supported by merchant efforts, but not so, apparently, the poorhouse.[66] In Rugao, the academy founded by the magistrate in 1724 never received merchant funds and was dependent on land rents alone.[67] The same was true of the Zhengxin Academy in Caoyan yard, near Dongtai.[68] In Yancheng, also located within the salt production zone, neither the orphanage nor the poorhouse nor any other benevolent institution owed anything to the generosity of the salt merchants.[69] Probably the Huizhou merchants on the coast wanted to move to Yangzhou at the earliest opportunity and did not feel at home in places like Dongtai and Yancheng.

The uneven distribution of resources within the prefecture was most obvious in education and its outcomes, the all-important examination successes. As Ho Ping-ti has shown, a survey of the distribution of *jinshi* degrees in Jiangsu province shows a fairly even spread over the prefectures with the exception of Suzhou and Changzhou, outstanding in their yield of *jinshi*, and of Haizhou, on the far north coast of Jiangsu, where the yield was very low.[70] When considered county by county, however, the Yangzhou figures tell a different story, with Jiangdu, Ganquan, and Yizheng yielding a high number of graduates and other counties in Yangzhou prefecture performing only modestly (see Table 10.2). Ho rightly regards Yizheng as a virtual appendage of Yangzhou. Not only was Yizheng a small county directly neighboring Yangzhou's home counties, but the county seat was close to Yangzhou and large numbers of people living in both the rural and urban areas of Yangzhou were registered in Yizheng. Aggregating these three counties for the period up to 1806—in Thomas

Table 10.2
Jinshi Graduates from Yangzhou Prefecture Under the Qing to 1806

Jurisdiction	Number of *jinshi*
Jiangdu/Ganquan/Yizheng	120
Taizhou/ Dongtai	29
Gaoyou	32
Baoying	22
Xinghua	23
TOTAL	226

NOTE: Excluded from this calculation are the county-level jurisdictions of Tongzhou, Haimen, Taixing, and Rugao, which were divided from Yangzhou prefecture with the creation of Tongzhou as an independent sub-prefecture in 1727.
SOURCE: *Ming Qing like jinshi timing beilu.*

Metzger's phrase, the year of "increasing difficulties" in the Lianghuai salt monopoly[71] — results in 120 graduates, more than half the total number for the prefecture.

In fact, these three counties did much better than the figures suggest, for the numbers do not include successful candidates who were long-term residents of Yangzhou but were registered either in other prefectures or in other Yangzhou counties. The editors of the 1810 prefectural gazetteer omitted the names of fourteen successful candidates in Jiangdu listed in the 1733 edition because they had subsequently been identified as registered elsewhere, most notably Shanxi. Forty-six of the 120 graduates from the three core Yangzhou districts in the period to 1806 are also listed in the Lianghuai salt gazetteer and hence were from currently registered salt families. To judge from their distinctive surnames, between ten and twenty others were from Huizhou or West (Shan-Shaan) families residing in Yangzhou. Perhaps no more than half the locally registered graduates from these three districts were of native stock.[72]

Whatever the situation elsewhere in Jiangsu, academic success was far from evenly spread either geographically or socially within Yangzhou prefecture. Rather, it was concentrated in the counties around the prefectural city and among the immigrant families of the salt merchants. Nor, on the basis of examination figures, can it be argued that the salt monopoly induced prosperity in "many a northern Kiangsu locality."[73] The salt-producing counties of Xinghua and Taizhou/Dongtai did noticeably worse than Gaoyou, where salt merchant money was negligible.

The Blurring of Social Boundaries?

The initiatives of Huizhou merchants in urban social management, combined with their prominence as owners of urban land and as candidates in the civil examinations, bring us to the issue of their relationship with the local society and to the problem of status differentiation in Chinese society during the late imperial period. The position of the salt merchants at the pinnacle of Yangzhou society contradicted the formal hierarchy of Chinese society, which was premised on the functional differentiation of the so-called gentry from other social groups—peasants, artisans, and merchants. The salt merchants' social mobility underpins an understanding of Chinese society in the eighteenth century that sees a transformation in the social basis of the gentry. As noted in the Introduction to this book, the preliminary evidence for this view was supplied by Ho Ping-ti, who argued that the social categories of merchant and scholar-official were beginning to blur in the Qing and that native-place particularism was losing its salience as geographic mobility led to interregional social and economic integration.

The blurring of social boundaries between literati and merchants in eighteenth-century Yangzhou is one of the most commonly mentioned changes in the Ming-Qing period. The Ma brothers Yueguan and Yuelu are often adduced as evidence for this thesis, not least because they actually emerge to view in a well-known painting, *The Ninth Day Literary Gathering in the Temporary Retreat* (*Jiuri xing'an wenyan*) (see Fig. 22). In a catalogue entry on this painting, Ho Wai-kam gives expression to a now-established view of Yangzhou, and even of Chinese society, in the eighteenth century. The painting is a group portrait and is of great interest because of the social prominence of a number of those portrayed. According to Ho:

Fig. 22 The Ma brothers and friends, depicted on the Double Ninth festival in 1743. This portion of the painting shows, from left to right: unnamed servant 1, Wang Zao, Hong Zhenke, Lu Zhonghui, Ma Yueguan, unnamed servant 2, Wang Yushu, Cheng Mengxing, Ma Yuelu, Fang Shijie, unnamed servant 3, Chen Zhang, Li E, unnamed servants 4 and 5. (Detail, Fang Shishu and Yeh Fanglin, Chinese, late 17th–early 18th century, Qing dynasty. *The Literary Gathering at a Yangzhou Garden*, 1743. Handscroll, ink and color on silk, 31.7 x 201 cm. © Cleveland Museum of Art, 2004; The Severance and Greta Millikin Purchase Fund, 1979.72).

With the exception of two retired officials, the rest of the group were primarily commoners; although a number of them were qualified candidates and were recommended for the highly esteemed special examination, the *po-hsüeh hung-tz'u*, most had declined the honor. The economic and social changes of the eighteenth century had made it possible for some of the selected members of the merchant class to be accepted by, and integrated with, the intellectual elite.[74]

Yangzhou, as Ho also notes, was then "a city of enormous economic and cultural vitality, the center of a prosperous salt trade, and the hub of the Grand Canal for south-north traffic."[75] Yet it is not only in the context of Yangzhou that the integration of "the merchant class" with "the intellectual elite" is discussed. Rather, Yangzhou is regularly cited as an "example" of what was happening throughout China—the richest city in the most prosperous century epitomizing trends within Chinese society at large.

Liang Qizi's work on philanthropy in late imperial China supports this view. Despite the prominence of salt merchants and officials in philanthropic activities in Yangzhou, Liang downplays the significance of both the extra-local origins and the occupational profile of salt merchants on the grounds that the merchants "had intimate dealings with local literati," were often themselves from literati or scholar-official families, and worked closely with local officials.[76] Her point is directed at the question of the social hierarchy in Chinese society; that is, she is interested more in the relationship between merchant and scholar than that between Huizhou and Yangzhou. She goes on to state that the two most prominent philanthropists of the early Qing were, like the majority of Yangzhou philanthropists, "originally of Anhui (i.e., Huizhou) registration," but at the same time she is at pains to emphasize the nexus in philanthropic undertakings of merchant wealth, the responsibilities of local officials, and the moral obligations shouldered by the gentry.[77]

Liang's conclusions are based on Yangzhou sources but are echoed in the assumptions of scholars exploring wider fields. Benjamin Elman, drawing on Silas Wu and Ho Ping-ti, observes: "It is nearly impossible to distinguish literati from merchants in eighteenth-century Yangchow, for example."[78] Martin Huang, while noting that "the rise of the merchant class and the elevation of their social status can be traced to the fifteenth and sixteenth centuries," cites Ho Ping-ti in arguing from "the most *representative* example" (italics added) of merchant success, Yangzhou, that the blurring of social distinctions and the consequent identity problems of literati were basically eighteenth-century phenomena.[79]

The "blurring" thesis, which has implications for the analysis of both occupational status and native-place identity, has acquired near-paradigmatic status in eighteenth-century studies and can shape even contrary evidence to its own ends. Tracing the vicissitudes of a prominent family in Shandong, Susan Mann suggested that "there can be little doubt that the Zhoucun Lis [in Shandong] harbored merchants in their midst," but she was able to identify only one, whose relationship to the other members of the family is unclear.[80] This family, she argues, "might have covered their commercial tracks," but if so, that would surely stand as evidence of the continued low status of commerce. Rowe observed the sustained importance of native place in the sojourner population of nineteenth-century Hankou but stressed "the extent to which organization along bonds of common origin was violated," agreeing with Ho Ping-ti and Tou Chi-liang "that sojourners' sentiment and sense of personal identity were deflected toward the host locality."[81] The present study supports this view but emphasizes that native place continued to have a high degree of salience within the "host locality." In other words, the fact of identification with the host locality does not alter the fact of native-place identity. The Chinese in Singapore offer a case in point.

Further, as discussed in greater detail in the following chapter, Elman found that classical scholars in the Qing "seem to have been very much a part of the traditional elite" but concluded in light of the "blurring of social boundaries" that his data do "not prove conclusively that evidential scholars were or were not increasingly from merchant families." Rather, "we *may* see a new social group that includes sons of merchants and literati carrying out evidential scholarship" (italics added).[82] Ho Wai-kam was less tentative, and interpreted the social group represented in the Temporary Retreat in precisely the terms Ho Ping-ti used for Yangzhou society: merchants and literati mixing easily. Correspondingly, Ginger Hsü notes that "this painting is evidence of the disappearing boundaries between merchants and the intellectual elite in late imperial Chinese society."[83]

Who Was Not at the Ma Brothers' Party?

Would an alternative description of Chinese society permit another way of seeing *The Ninth Day Literary Gathering in the Temporary Retreat*? Let us review the evidence it offers for the "blurring of social boundaries." Executed in 1743 by Fang Shishu (1692–1751) and Ye Fanglin, the painting depicts a party hosted by the Ma brothers in their Temporary Retreat garden (see Fig. 22). The Ma brothers were slightly anomalous among the Huizhou sojourners in Yangzhou in that their ancestral place was not Shexian but the poorer, outlying county of Qimen, noted more for its tea merchants than for its salt merchants.[84] But their profile is typical of the Huizhou philanthropist as identified by Liang Qixi, and as patrons of the arts and letters in Yangzhou, they outdid most of their Shexian contemporaries.[85] The painting of the gathering at the Double Ninth festival is a fitting testament to their role as cultural benefactors.

The Double Ninth festival, held on the ninth day of the ninth lunar month, was popularly celebrated by climbing a hill and imbibing drinks made with chrysanthemums, ritual practices undertaken in imitation of Huan Jing of the Eastern Han, who was advised to take refuge on high ground and drink chrysanthemum wine in order to avoid imminent catastrophe.[86] A reference to high ground is provided in this painting by the depiction of rocky outcrops, typical features of a Yangzhou garden and suggestive, in the context of the Double Ninth festival, of a mountainous setting. Chrysanthemums are being planted by servant boys in the foreground, and a chrysanthemum flower is clutched in the hand of one of the guests—the poet and mathematician Li E (1692–1752). Other guests are absorbed in examining a scroll, which can be presumed to be the painting of Tao Yuanming (365–427) by Qiu Ying (1494/95–1522), which had been brought out to serve as the focus of the afternoon's poetry games. When Tao Yuanming "plucked chrysanthemums by the eastern fence," he established an enduring association between the flower and himself as poet; both came to signify retreat from the humdrum world.

The guest-of-honor at the Double Ninth party was Cheng Mengxing, pictured playing a zither. "Cheng" is a typical Shexian surname, and it was ubiquitous in Yangzhou at this time.[87] Cheng Mengxing was the pride of the Cheng clan in his time, having been appointed to a position as a Hanlin compiler after passing the metropolitan examination.[88] His status in this gathering is apparent from the central positioning of his figure behind the table on which the instrument has been placed. The host, Ma Yuelu, is standing beside him in a politely deferential posture. The other Ma brother is one of the two men holding the scroll, seated on the left. Li E obligingly left a full description of the party in a colophon on the painting identifying each guest (for a list of those in attendance, see Table 10.3).

In light of evidence from the late Ming, it may be questioned whether the mixture of guests at the gathering was peculiarly a product of economic and social change in the eighteenth century. A comparable portrait of Zheng Yuanxun and his guests in the Garden of Reflections would surely have shown a similar social composition.[89] The timing of such developments aside, Fang and Ye produced a fair enough portrait of Yangzhou society in the eighteenth century. A mixture of salt merchants, merchant-gentry,

Table 10.3
Ancestral Places of Men Portrayed in
The Ninth Day Literary Gathering

Name	Ancestral place
Chen Zheng	Hangzhou, Zhejiang
Cheng Mengxing*	Shexian, Anhui
Fang Shijie*	Shexian, Anhui
Fang Shishu*	Shexian, Anhui
Hong Zhenke*	Shexian, Anhui
Hu Qiheng	Wuling, Hunan
Li E	Qiantang, Zhejiang
Lu Zhonghui*	Shexian, Anhui
Ma Yueguan*	Qimen, Anhui
Ma Yuelu*	Qimen, Anhui
Min Hua	Shexian, Anhui
Quan Zuwang	Ningbo, Zhejiang
Tang Jianzhong	Tianmen, Hubei
Wang Yushu*	Shexian, Anhui
Wang Zao	Wujiang, Jiangsu
Zhang Shike*	Lintong, Shaanxi

*From a salt merchant family.
SOURCES: Li E, colophon to *Jiuri xing'an wenyan*, trans. in Chou and Brown, *The Elegant Brush*, pp. 133–34; Ho Wai-kam, "The Literary Gathering at a Yangzhou Garden," pp. 372–75; YZHFL, juan 4, pp. 85-88.

and sojourning officials and literati engaging in literary activities in a garden says much about Yangzhou in this period: the composition of the elite; the leisure activities enjoyed by its members; the devotion of urban space to the development of gardens; the careful crafting of a high culture ethos that helped counter the stereotype of the merchant as low, vulgar, and grasping.

Given the centrality of the salt merchants to Yangzhou society, this portrait might be considered less remarkable for documenting the blurring of social divisions than for its confirmation of the unchallenged place of literati culture. But the portrait holds another matter of sociological interest. Li E stated in his colophon that "of the sixteen of us, some are native (to Yangzhou), and others are only visitors." Of those described by Li E as being of Jiangdu, that is, natives of Yangzhou's home county, Lu Zhonghui, Cheng

Mengxing, and Min Hua were actually of Shexian origin.[90] The salt connection is overwhelmingly obvious: the Ma brothers, the Fang brothers, Wang Yushu, and Cheng Mengxing were from salt merchant families. Hong Zhenke "lived by the sea" (that is, on the coast, which was devoted to salt production), and his mother was from the Ma family; so he may safely be reckoned among the merchants as well.[91] Zhang Shike, like the Huizhou men, was from a place that typically produced salt merchants—in his case, Lintong in Shaanxi. He may well have been from "a family of poets" as Ju-hsi Chou and Claudia Brown suggest, but even poets need something to live on. Yangzhou had a large number of people from Lintong bearing the surname Zhang, and all of them were probably involved in the salt trade or living off the inherited proceeds.[92]

At least two social categories are missing from this painting. Women—courtesans and entertainers—were regularly present at literary gatherings in Yangzhou gardens, and their absence here suggests a particular construction of the scholarly community by the designers of the painting. By contrast, paintings of literary gatherings from earlier dynasties, with which these Qing literati were no doubt acquainted, commonly featured female attendants.[93] Fang Shishu's contemporary in Yangzhou, the Fujianese painter Hua Yan, produced a fanciful representation of the famed Jingu Garden of the Three Kingdoms period, depicting the owner Shi Chong with both male and female attendants, the latter apparently including Shi's concubine in the person of the flautist. Hua Yan was probably drawing on earlier paintings, but as a resident in a salt merchant household, he was no doubt familiar with garden scenes and may have depicted a common occurrence in the Yangzhou literary garden.

Also conspicuous by their absence from the portrait are men of Yangzhou ancestry. Did Huizhou and Yangzhou people move, then, in different social spheres? Li Dou suggested as much with his list of scholars in the Ma brothers' circle. Local people are not absent from the list, but they are few: the painter Gao Xiang, a neighbor and old friend of the Ma brothers; the Hanlin academician Wang Wenchong (js1733); one guest from Taizhou; and another from Baoying.[94] The same pattern can be observed in literary circles forming around other merchant families. In 1746, a literary gathering was held at He Junzhao's Eastern Garden, discussed in Chapter 8. Li Dou gave the native-place origins of some 40 of the 64 scholars either connected with this gathering or known through their writings to have visited the garden. Two are described as being from Yangzhou, two as being from Jiangdu county (one of whom has the highly distinctive Shexian surname Min), and another as a native of Xinghua county. The remainder were variously from southern Jiangsu, Zhejiang, Anhui, or the western provinces.[95]

The salt merchants evidently enjoyed the company of men from far and wide. To judge from their circles, however, they counted few local people among their close acquaintances. The Double Ninth painting in fact supplies an unusually clear illustration of an aspect of Yangzhou society on which Ho Wai-kam did not comment: the bifurcation of the population along native-place lines. The Ma brothers' garden was one sign of this divide: the great gardens of Yangzhou were concentrated in the hands of Huizhou sojourners. Among the Double Ninth party guests, Cheng Mengxing, Wang Yushu, Zhang Shike, and Lu Zhonghui were garden owners, and all except Zhang were from Huizhou.[96]

Considered from this perspective, the Double Ninth painting can be seen to offer other criteria by which to consider the question of relations between social classes in Yangzhou, criteria arguably better than the mixing of men who, whatever the source of their wealth, were in the final analysis all *wenren*, men of letters. We might well ask what were the native places of the unnamed servant boys in the painting. Since the Ma family had been in Yangzhou for at least three generations, it may be presumed that the servants were locals. If so, we have in this portrait a representation of Yangzhou society as masters and servants demarcated along sojourner-local lines.

The Double Ninth painting went missing at some point. Li Dou, at least, did not know where it was in the 1790s.[97] By this time, the Temporary Retreat garden had been absorbed into the grounds of the palace built for the Qianlong emperor's visit in 1784, and the Little Translucent Mountain Lodge—the Ma brothers' other garden—had been sold.[98] The painting was still in Yangzhou, however, because in 1808 Ruan Yuan added a colophon to it.[99] The Huizhou merchants were fading from the Yangzhou scene in the early Jiaqing reign, and the salt trade was in the doldrums. Ruan Yuan's colophon might be read as a local scholar's salute to Yangzhou's fading Huizhou heritage.

Redefining Salt Merchants

The Huizhou sojourner and immigrant families in Yangzhou demonstrably made up the city's elite at least from the late sixteenth century to the early nineteenth. We are left with the problem of how to understand and conceptualize this phenomenon in the context of a literature that emphasizes merchant mobility and the "blurring of social boundaries." One issue that needs to be clarified in this respect is the term "gentry-merchant" (*shenshang*), which has been advanced as evidence of the blurring process.[100] The terms *shen* and *shang* do occur in conjunction in the Yangzhou records, but they prove to indicate a plural rather than a compound form, "gentry *and* merchants," not gentry-merchants.

Considerable ambiguities are attendant on the translation of such terms, but the different histories of the Yangzhou orphanage offer good material for validating the first reading. In the prefectural gazetteer, the contributions of "*shen shang* Li Zongkong, Min Shizhang *deng* [et alia]" are noted, which could be read as a reference to gentry-merchants. But the longer list of names in the 1806 salt gazetteer commences: "*junshen* [prefectural gentryman] Li Zongkong, *shangren* [merchant] Min Shizhang."[101] This compels a rereading of the earlier entry as "gentrymen and merchants," especially since Li Zongkong held a *jinshi* degree, and Min Shizhang was plainly a merchant and had no degree.[102] The same reading is certainly the most plausible for another instance of the term, a list of those *shen shang* who supplied fire-fighting equipment to the Hall of Benevolence.[103] The term does not seem to occur in discussions of a single individual.

In her discussion of the term *shenshang*, Joanna Handlin Smith differentiated developments in the late Ming and the Qing. As she noted, late Ming references to disappearing status distinctions are rare, and the same sources are cited over and over again in works making claims in this respect for this period.[104] (The same, of course, can be said of Qing studies, which refer repeatedly to the example of eighteenth-century Yangzhou.) Smith went on to cite the appearance of the term in Qing documents as evidence of increased visibility and respectability "of merchants *as merchants*" in the early Qing. Of her two references in support of this claim, one is to Li Zongkong and Min Shizhang, discussed above, and the other is Susan Mann's discussion of the term, which makes it fairly plain that it was a nineteenth- and early twentieth-century phenomenon. As Mann noted, Chang Chung-li in *The Chinese Gentry* found relatively few examples of this term in the local gazetteers, and most were late examples, i.e., from the nineteenth century, when the historical terrain was beginning to change considerably.[105] Mann herself wrote that "in late imperial times, the line between 'merchants' (*shang*) and 'gentry' (*shen* or *shi*) became impossible to draw" and that "the blurring of gentry and merchant roles [evident in the late nineteenth century] had deep historical roots in local society."[106] But in Yangzhou, it would appear that the distinctions were steadily drawn. There were *shen* such as Li Zongkong and *shang* such as Min Shizhang.

With the rise of the salt merchants under the Qing, the technical distinction between salt gentry and the landed gentry in the examination system lost much of its formal meaning. The sons of the salt merchants increasingly competed in the examination system as residents of Yangzhou rather than the scions of salt merchants: in other words, increasing numbers of them had changed their household registration to the locality and did not have to utilize merchant registration. A quota of fourteen *jinshi* was set for merchant students from the Lianghuai region, but in 1779 it was found that there was only one merchant candidate.[107] The quota was accordingly reduced from fourteen to four. Comparable adjustments were taking place in other sectors.[108]

Such developments can be interpreted, and usually are, as qualitative shifts in the status of merchants during the late imperial period, but it is far from clear that what might be said about merchant status and merchant-gentry relations on the basis of the salt merchants is applicable to other sorts of merchant. Attention must always be paid to context. When we can show that tea merchants, timber merchants, cotton merchants, enjoyed the sorts of social mobility attributed to salt merchants, the blurring of social boundaries will have a surer foundation. As for the salt merchants themselves, if we consider them in the light not of their profligacy but of their steady attention to the fabric of society, their undertakings in Yangzhou seem less like a departure from old norms and conventional practices based on the ascendancy of the landed gentry than their replication, *mutatis mutandis*. It is not so much that the salt merchants overcame social barriers as that they were incorporated into a particular social stratum.

Rather than being changes in the position of salt merchant *qua* merchant, these various initiatives might be interpreted as redefining the salt merchants, in the context of the territorially defined salt monopolies, out of trader status into a social stratum coeval with that of the great landed families of the provinces. In other words, the boundaries did not blur, but they shifted. The salt monopolies, with their elaborate administrations, fixed borders, and revenue quotas, were constructed in parallel with the provinces, and bore a similar relationship to the imperial government: they provided revenue in the form of taxes, and they produced scholar-officials. Moreover, salt could be used, like land, as the basis for changing one's household registration: an investment in so many lots of salt within a given salt administration was clearly the equivalent of procuring land in a particular county. Instead of crops, salt merchants can be seen as farming salt. Their position with respect to the salters was not much different from that of the absentee landlord to his peasant tenants, except that they amassed greater fortunes. This comparison occurred to Kong Shangren, who wrote whimsically that "the salt yard is like a prefecture, and the salter groups and salterns are like departments and counties. As for the grasslands [used for fuel], they are just like fields [for tilling]."[109]

The claims made by salt merchants and salt officials alike for the right and proper place of the salt trade and its practitioners in the workings of the empire were only partly successful. None of the achievements of salt merchants in the examination arena prevented them from continuing to be the butt of derogatory remarks and sarcastic comments. Chinese society continued to distinguish between the salt merchant as merchant and the literati, even if there was actual genealogical confusion between the two categories. On the other hand, the salt merchants' continuing aspirations to literati status confirm the significance of the scholarly class in the Chinese social hierarchy. Had the Huizhou salt merchants been portrayed in their counting houses rather than in their gardens, then indeed they could be seen as having challenged the established order.

The history of Yangzhou, a place where this order appears to have been subverted more than in most other places, offers no example of such a challenge, because by the end of the eighteenth century the salt merchants were declining in number and importance. They were, after all, "official merchants" (*guanshang*). Their ranks thinned as the central government weakened in China, for not unrelated reasons, and the foundations of their extraordinary wealth were destroyed as the government monopoly was restructured in successive phases during the nineteenth century.[110]

ELEVEN

Native Place in an Immigrant City

"Yangzhou is divided between the Dipper and the Ox," wrote Zhou Sheng. "Its people are literary and refined."[1] Places, like people, in late imperial China were differentiated by their literary attainments. Zhou Sheng's characterization of the people of Yangzhou identified the city as occupying a prestigious place in the cultural geography of Qing China. The academies established in Yangzhou in the first half of the eighteenth century declared the city's importance more solidly if less noisily than paintings, gardens, and theater. In the second half, a group of local scholars, some of whom had been students at these same academies, began to form. Collectively known as the "Yangzhou school" (*Yangzhou xuepai*), they had varied levels of success in the examination system, but together brought glory to their native place with their learning.

The Yangzhou school was named by the Tongcheng scholar Fang Dongshu (1772–1851), who was for some time employed by Ruan Yuan. Fang was an arch defender of Song learning and deployed this term in a general attack on Han learning, as manifested by the Yangzhou scholars among others, but he deserves credit for giving a name to the last great cultural product of Qing Yangzhou.[2] Like "the eight eccentrics of Yangzhou," it is a problematic term. Scholars who have been grouped under this rubric range from Wang Maohong (1668–1741), who studied the works of Zhu Xi, to the early twentieth-century anarchist Liu Shipei (1884–1919), and the two had little in common apart from a good command of the classics.[3] But some fifteen scholars active in the late eighteenth and early nineteenth centuries were demonstrably linked by personal and scholarly ties (see Appendix G). Of these, the middle cohort—Ruan Yuan, Jiao Xun, and Ling Tingkan (1757–1809), among others—played an integrating role, feeling the influence of older scholars such as Ren Dachun (1738–89), Wang Niansun (1744–1832), and Wang Zhong (1745–94) and influencing the younger generation in turn.

The timing of their emergence deserves attention. Most of the artists whose names collectively became a byword for eighteenth-century Yangzhou were dead by the time the scholars emerged. Jin Nong died in 1764, and Zheng Banqiao a year later. Ruan Yuan, the pivotal figure in the Yangzhou school (see Fig. 23), was born in the year of Jin Nong's death. In the same year Li Dou began compiling his famous guidebook, to which Ruan later contributed a preface. The second half of the Qianlong reign, during which Li's work was slowly taking shape, was a period of transition for Yangzhou. "Yangzhou's fullest prosperity," wrote Ruan, "was reached between the fortieth and the fiftieth years of the Qianlong reign [roughly 1775–85]. I witnessed it with my own eyes when I was a child."[4] After this time, the position of Huizhou merchants in Yangzhou gradually weakened, and Ruan Yuan and his contemporaries began to write. The literary productivity of Yangzhou scholars reached a critical mass in the early decades of the nineteenth century. Liu Shouzeng (1838–82), singing the praises of his own locality, was hardly exaggerating when he stated that in Yangzhou between the Qianlong and Xianfeng (1851–61) reigns, "the flourishing of classical learning was without compare in the southeast, save for Suzhou and Changzhou."[5]

Like the painters active in earlier decades, the Yangzhou scholars attended to a great variety of subjects and enlarged the domain of literati interests. They combined research on the his-

Fig. 23 Portrait of Ruan Yuan (anon.)
(SOURCE: Ren Zhongyi, *Guangdong bainian
tulu*, p. 31, Fig. 108).

torical and philological interests that dominated Qing learning
with less orthodox studies—astronomy, mathematics, music,
drama. From the perspective of the locality, however, it is the research on Yangzhou and its region by a number of these scholars
that invites particular attention. Their interest in such issues
points to an identification with native place that has historical
significance in the context of the reduced presence of Huizhou
merchants in the early nineteenth century.

Merchants and Scholars

Not surprisingly, the burgeoning of scholarship in Yangzhou in
the late eighteenth and first half of the nineteenth centuries is attributed to the economic and cultural capital amassed by the salt
merchants over the preceding century. These merchants, in Liang
Qichao's view, "played the same role as the great families and

wealthy merchants of Italy during the Renaissance."⁶ Liang's statement was made in the context of a survey of Qing scholarship at large, but he has since been echoed by commentators on scholarship in Yangzhou.⁷

The relationship between the merchant culture of Yangzhou and the rise of the Yangzhou school was complex. The Huizhou immigrants contributed richly to learning in Yangzhou. Their literary parties, their examination successes, and their support for schools and academies in the city imparted an extraordinary vitality to literary life in the city. The establishment of the Anding and Meihua teaching academies led to an influx of established scholars who taught at the academies or participated in their activities and of talented students who would be the next generation of classicists. Notable among the teachers was the "Tongcheng school" scholar Yao Nai, sometime director of the Meihua Academy. The famous scholars Wang Niansun, Wang Zhong, Jiao Xun, Duan Yucai (1735–1815), and Hong Liangji (1746–1809) were among the students.⁸ In the late nineteenth century, local scholar Xue Shou, after reading Li Dou's *Painted Barques*, wrote wistfully:

Among the scholars [in my county] were some of great talent and learning. Men came and went among them from distances of thousands of *li*. There were great discussions and enormous gatherings, each [participant] wrangling with the visitor for the sake of esteem and glory. There were also teachers and scholars who appreciated talent and advocated refinement. In collections of the works of antiquity, [Yangzhou] was superior to other prefectures.⁹

The salt administration had a significant role to play in the creation of this environment, both through its support of the academies and through the participation of salt officials in literary life. As salt controller in Yangzhou, Lu Jianzeng emulated the example of Wang Shizhen in the early Qing by encouraging a lively scholarly community. At the time of the spring cleansing ritual in 1757, he hosted one of the great literary gatherings of the age, one conducted — again in emulation of Wang — by the Rainbow (Red) Bridge.¹⁰ Both the "Wan school" scholar Dai Zhen (1724–77) and his "Wu school" counterpart, Hui Dong (1697–1758), enjoyed Lu's patronage: indeed, it appears that these two defining figures of Qing scholarship met in Yangzhou, under his roof.¹¹

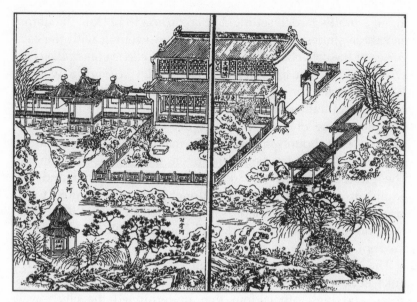

Fig. 24 The Fountain of Letters Pavilion depicted in an early nineteenth-century woodblock print (SOURCE: prefatory illustrations in *JQLHYFZ*).

To its position as salt capital of the empire, Yangzhou owed its greatest scholarly acquisition in the eighteenth century: a branch of the Four Treasuries library. The Four Treasuries collection was the most ambitious of the many bibliographical projects launched in the Qing dynasty and was directed at establishing the definitive Chinese canon. The original collection was housed in a library in the imperial palace, but six other sets were prepared and housed in branch libraries. Three of these were located in the Lower Yangzi region: Yangzhou, the Zhenjiang island county of Jinshan (Gold Mountain), and Hangzhou. These locations testify to the extraordinary significance of the Lower Yangzi Delta as the heart of Qing cultural productivity, and they were justified by reference to the many scholars active in Jiangsu and Zhejiang.[12] The southern libraries were distinguished from their northern counterparts by being readily accessible to accredited scholars.[13]

The Yangzhou library building, known as the Fountain of Letters Pavilion (Wenhuige; see Fig. 24), was constructed in the grounds of the Tianning Temple palace. A woodblock print published in the 1806 salt gazetteer shows an elegant two-story building in a pretty garden setting, but according to Li Dou, the library had three stories, functionally divided in keeping with the categories of the collection. The encyclopedic *Collected Illustrations and Writings from Past and Present* (*Gujin tushu jicheng*), a work in 10,000 fascicles compiled during the reign of the Qianlong emperor's grandfather and printed during the reign of his father,[14] occupied pride of place in the center of the ground floor. The volumes were bound in yellow silk, the imperial color. On either side were the classics, bound in green, and the histories, in red. The philosophy collection occupied the middle floor and was bound in jade-colored silk. The upper floor was devoted to the largest collection, belles-lettres, represented by 1,229 titles copied by hand and 43 printed works, all bound in silk the color of arrowroot, a pale lilac.[15]

The library confirmed Yangzhou's standing in the empire. The links between the salt monopoly, the merchant-gentry of Huizhou, and high officialdom that underpinned the city's well-being were personified in the man appointed to the directorship of the Yangzhou library: the well-connected Xie Shisong. Xie was registered in Yizheng but lived in Yangzhou and, like so many prominent representatives of either place, was the offspring of a merchant family from Shexian. His father, Xie Rongsheng, was a *jinshi* graduate of 1745 who rose to the position of senior vice-president of the Board of Rites. His mother was a daughter of Grand Secretary Chen Hongmou, who had served as prefect of Yangzhou in the years 1729–31 and betrothed his daughter to Rongsheng at that time.[16] The salt controller in Yangzhou was charged with searching out books for the collection, as were the governors and governor-general of Jiangnan.[17] The Yangzhou salt merchants memorialized the throne asking to be allowed to support the Yangzhou and Jinshan libraries, although the offer was declined.[18]

The Literary Inquisition

The Four Treasuries project involved collecting books both for inclusion in the library and for destruction. A literary inquisition was under way while the middle cohort of Yangzhou scholars was growing up. Frederick Mote has written evocatively of the atmosphere of dread in which the literati of the eighteenth century functioned, an outcome primarily of literary witchhunts conducted sporadically during the earlier decades of the century and intensifying at the height of the Qianlong reign.[19] To this atmosphere has been attributed the ossification of Qing learning, a characterization of eighteenth-century scholarship that detracts from its achievements but certainly sums up the constraints within which it was pursued.

Two cases of literary inquisition under the Qianlong emperor heightened this atmosphere in Yangzhou. The first occurred in 1768, the year of the Lu Jianzeng scandal. It erupted in the dead of winter, on the third day of the first lunar month, in the midst of the New Year's celebrations. The salt controller Zhao Zhibi was sitting in the main hall of his yamen to receive salt tax payments for the coming year, when a man entered the hall and began loudly declaiming against the imperial government. He presented Zhao Zhibi with a red envelope that contained three folded red bills, each containing three white sheets of paper. On each sheet was written a seditious poem. On investigation, it was found that the man was one Chai Shijin, also known as Jiang Kui, forty-seven years old and originally a boatbuilder from Huai'an. Shortly before, he had lost both his wife and his son; according to the testimony of his younger brother, also resident in Yangzhou, he had become mentally unbalanced in consequence. The high officials looking into the case were inclined to agree with this diagnosis, and the emperor took the same view, exempting the man's relatives from association with the crime. Chai himself, however, was condemned to death by beating, "as a warning through punishment."[20]

The second case unfolded ten years later in the Yangzhou county of Dongtai and ended in the arrest and punishment of Yangzhou prefect Xie Qikun (1737–1802), among others. It centered on Xu Shukui, a 1738 *juren* of Dongtai county, whose works were posthumously printed on the initiative of his son in 1763. A dispute over land involving the Xu family led the other party to denounce these works. The case finally led in late 1778 to the trial in Beijing of various members of Xu's circle, the informer, and a number of high-ranking officials accused of covering up the crime, including Xie Qikun and the Dongtai county magistrate.

Xie was ordered to redeem his crime through military service; the magistrate was sentenced to 100 blows and three years' banishment. Xu Shukui's grandsons and two of his students, along with Jiangsu financial commissioner Tao Yi and his assistant, were condemned to death without delay, but their sentences were deferred until the autumn of 1779. The informer was allowed to return home to Dongtai.[21]

These two cases differed in their implications for local literati. Chai Shijin's trial and punishment, involving a commoner of obscure origins, could have been of only remote interest to the scholarly class. By contrast, the Xu Shukui case embroiled minor gentry and officials of the field administration and involved things dear to the interests of the gentry: literature and land. Its importance to the throne was made public in the most horrific manner. The court ordered the Jiangsu governor to go to Dongtai and supervise the disinterment of the bodies of Xu Shukui and his son, Xu Huaizu (d. 1777). Their bodies were chopped into pieces and scattered in the wilderness, and their skulls were exposed in public for the edification of the masses. The belongings of the Xu family were confiscated in their entirety.[22] This was the show trial of the inquisition for Jiangsu province, as the Wang Xihou case had been in Jiangxi,[23] and it made clear to the provincial scholar-gentry that they needed to exercise caution in what they wrote and published.

The inquisition removed a large number of literary works from circulation, in many cases obliterating them altogether. Qing literature was thinned out, and history along with it. Popular culture, too, was affected. In Yangzhou the *luantan*, or local forms of theater, came under investigation. In the late Qianlong and again in the early Jiaqing reign, stern prohibitions of local drama were promulgated.[24] The prominent Yangzhou scholar Jiao Xun would later recall the pleasure of listening to local opera in his youth.[25] It is probable that this was a rare pleasure in his old age.

The inquisition and the compilation of the Four Treasuries inevitably touched the lives of Yangzhou scholars. At the time of the Xu Shukui case, Ren Dachun, from the Yangzhou county of Xinghua, was employed in the Imperial Library, where the collation and storage of the Four Treasuries was in progress.[26] He was in Beijing for many years, long enough to welcome and provide guidance to a youthful Ruan Yuan when the latter arrived there in 1786.[27] In 1777, Ren was working alongside Liu Yong (1720–1805), who as Jiangsu commissioner of education would play a central role in the Xu Shukui case the following year. Their superior was Ji Yun (1724–1805), who, as the maternal grandfather of Lu Jianzeng's son, had been caught up in the salt scandal of 1768.[28] Wang Chang (1725–1806) had tutored Lu Jianzeng's children in Yangzhou and was caught up in the same scandal. Restored to favor, he was one of three high-ranking scholar officials who recommended Wang Zhong as corrector of the Four Treasuries collection at Jinshan.[29] Ruan Yuan later presented 60 rare books omitted from the collection,[30] and Ling Tingkan was appointed to the Drama Bureau in Yangzhou.[31]

Other Yangzhou scholars were caught up in the inquisition in lesser ways. Both Jiao Xun and Jiang Fan (1761–1831) were employed in later years on the compilation of the prefectural gazetteer, the groundwork for which was laid by Ruan Yuan and Jiao Xun when they began compiling a local historical chronology, *Yangzhou tujing*. In compiling the gazetteer, Jiao and Jiang naturally had to make decisions about entries. Their self-censorship is evident in the absence of Xu Shukui's name from the list of local degree holders. Xie Qikun, prefect of Yangzhou at the time of the Xu Shukui case, appears only briefly in the listing of officials (he was still alive when the gazetteer was being compiled, which may have made a biographical entry inappropriate). But Tu Yuelong (d. 1798), the magistrate of Dongtai who had been punished for his role in the Xu Shukui case, was honored with a short biography that recorded his contributions to local education.[32] He had survived the ordeal of the inquisition and was restored to office, ending his days twenty years later as a prefect, respected for his tireless commitment to his official duties and his peerless erudition.[33] Rehabilitation of offending officials was, of course, common, but the compilers of the gazetteer could not but have been conscious as they entered the details of Tu's magistracy that he had been disgraced. They may have been making a statement in their quiet affirmation of his standing as a respected scholar-official in the prefecture.

Pamela Crossley has discussed the Four Treasuries project in the context of the Qianlong emperor's concern to provide a stable history of Manchu and Mongol origins. "For the Chinese," writes Crossley, "no project of this kind was necessary. The ancient classics contained all that was supposed to be knowable about Chinese origins."[34] But as Crossley notes, the project was simultaneous with the expulsion of Chinese bannermen from the banner system, which amounted to an ethnic cleansing of the garrison towns in the provinces. The implications of these developments are most obvious for issues of Manchu identity and the boundary between Manchu and Han in China. Yet, as discussed further below, local histories of one sort or another proliferated in Yangzhou in the half-century after the Qianlong reign. The identity politics of historical documentation may, then, have been infective. An early and famous example of the local history genre was an essay by Wang Zhong, "Answering Questions About Guangling" ("Guangling dui"), written in 1787.[35] The themes of such works were consistent with some of the concerns evident in the development of a corpus of works on the Manchus: lineages, origins, loyalty, language, customs.

Genealogies of Scholarship

Another context for Yangzhou scholarship is provided by the rise of evidential scholarship, which again has links with salt merchants and the Huizhou diaspora. An impressionistic overview of developments in Qing studies shows many points of connection between the salt trade and the scholars of the evidential school. Yan Ruoju (1636–1704), perhaps the greatest classicist of his time, was born in Huai'an to a salt merchant–gentry family that during his lifetime was still registered in Taiyuan, Shanxi.[36] The even more famous Dai Zhen was the son of a cloth merchant, not a salt merchant, but he was from the Huizhou county of Xiuning, which, like Shexian, prospered from the salt trade.[37]

The Huizhou diaspora had a foundational importance in Qing philosophy, beginning with Yan Ruoju and confirmed in the influence of Dai Zhen. Liu Shouzeng, the grandson of prominent Yangzhou scholar Liu Wenqi (1789–1856), described his father and grandfather as intellectual heirs to a tradition beginning with Gu Yanwu (1613–82):

On coming to the beginning of the Qianlong reign, we find that [worthy] teachers were few, and scholarship was in decline. Then Jiang [Yong, 1681–1762] of Wuyuan [in Huizhou] rose to eminence from a lowly county: his achievements in revision and transmission [of the classics] were great. His learning spread to Dai [Zhen] of Xiuning. Dai and his disciples flourished in Yangzhou.

According to Liu, Wang Niansun of Gaoyou and his son Yinzhi (1766–1834) took up Dai Zhen's study of phonological exegesis, Ren Dachun of Xinghua did the same with respect to ancient regulations and institutions, and Ruan Yuan, who befriended the Wangs and Ren alike, learned from them what their teacher had said. Liu Wenqi worked closely together with Liu Baonan (1791–1855) of Baoying, and these "two Lius" were among another seven or eight scholars of Yangzhou who also propagated Jiang Yong's learning.[38]

Not everyone would agree with this lineage. Liang Qichao, for example, differentiated Ruan Yuan quite sharply from the "two Wangs" of Gaoyou. The place of the Wangs in Qing intellectual history is not in dispute. They are universally regarded, along with Duan Yucai of Jintan in southern Jiangsu, as Dai Zhen's major disciples. Wang Niansun was personally tutored by Dai. Another disciple was Jiao Xun, who was inspired by Wang Zhong to an interest in Dai's work and showed its influence in the breadth and originality of his own research into subjects such as the *Book of Changes*, mathematics, and ontology. "I read Dongyuan [Dai Zhen]," recorded Jiao, "and I subjected my mind to him."[39] Dai Zhen, as noted above, was from the Huizhou county of Xiuning. He and his teacher Jiang Yong were thus from the same prefecture that supplied Yangzhou with its salt merchants. His scholarly lineage is referred to as the "Wan school," Wan being the conventional short term for reference to Anhui. But "Huizhou school" would seem a more appropriate designation.

The Yangzhou scholars were divided in terms of their relationship to the two major schools of the eighteenth century — the Wu and Wan schools. This variation in orientation is no doubt responsible for the relatively low level of recognition accorded them as a group. Zhang Shunhui, the foremost and for some decades virtually the only Chinese authority on the school, regards intellectual diversity as characteristic of the school as a whole: Yangzhou scholars "energetically debated one another; yet although each continued to defend his own position, they never slandered one another."[40] The grounds for classifying these scholars as a "school," then, lie not in their agreement with one another or in their pursuit of similar lines of study but in the community of scholarship produced by native-place, personal, and professional ties.

The Social Origins of the Yangzhou School

The role of the salt trade, the Huizhou immigrants, and Huizhou scholarship in the Yangzhou school returns us to the issue of the "blurring of social boundaries" discussed in the preceding chapter and tentatively raised by Elman with respect to the social origins of classical scholars of the Qing dynasty. The effects of the "blurring" thesis are particularly evident in the tensions between Elman's speculations and his conclusions. Among Ming-Qing scholars, those from Yangzhou, above all, should show evidence of merchant origins. Indeed, in his study of Qing scholarship, Elman provided only three examples of scholars with merchant origins from the empirewide cohort of prominent scholars born between 1746 and 1765, and all are members of the so-called Yangzhou school.

As indicated in the preceding chapter, it is almost impossible to separate the issue of occupation or status from that of native place in the context of Yangzhou, but status has been the main subject of historical analysis. In discussing the social origins of Yangzhou scholars, Ōtani Toshio commented particularly on the self-educated Ling Shu (1775–1829) and his nephew Liu Wenqi, whose father was a physician.[41] The statecraft scholar Bao Shichen (1775–1855), who knew Ling Shu, left a detailed account of his circumstances:

Mr. Ling was a poor man and lived in the city. At ten years of age he had attended primary school [in his home village]. When older, he had not yet completed reading the Four Books when he again went to primary school in the role of all-purpose guardian. Unhappily, he was unable to understand the ancient commentaries because his studies had come to an abrupt halt. A rich man lived next door to him and had engaged a teacher of the classics for his sons and younger brothers. For many months Mr. Ling took advantage of the night to listen to his teaching from outside the verandah. Becoming aware of this, the teacher shut the window, not accepting Ling; Ling was really angry and sought to leave the old place and go to the city, where he might by day perform services on commission as of old while studying in private till dawn. When he was twenty, he abandoned his original profession, gathered together some pupils, and became a primary teacher, little by little turning himself into a scholar. This was, however, a lowly position, not appropriate to his ambitions, for Ling's learning lifted scholarship to an incomparable degree, and his peers did not dare dispute him.[42]

Bao commented in the same passage on the impoverished youth of Wang Zhong, whom he regarded as more talented than Ling but less hardworking, and also on the humble circumstances of Liu Wenqi, whose "fame in Jiang-Huai for knowledge of the classics was all due to the learning he received from [his uncle] Ling." Ling, clearly, was a self-made man, as opposed to Wang Zhong, whose educated mother instructed him in the classics, and Liu Wenqi, who benefited from an uncle "who loved him dearly and did not want to abandon him to a life of trade."[43]

Was the ability of these men to transcend their humble backgrounds due to circumstances peculiar to the eighteenth century? A comparison of the mid-Qing Yangzhou school with the mid-Ming Taizhou school would suggest not. The social origins of Taizhou school scholars from the greater Yangzhou area were quite diverse. Commentators who have noted this phenomenon have advanced the same hypotheses and drawn the same conclusions as historians studying the Yangzhou school, which flourished two to three centuries later. Wang Gen (1483–1541), leader of the Taizhou school, was an independent salt trader, and all eleven of his local disciples were from commoner families, three of them from impoverished backgrounds.[44] William Theodore De Bary long ago pointed out the numerical dominance of men of scholar-gentry connections among thinkers associated with the Taizhou school—17 out of 25, of whom 11 also had *jinshi* degrees.[45] The proportions are more or less comparable to the gentry-commoner ratio in the Yangzhou school, as are the key roles played by the gentry members. If the Yangzhou school featured dynamic class relationships, as Ōtani suggests, these were perhaps not long-term, secular changes but simply the vicissitudes of status made possible by long-standing socioeconomic circumstances. Otherwise a tendency toward humble professional origins for ever greater numbers of scholars in the late Qing would be obvious. Elman's examination of the social origins of Qing scholars, despite his speculations on a shift in the social base, suggests the contrary.

Elman's conclusions, insofar as they concern Yangzhou scholars, bear directly on the Huizhou context for these scholars and deserve scrutiny. The three examples of merchant origins he offers for the cohort of prominent scholars born between 1746 and 1765 are Ruan Yuan, Jiao Xun, and Ling Tingkan. Of these, Ruan Yuan was the best known in his lifetime. Raised in genteel poverty in a city where wealth and power mattered, a Yangzhou native in a city dominated by Huizhou immigrants, the scion of a military family in a culture that "valued the literary and made light of the military," Ruan achieved the highest successes in his scholarly and official career. Well regarded by three emperors, he mixed with great statesmen, helped able scholars, and expended enormous energy in the promotion of serious literati culture.

As Elman points out, Ruan had a family connection with the salt trade. His father's birth mother, the concubine of his grandfather Ruan Yutang (1695–1759), was a cousin of Jiang Chun, and his own first wife was from the same family. His father found occasional employment in the salt trade and Ruan himself was tutored by Li Daonan (*js* 1759), son of a salt merchant. He was friends with, and subsequently a patron of, Ling Tingkan, also the son of a salt merchant.[46]

The significance of these connections is open to question. Ruan's paternal grandfather was a military *jinshi*, as Elman also notes, and moreover from a local family, established in Beihu, north of Yangzhou, since the end of the Ming. Elman errs in stating that his mother was the daughter of a merchant, perhaps mixing the social origins of the mother and paternal grandmother. Ruan Yuan's maternal grandfather received the *juren* degree and served as magistrate of Datian county in Fujian.[47] As for the paternal grandmother, the fact that she was one of a number of concubines rather than a wife diminishes the significance of her Huizhou mer chant connections.[48] She may herself have been one of any number of superfluous daughters of her father's concubines.

Ruan Yuan's two marriages deepen an impression that his connections with the salt trade were marginal. His first and rather brief alliance with a daughter of the Jiang clan was no doubt a consequence of family connections through his paternal grandmother; in later life he effectively distanced himself from this connection. When this wife died in 1791, the family could not afford extra land for burial near the clan tombs; she was buried some distance away. After Ruan Yuan had become a prominent official and was wealthy enough to buy extra land in the cemetery, he was prepared to move his mother's coffin to the new site but not his wife's.[49] His second wife was chosen from outside the confines of Yangzhou society, not to mention of merchant circles. She was a daughter of the clan of Confucius, from Shandong province. His social circle in his adult years did not feature merchants. In Ruan's biography of his famous relative Jiang Chun, he described Jiang as a man of Shexian, Anhui, and the Yangzhou prefectural gazetteer of 1863 lists this biography in the category of sojourners (*liuyu*).[50]

Jiao Xun, Ruan's relative by marriage, never attained the *jinshi* degree and had no official career of note, but his scholarship exceeded Ruan's. His studies of the *Book of Changes* ensured him a significant place in the Chinese canon. Jiao boasted few connections with merchant circles, apart from indirect links provided by his association with Ling Tingkan and his family relationship with Ruan Yuan. He grew up in Beihu, where the Ruan clan was also based as of 1644, and his father was a landowner with 800–900 *mu* of land to his name.[51] Both his father and his grandfather had a deep interest in the *Book of Changes*. Jiao clan members in the early Qing shifted between farming and trade (*huo nong huo gu*) to make a living,[52] but the steady entry of members of this clan into the ranks of military graduates and the fact of their sustained ties with farming suggests merely that they were struggling with economic exigencies, not determinedly entering a new social stratum. Elman includes Jiao Xun in his little group of scholars of merchant origin because he attended the Anding Academy, which according to Ho Ping-ti was "reserved for children of merchant families in Yangzhou."[53] But Ho's statement was incorrect. Founded expressly as a "place for training scholars of the prefecture," the Anding Academy eventually attracted students from far and wide.[54]

The Yangzhou school's strongest link with the Huizhou merchants, and a clear case of merchant origins for a prominent Qing scholar, was Ling Tingkan, the studious progeny of a Shexian merchant of Haizhou. Ling Tingkan was born in Banpu salt yard in Haizhou, on the far north coast of Jiangsu province. His father had moved to Haizhou from Shexian to make a living in trade. He must have been poorly connected to go so far afield: this was an "out-of-way, distant place," as Ruan Yuan commented. He married a local woman, surnamed Wang (1713–1805), although not the Wang with a water radical which would have definitively signified Huizhou origin.[55] He died when Tingkan was only six.

Ling paid due observance to his Huizhou heritage. He "returned" to Shexian in 1805 to conduct mourning rites in the wake of the deaths of his brother, mother, and wife in that year, and it was in Shexian that he died four years later. All the same, there was a Haizhou side as well as a Huizhou side to Ling, and since his father died when he was only six years old, his Haizhou mother probably had more influence on him than might otherwise have been the case. According to Ruan Yuan—who echoed the old tale about the mother of Mencius choosing a neighborhood appropriate for his upbringing—Tingkan's mother was not at all keen about his engaging in trade and urged him to travel so as to make the acquaintance of scholars.[56] Although he gained the *jinshi* degree, Ling Tingkan pursued his scholarship almost as a private matter. He resisted having his works printed, and they were published only posthumously, largely on the initiative of Ruan Yuan.[57]

Surnames point to other candidates among the Yangzhou scholars for Huizhou, and hence possibly merchant, ancestry, most notably Wang Zhong.[58] Achieving only the first degree, abused in some quarters for his unorthodox opinions, short-lived, having few material resources, and producing a relatively small literary opus, Wang was a renowned scholar in his own time and even more so after his death. He is regarded by Shimada Kenji as the pioneer of *zhuzi* studies—research on pre-Han philosophers other than Confucius and his disciples—which came into their own during a period of deep self-questioning by Qing literati toward the end of the dynasty.[59] Ruan Yuan was proud to be regarded in the same light as Wang Zhong:

Mr. Xie Jinpu [Yong, Shaozai] of Jiashan, vice president of the Board of Civil Appointments, twice served as examiner in Jiangsu. I, Yuan, sat for examination in the year [1785]. Shaozai selected me as the first in exegesis of the classics and also as the best in poetry and prose. Shaozai stated: "When I served as examiner [here] earlier, I had Wang Zhong; this time, I have Ruan Yuan. Both are scholarly men."[60]

Wang Zhong is not commonly described as being of Huizhou origins, but two elements of his biography by Tu Lien-che suggest such origins: first, he had a typical Huizhou surname (Wang with the water radical), and second, his great-grandfather was an expert in the making (or theory of making) of red seal ink—an interest consistent with Huizhou expertise in inks and printing generally. But unlike Ling Tingkan, he did not acknowledge his Huizhou origins, regarding himself as a native of Yangzhou. According to a later commentator, Chen Qubing: "Although Wang was a man of Shexian, he was brought up in very humble circumstances, and all his life harbored a loathing for the wealthy. He used to take up a position by the stone lions outside the Meihua Academy and strike salt merchants on the nape of the neck and scold them to their faces."[61]

The biographies of Yangzhou scholars suggest the advisability of also looking at the maternal line in assessing social origins. Wang Zhong, his son Wang Xisun (1786–1847), Liu Baonan, and Ling Tingkan were each raised by his mother after the early death of his father.[62] In Ling Tingkan's case, his mother must have been instrumental in inducing the sense of filial piety that led to his affirmation of his paternal ancestry. Wang Zhong's mother, Zou Weizhen (d. 1787), left a different legacy. The daughter of a humble local scholar, Zou gained an education by listening to her father teaching his pupils. Later, as a widow, she accepted female students to help support her four children.[63] In "Guangling dui," Wang Zhong begins with conventional apologies for the slightness of his learning, explaining that he had "lost [his] father in childhood and did not have the experience of being trained by father or elder brother."[64] Tu Lien-che surmises that employment in a bookstore may have helped him overcome the disadvantage of a lack of formal schooling, but some credit must go to his educated mother.[65] As for native place, the Zou family can be traced back to Wuxi, but the family had been in Yangzhou for seven generations by the time of Wang's maternal grandfather, and knowledge of the line had virtually been lost, suggesting that any possible ties with alternative native places had been sundered. Wang's obituary for his mother shows that he had few grounds for feeling attached to his father's family.[66]

An intimate glimpse of the significance of the mother in the transmission of literary skills to sons is provided by Jiao Xun's account of his talented great-grandmother. Her surname was Bian. Her father had no sons, and he doted on this daughter: rather than having her taught women's crafts, he instructed her in poetry and painting. When she married, her dowry boxes were covered with paintings and poems from her own hand. When Xun was a small boy, he often saw these boxes: his mother would take his hand and guide his finger along the characters of the poems, teaching him to recite them.[67] Ruan Yuan, too, was indebted to his mother for much of his early training, because his father was frequently absent.[68] The fact that his mother's line achieved higher attainments than his paternal ancestors in scholar-officialdom suggests the social dynamics by which the matriline might assume an unstated but profound importance in an aspiring scholar's view of his place in the world. In Liu Wenqi's case, for example, his maternal uncle Ling Shu appears to have been a much more important figure in his life than his own father.

The fact remains that a simple survey of lineage among the Yangzhou scholars shows more rather than fewer connections with salt merchants and/or Huizhou families than were evident to Elman. Nonetheless, the Yangzhou school as a whole basically illustrates Elman's initial finding that Qing scholars were "very much a part of the traditional elite." With all its diversity of native place and socioeconomic origins, the community of Yangzhou scholars owed its identity and characteristics to ties formed by intermarriage, education, and patronage within family and official circles. These ties are obvious in the biographies of these scholars in *Eminent Chinese of the Qing Period*, are evident in their writings, and have been documented by later scholars writing on the "Yangzhou school."[69]

To summarize, in contrast to their relationship with the salt merchant community, the Yangzhou scholars as a group exhibit strong transprefectural ties. City-bred men such as Wang Zhong and Ruan Yuan enjoyed close relations with the landed gentry from outlying counties such as Gaoyou, the native place of father-and-son team Wang Niansun and Wang Yinzhi; Xinghua, the native place of Ren Dachun; and Baoying, which yielded Liu Tai-gong (1751–1805) and Liu Baonan.[70] Beijing was as important as Yangzhou in the development of these ties, for it was in the imperial capital that some of these men first met. Shared native-place roots in Yangzhou prefecture undoubtedly provided the grounds for such meetings, and the Yangzhou *huiguan* in Beijing, where Ruan Yuan stayed, was an obvious place of rendezvous.[71] The importance of Beijing points away from salt merchants and toward scholars and officials as the natural social world of the Yangzhou scholars. As for the "Yangzhou school" scholars of Huizhou patrilineage, none came from a well-placed, socially successful salt-merchant family, none attained significant positions in the regular bureaucracy, and only Ling Tingkan achieved the *jinshi* degree. By and large, the Yangzhou scholars were strongly based in the landowning gentry, although including scholars of relatively humble origins such as Ling Shu, Liu Wenqi, and Ling Tingkan.

Native Place on the Agenda

It is common for studies of the Yangzhou school to focus on its diverse orientations, even when the relationships between them are also discussed. Zhang Shunhui, however, did attempt to offer an integrated analysis. Among the characteristics he attributed to the Yangzhou scholars were a flexible and developmental analytical viewpoint, particularly evident in Jiao Xun, who rejected categorization of scholarship into "Song Learning," "Han learning," and "evidential research"; an unusual range of research interests—including Wang Zhong's investigation of Zhou and Qin philosophers, Ruan Yuan's research into inscriptions on bronze

vessels, Jiang Fan's large-scale historical work, Jiao Xun's research into dictionaries and drama, and Liu Yusong's collection
of folktales and proverbs; a willingness to depart from Song and
Han commentaries on the Confucian canon for independent textual research into earlier works; an integrity of character that
paired deep scholarship with upright behavior; and a willingness
to recognize their own shortcomings. This last characteristic he illustrates with some disarming anecdotes: Jiao Xun collected letters critical of his work and published them as a supplement; at
eighty years of age Wang Niansun admitted that in compiling his
edition of the *Huainan zi*, he had not had access to Song works on
the subject; he viewed this as regrettable and "humbly went a
thousand *li* to be taught by Gu [Guangqi; 1776–1835]," who was
his junior by more than thirty years.[72]

One feature of Yangzhou scholarship not specially noted by
Zhang is a high degree of interest in the locality. A number of
works on the city and prefecture by Yangzhou scholars are cited
as sources in the present book. The majority appeared in the decades immediately following publication of Li Dou's *Painted
Barques*: Ruan Yuan's works on poetry and the historical circumstances surrounding its composition in Yangzhou, Jiao Xun's collection of biographies and writings by residents of Yangzhou
during the Qing dynasty, Liu Baonan's collection of biographies
of Yangzhou's Ming loyalists, Liu Wenqi's historical study of the
waterway system, Jiao Xun and Jiang Fan's topographical history,
and Jiao's little gazetteer of the North Lakes.[73] Alongside the
works of these acclaimed scholars were others. Ruan Xian, a
cousin to Ruan Yuan, edited a sequel to Jiao Xun's gazetteer of
this rural district, to which Ruan contributed a preface. Another
cousin, Ruan Heng, compiled a collection of woodblock prints of
Yangzhou's famous sites. Lin Sumen, Ruan Yuan's maternal uncle, published two works of verses on Yangzhou customs, one a
collection of local "bamboo verses," the other his own compositions.[74] To this list we could add works of fiction produced in the
first half of the nineteenth century (see below).

Taken together, these works suggest an insistence on the importance of place for defining community. This has implications
for an understanding of Yangzhou as a "melting pot," a place
characterized by a "tolerance of difference,"[75] a view more defen-

sible for some periods of the Ming-Qing era than others. Certainly by the early nineteenth century, a case could be made that difference was becoming less tolerable. In the fifth edition of the Jiangdu county gazetteer, published in 1811, the compilers complained about the number of long-term sojourner families in Yangzhou.

The Liu family of Lanzhou, the Qiao and Gao families of Xiangling, the Zhang and Guo families of Jingyang, the Shen family of Xi'an, and the Zhang family of Lintong are all registered in their ancestral places, but in reality all live in Yangzhou. . . . Moreover, the great Shexian clans of Cheng, Wang, Fang, and Wu have been in Yangzhou for successive generations, and those who still retain their original registration simply cannot be counted. . . . In listing names on the [ancestral] tablets, what should be conformed to has long been abandoned. The father might be recorded and not the son, the younger brother included and not the elder, violating proper feeling. This cannot be approved.[76]

The insertion in this gazetteer and not in earlier editions[77] of such a bald statement of discontent with the composition of Yangzhou's population suggests that the times were out of joint in Yangzhou, and commentators were in search of an explanation.

An awareness of difference is not the same as the fact of difference. The preceding chapters have distinguished Yangzhou and Huizhou people on the basis of observed social phenomena such as garden ownership, philanthropic initiatives, residential patterns, and occupational structure. Whether such differentiation was consciously felt by local people living in the lanes of Yangzhou is another matter, and the fact that the literary sources on such matters were written by educated men complicates analysis of their historical significance. In other words, the proliferation of works on Yangzhou in the early nineteenth century attests to a growing consciousness of the locality among local elites, but it does not necessarily tell us much about the quotidian experience of those people whose names will never be known to us.

Lin Sumen's collection of 300 verses stands as one piece of evidence of this growing consciousness and provides insights into the mechanisms by which a sense of difference was maintained. Lin was not only a relative of Ruan Yuan on the maternal side but also an old associate of Ling Tingkan, who wrote a preface for the collection. Imprinting himself on his times by virtue of his writings on popular culture, he provides an insight into the wider world of the Yangzhou scholars, who mixed with people other than classicists and philosophers and who trod the same streets as Lin and saw the same sights. His verses cover a range of topics on Yangzhou local culture, from commodities available in the markets to rites of passage to literary and artistic achievements. Equally valuable are his commentaries on his own verses, which reflect on changes in Yangzhou over recent times.

Among the many local cultural practices noted by Lin is musical drama, or "opera," as it is often termed. Under this heading, Lin began by noting the pre-eminence of Beijing as a site of operatic performances, with Suzhou ranking second. In Yangzhou, he noted, venues for opera had been established only shortly before, in imitation of the Beijing model. But Lin was also aware that Yangzhou had been a flourishing center for musical drama in former times.[78] The gap between the heady days of the Qianlong tours, when Huizhou merchant troupes set the pace in opera, and the first half of the Jiaqing reign when Yangzhou was producing pale imitations of Beijing styles, is clear.

The history of opera in Yangzhou suggests that local and immigrant culture may have been distinguished over a longer period than the sudden flurry of writings on the Yangzhou region in the early nineteenth century might indicate. The development of this form of theater in Yangzhou followed two main trajectories. One was the rise of Kunshan or Kun Opera (*Kunshan qiang*, *Kunqu*), which evolved during the sixteenth century in the Wu-speaking area of Jiangnan, began declining after the middle of the seventeenth century, and then was jolted back into prominence by Kong Shangren's *Peach Blossom Fan*.[79] In Yangzhou, as Colin Mackerras has shown, Kun Opera flourished under the patronage of the salt merchants, at least seven of whom owned opera troupes. This, like most other things in Yangzhou, was predominantly a Huizhou enterprise: the only non-Hui personage among the owners was surnamed Zhang, who was probably a native of Lintong in Shaanxi.[80]

The major stimulus behind the development of the opera troupes was probably the Qianlong emperor's first southern tour, in 1751. The emperor had already manifested considerable interest in the performance of musical dramas at court, and his exposure to opera in the south resolved him to import actors from Jiangnan.[81] It can be no coincidence that the opera troupes recorded by Li Dou developed after this period. Privately owned opera companies in Yangzhou flourished in the late seventeenth century as well, but it was only in the second half of the eighteenth century that they proliferated.[82]

The nonlocal origins of Yangzhou's Kun Opera are evident in the origins of the art form itself, in the native-place origins of the company owners, and in the composition of the companies. In Yangzhou, the temple where actors registered was located in Suzhou Singers Street (Suchangjie), and the best actors used in the salt merchant companies were from the south.[83] But there was also a local tradition of musical drama, known as "Yangzhou clapper" (*Yangzhou bangzi*) or "clamorous performance" (*luantan*), belonging to the "flowery class" (*huabu*) of opera, a category that incorporated all non-Kun forms under the various names by which they were severally known.[84]

Local opera, in contrast to Kun Opera, "involved local people."[85] Only one salt merchant owned a "flowery" troupe — Ruan Yuan's maternal relative Principal Merchant Jiang Chun.[86] Given the prestige of Kun Opera at this time, Jiang's formation of this troupe may be read as a gesture toward the local populace. Ginger Hsü interprets his initiative as evidence of the Yangzhou "melting pot," but she also notes that it was "most unusual." As such, it seems to be evidence only of Jiang Chun's idiosyncrasies.[87] The company was soon dominated by famous performers from outside Yangzhou — notably Suzhou and Anqing — and ended by offering a blended repertoire.[88]

Local opera had its roots outside the city. The performers sprang from Shaobo to the north, Yiling to the east, the North Lakes market towns, and others.[89] The repertoire was taken largely from Yuan drama, and "the musical rhythms and costumes were exceedingly rustic."[90] As noted above, Jiao Xun, a resident of the countryside north of Yangzhou, was a great aficionado of local opera. He wrote extensively on the subject, nostalgically recalling his first exposure to it as a child and affectionately writing of men "taking their old wives and young grandsons on cart or boat to watch [the opera] by the lakeside."[91]

Unlike Kun opera, local opera was performed in the open. Troupes played to rural audiences in the spring and then moved into the city in the heat of summer, when the Kun troupes were not performing. (The busy season for Kun opera appears to have been between the Qingming festival in the third month and the Dragon Boat festival in the fifth month.)[92] Traveling "flowery class" troupes competed with local opera: clapper opera troupes from Jurong in southern Jiangsu, Anqing "double reed" (*erhuang*) troupes from Anhui, Yiyang-style performers from Jiangxi, and the Luoluo troupes from Huguang. These, like the local troupes, made their way into the city during the hot summer months, when the Kun opera troupes disbanded.[93] It is unclear who dominated the audiences for these visiting troupes, but it is known that Yiyang opera was popular with Huizhou people.[94] Of local opera, Li Dou stated: "The local opera of my district (*wu xiang*) gives pleasure only to the local people (*xiangmin*)."[95]

Three social distinctions are suggested in this sketchy outline of opera in late eighteenth-century Yangzhou. One is between the city and the surrounding towns or rural areas; a second is between the moneyed elite and the common people; and a third is between the salt merchants (mostly from Huizhou) and the natives. The three can with some license be collapsed into one. Almost all Huizhou people were city dwellers, engaged mostly in the salt trade, and rarely owning land in the Yangzhou region apart from what was needed to support the clan shrines. Yangzhou natives had connections to the land, were rarely engaged in the salt trade except when in the service of the Huizhou merchant families, and probably accounted for the bulk of the common people in the city. Kun opera, clearly, was not indigenous to Huizhou but was developed in Yangzhou by Huizhou people, who regarded Suzhou culture as a cut above the local culture of their adopted city. The thriving *luantan* troupes are evidence of a cheeky cultural vitality in the city's hinterland, from which urban society in the marketplaces and on the streets drew sustenance.

Opera draws attention to an aspect of local culture that may have helped sustain boundaries between local and immigrant groups in Yangzhou. Language would appear to have been a key difference between the two forms of opera, with Kun opera using

Wu dialect and local opera being performed in the vernacular.[96] Because of the sustained influx of Huizhou people into Yangzhou throughout the seventeenth and eighteenth centuries, the distinctive Huizhou dialect (Wan'an) must have had a pronounced presence.[97] It was quite different from the local patois, which is classed among the Eastern Mandarin dialects but not readily intelligible to a standard Mandarin speaker. Immigrants from the West provinces brought to the city another set of dialects, all from the Mandarin family but exhibiting great diversity. In Yangzhou, the marketplaces must have echoed with the sound of various languages and accents, with buyers and sellers sometimes struggling to reach an understanding. Li E's miswriting of the names of two of the guests at the Ma brother's Double Ninth literary gathering may well have been due to his misunderstanding the pronunciation of a name.[98] Zhou Sheng, who commented at some length on dialect differences in Yangzhou, confessed himself puzzled as to how to write the sound *jia* (first tone) in the Yangzhou dialect word *yajia* (cunning).[99]

To reiterate an earlier point, the fact of social differentiation has to be distinguished from the significance attributed to it. It would overstate the case to claim that Hui and Yang people never sat in the same theater, metaphorically speaking. Two centuries of close association led to cultural exchange between the communities. Lin Sumen provided good documentation of cultural change among Yangzhou natives in the late eighteenth and early nineteenth centuries. Weddings were one practice in which Huizhou influence on Yangzhou customs was to be observed:

When the woman being married enters the sedan chair, she does not wear hair ornaments but is covered with a red veil. Over this she wears an ornamental hat made of bamboo threads, at least one *chi* high, onto which are basted figurines of silk. When the time comes for her to leave the sedan chair at the husband's house, one of the assembled relatives lifts the top off the sedan chair, and the ornamental hat bursts into view. Because there is a struggle [over who will lift the top], this signifies congratulations on the wedding. This is a Huizhou ritual. Recently Yangzhou practice has been following suit. [Further], it is said that in Huizhou there are many tigers. If the woman being married encounters a tiger, the sedan chair bearers all take flight, and the tiger captures the woman. Since, like the ornamental hat, this can hardly harm the bride, in Yangzhou city there are now also "tigers."[100]

This emulation of Huizhou practice suggests not only the number of Huizhou-style weddings seen by Yangzhou people over the years but also the prestige of Huizhou culture in Yangzhou. What the rich and mighty did, the lesser were keen to emulate. Although these features of the wedding journey failed to survive the demise of the Huizhou community in Yangzhou (contemporary accounts of traditional wedding practices in Yangzhou fail to mention them),[101] they substantiate the view that local cultural practices in this period were a mix of Hui and Yang customs.

Food was another domain in which Huizhou influence was evident. Lin Sumen noted the *san xian da lian* (literally "three fresh, great-joined-together") noodles served in the Yangzhou markets. This was a Huizhou dish.

Yangzhou is known for its Hui noodles. The *san xian* are chicken, fish, and pork. The *dalian* is a large bowl of noodles. When people from other provinces first come to Yangzhou city and eat noodles at the markets, they see a great bowl of soup like a basin of water. Some do not dare dip their chopsticks in, but one bite and the mouth will be watering.[102]

It is possible that wheat noodles were relatively rare in Yangzhou before the rise of the sojourner communities. This was rice territory, since the wet conditions in the city's hinterland inhibited the production of wheat. Hui noodles did not survive as a dish associated with Yangzhou. What did survive was fried rice (*chao fan*), still sold under the name "Yangzhou fried rice" in Chinese restaurants around the world, which may have been a product of the Huizhou presence. According to Lin, people in Yangzhou commonly ate rice at midday; the leftover rice was boiled up as rice porridge and consumed in the morning and at night. Servants in the houses of the rich, however, were distinguished from their counterparts in middle-ranking households by eating rice cooked in oil—a commodity used frugally by those of modest resources.[103]

While describing Huizhou influence on Yangzhou culture, Lin nonetheless distinguished clearly between the two cultural traditions. His attention to such differences demonstrates the consciousness of local and immigrant that appears to have been characteristic of Yangzhou scholars in the first half of the nineteenth century. He had a native's keen eye for the origins of different practices and was able to document sustained differences as well as changes over time. Fine detail is apparent in his account of the winter sacrifice to the Kitchen God, which to an outsider might seem marked by the preparation of rice balls by all families, of whatever origin. To someone in Yangzhou, the differences were apparent. Yangzhou people made their rice balls out of glutinous rice flour. Huizhou rice balls were made from plain rice flour, and both large and small balls were prepared.[104] The distinction is reminiscent of one recorded by Emily Honig in her study of Shanghai prejudices against Subei (Jiangbei) people: Jiangbei culinary practices were regarded as vulgar in Shanghai, and one of the signs of this vulgarity was the size of these glutinous rice balls.[105]

Perhaps the most interesting of Lin's observations on the cultural specificities of daily life in Yangzhou concerns Huizhou people's homes, which were literally signposted:

The families of the transport merchants in Yangzhou, even if they have been local people for a hundred years and are now just menials, all have banners (outside their houses) saying "So-and-so's shop." Thus great households, even though they are not shops, are also called "shops." If impoverished due to idleness, the great family will scatter and the house will be sold and divided up for dozens of families to live there, a passageway being created between the front and rear gates. This is then called a "through shop."[106]

Lin was laughing at the pretensions of merchant families and their descendants, but his account is weighty with significance for the issue of native-place identity in Yangzhou. Huizhou people who had been in Yangzhou for generations no doubt spoke the Yangzhou dialect, probably ate rice porridge at breakfast and dinner, and perhaps went to the outdoor theater in the hot summer months. But generations after moving to Yangzhou, they were still adhering to practices that allowed them to be seen as Huizhou people. The sign saying "shop" on the door allowed those who had suffered downward social mobility to cling to the veneer of respectability that came from identification with the elite of local society, the transport merchants from Huizhou.

Again, the significance of the early nineteenth century for a developing sensitivity to native-place origins is evident. "Shop," it would seem, had always proclaimed a Huizhou residence, but the social significance of the "shop" paradoxically became most apparent with the emergence of the "through shop." One is reminded here of Crossley's suggestion that the term "ethnic" can meaningfully be applied to the Manchus only beginning in the early nineteenth century. The early nineteenth century deserves consideration as a time when throughout China, communities may have been in the process of being redefined in more local and historically specific terms.

Home and the World

Lin Sumen's documentation of commodities available in Yangzhou suggests that one context for the development of local-place consciousness among scholars of the prefecture may have been the wider world into which China as a whole was gradually being incorporated. In other words, the movement or threatened collapse of familiar extra-local boundaries resulted in anxieties about community and a sharpening of concerns about local identity.

In the early decades of the nineteenth century, Yangzhou enjoyed a surprisingly intimate contact with the places outside the empire. Lin Sumen's itemization of commodities shows that in the first decade of the nineteenth century people in Yangzhou were experiencing a heady exposure to international trade. The popularity in the city of "Western" rats and Cantonese chickens as household pets, along with watches and alarm clocks as personal and household possessions, indicates the prestige of goods from distant Canton, the flourishing center of Sino-Western trade at the time.[107] Vernacular expressions also show Yangzhou's increasing contact with Canton. Lin noted the sudden rise of the term *xiyanghua* (Western language) for "gobbledygook"[108] and the saying *xiang fa Guangdong cai*, or "seeking a Cantonese fortune," a phrase used of a family that had lost its wealth.[109] That expression no doubt became increasingly common in Lin

Sumen's old age, when wealthy families lost their fortunes as the salt trade entered the doldrums.

Foreign commodities and foreign knowledge were part of the world of the Yangzhou scholars. Stephen Roddy has observed of Ling Tingkan that his "appreciation for Western science and technology . . . appears to have allowed for a certain openness toward the outside world, even perhaps a cosmopolitan outlook that recognized the significance of events in the world beyond China's own cultural sphere."[110] One piece of evidence advanced for this speculation is the enthusiasm Ling expressed to Ruan Yuan for foreign inventions he had seen in Yangzhou.[111] Ling's appreciation of science and technology was otherwise evident in his interest in mathematics. This he shared with Jiao Xun, who used math to original effect in his study of the *Book of Changes*, and with Ruan Yuan, whom Ling assisted in the compilation of a collection of biographies of mathematicians. Neither the study of mathematics nor the compilation of biographies was an exceptional undertaking in the world of Qing scholarship, but as Zhang Shunhui has noted, Ruan Yuan's biographical project was a first in China for its systematic presentation of the writings of scientists. Among Ruan's subjects were 35 foreigners, certainly an unusual feature for such a well-established literary genre as this one.

Contemporary works of fiction associated with Yangzhou show a comparable originality. *Destinies of the Flowers in the Mirror* (*Jing hua yuan*), written by Ling Tingkan's student Li Ruzhen (1763–ca. 1830),[112] is notable for its effective decentering of the Middle Kingdom through the narrative device of having the central protagonists travel to lands where social criteria were quite different from those operative in late imperial China. Roddy attributes Li Ruzhen's capacity for cultural relativism to the influence of Ling Tingkan, confirming Ono Kazuko's contextualization of Li in Yangzhou scholarship and incidentally implying Yangzhou roots for Li's worldview.[113]

Dreams of Wind and Moon, written in the 1840s, offers an intellectually appealing counterpart to *Flowers in the Mirror*. Whereas Li Ruzhen's novel gazes out from the Middle Kingdom to other places, *Dreams* looks intensely inward to provide the first fictional treatment of Yangzhou itself. In Patrick Hanan's view, this was "the first Chinese 'city novel' in any meaningful sense," and Yangzhou "the first city to be so treated."[114] In striking contrast to *Flowers in the Mirror*, this novel takes an utterly familiar world and throws it into sharp relief. The early date of composition of this quintessentially urban novel suggests a high degree of sensitivity in Yangzhou to the changing nature of the world around it.[115]

Considered in light of these novels, the outpouring of scholarly works on local history and culture from the younger members of the Yangzhou school assumes historical significance as a product of a time of profound change. Like the owl of Minerva taking off at dusk, the Yangzhou scholars of the late eighteenth and early nineteenth centuries perhaps saw what had happened in the day just finished. Their consciousness of their own local place in the world was arguably enhanced by the changing parameters of the world itself. Certainly they were writing Yangzhou into Qing history, providing their native place with a line of descent that linked the present to the Hancheng of the Spring and Autumn period, before the time of Confucius.

In 1804 Ruan established a family temple in the Old City, installing ancestral tablets recording his own patriline.[116] In doing so, he completed a process of native-place cultivation initiated two centuries earlier, when the Huizhou gentryman Zheng Yuanxun busied himself with the creation of a moral community in Yangzhou by encouraging female chastity and filial piety and by rallying his fellow townsmen to philanthropic causes. At the same time, the creation of the family temple in the heart of Yangzhou can be read as asserting local ownership of an immigrant city. Such an assertion was implicit in Ruan's writings and those of his circle. Together the later Yangzhou scholars did for the city what Ruan Yuan did for his own family: they supplied it with a good genealogical record.

Postscript

At the end of the Qianlong reign, Li Dou's Chronicle of the Painted Barques of Yangzhou *made its appearance after three decades of gestation. Li's work profoundly influenced the way in which the eighteenth-century city would be viewed by later generations. The parade of scholars, officials, merchants, and artists, the exhaustively itemized sites of leisure, the poems quoted and the scholarly works named, form a dizzying array of impressions that can too readily blur into a generalized idea of an eighteenth-century boom.*

As suggested in the preceding chapters, if we attend to the timing of these various developments, a more complex view of Yangzhou's eighteenth century emerges. The Yongzheng era was important as the time when major urban institutions took form. The critically acclaimed artists of the century were active mostly in the period up to the 1760s. The great period of garden building commenced in the 1750s and continued to the 1780s. Opera and the urban economy flourished in these same decades. After this time, local scholars became prominent, and from quite early in the nineteenth century, a great burgeoning of local scholarship was in evidence.

Nothing comparable to the Painted Barques *exists for the nineteenth century, a period that in some respects seems like the seventeenth century run in reverse, with a partial unraveling of much that had been achieved in preceding decades. For a range of complex reasons, the city never recovered from the devastations of the Taiping Rebellion. Although the city's nineteenth-century history deserves to be explained on its own terms, the eighteenth century casts a long shadow. It established a way of thinking about Yangzhou that would forever complicate how outsiders viewed the city and how local people saw themselves.*

Toward the end of the nineteenth century a peculiar form of slang arose in the city. Known as "duck talk" (yayu), it originated as a code for financial transactions in a local bank. In this argot, the word for "one" became "night's bright pearl" (yemingzhu), the word for "two" "ear's edge" (erduobian), the word for "is" or "yes" (shi) "silken melon" (sigua), the word for "not" or "without" (wu) "natural color" (bense), and so forth. It was gradually elaborated to a point that local people could abuse uninformed outsiders without the latter having the faintest idea that they were being maligned.[1]

Although we cannot know how generally these terms were deployed in the city, the phenomenon suggests a society that was suspicious of outsiders and constructing boundaries to define itself. The trajectory of events in the nineteenth century, outlined in the concluding chapter, offers a partial explanation for why this happened.

TWELVE

Rather Like a Dream

I n 1817, Qian Yong (1759–1844), travel writer and social commentator, was dismayed to observe the dilapidated state of
Yangzhou's famous gardens. The Twin Tree Bookroom Garden (Shuangtong shushi) of the Zhang family of Lintong, Shaanxi,
had been reduced to ruins, "the pavilions and terraces desolate,
the grass and trees forming a wasteland." The Stone Mountain
Dwelling (Piannshi shanfang) of the Wu family of Huizhou had
been sold to an old woman, who had opened a noodle shop on
the site. Of the nine famous rocks in the Garden of the Nine Peaks,
only four or five remained. The once-glorious vista along Slender
West Lake Qian likened to a "yardful of broken pottery," conjur-
ing up images of ruined buildings, collapsed rockeries, weeds
running wild. Moved to verse, he wrote:

> Half the men of the *Painted Barques* gone now;
> Willows outside the Yihong Garden thick as smoke.
> Comparing past and present
> Is rather like a dream.[1]

Nostalgia is common in nineteenth-century literary works on
Yangzhou. In 1844 Ruan Yuan wrote an afterword to Li Dou's
Painted Barques and commented on the decline of his native place

in similar tones. In the last year of the Qianlong reign, 1795, he had found "Yangzhou still a thriving and bustling place as of old." Eight years later, in 1803, he passed through Yangzhou while on leave from office and went on an excursion with some friends. "From this time on," he wrote, "there was a gradual decline. The terraces crumbled into ruins, and the flowers fell from the trees."[2] On a visit home in 1834, he observed merchant families and scholars reduced to poverty and impecunious officials turning to service and trade to make ends meet. And in 1839, when he retired after a long and distinguished official career in the provinces, he found the gardens and libraries of his hometown neglected or in ruins. The city of his boyhood was hardly recognizable.[3]

Regret for the past and apprehension about the future no doubt played equal parts in Ruan's observations. The Opium War had erupted in 1840, and in 1842 British gunboats blocked the Grand Canal at the Yangzi River junction while the Treaty of Nanjing was being negotiated. Liang Zhangju (1775–1849), acting governor-general of Liangjiang during the Opium War, was trapped in Yangzhou for the duration, and Ruan Yuan kept him company. With three other scholar-officials who had taken refuge in Yangzhou, they formed a "club of five elders."[4] Ruan introduced Liang to a number of the city's famous sites; he must have wished that his guest could have seen them as they had been.

On the basis of Ruan's observations, one could easily write a history of Yangzhou in the early nineteenth century as a story of decline and fall, to mirror the conventional history of the empire as a whole. Such a history would tell only of absences: what used to be and was now no longer. An alternative history is possible. As we have seen, Ruan Yuan's kinsman Lin Sumen wrote in 1808 of new commodities, new fashions. In the 1840s, the characters in *Dreams of Wind and Moon* stroll the city streets clad in fashionable clothing (*xinshishi*) fabricated from Western cloth (*yangbu*), sporting watches, carrying tobacco pouches, the girls made up and dressed to kill.[5] To judge from this novel, Yangzhou was still a busy city in the 1840s. Consider the scene at the southern gate of the New City, where the customs house was situated:

An unbroken line of pedestrians, densely packed houses; the revenue bureau searching out smuggled goods; the county tiger-head sign hung up to demonstrate authority in Yangzhou, the gate guards sternly apprehending villains, their hanging wolf's-teeth quivers [showing them off] in all their military glory; the lanterns of the guesthouses summoning visitors as they come and go; the notices on the shopfronts advertising the city traders with business monopolies; people entering the city, people leaving the city, their breath forming clouds; fellows humping things on their backs, fellows carrying things on shoulder [poles], the sweat flying like rain.[6]

Gong Zizhen (1792–1841), who visited Yangzhou in 1839, the same year Ruan Yuan retired, was pleasantly surprised to find it still a gracious city after the reports he had heard of its reduced circumstances. "Among the several tens of department and district cities between the Yangzi and the Huai," he wrote, "there are none of [comparable] charm and splendor."[7] On the basis of these observations, Yangzhou could be viewed as a city in transition rather than as one in decline in the early nineteenth century, when it was adjusting to a more modest but still considerable place within the urban system of the Lower Yangzi Delta.

On the other hand, both the salt trade and the hydraulic system in Jiangbei entered a period of crisis in the first half of the nineteenth century. The downturn in hydraulic management occurred in tandem with, and was linked to, problems in the salt trade. Thomas Metzger has described the years 1805–30 as a period of "increasing difficulty" in the salt monopoly.[8] These were also the years in which key installations in the water-control system in Jiangbei were collapsing, and the dynasty was struggling to recover from the effects of first the Miao and then the White Lotus Rebellion while simultaneously coping with piracy on the coast and the rising pressures of the opium trade.

One could interpret these developments together with the Opium War of 1840–42 and the Taiping Rebellion of 1850–64 as a series of unconnected events that combined to undermine the foundations of the city's wealth and its administrative importance. But they were, of course, related by administrative and economic factors. In the end, the city was affected by the adversities that afflicted the dynasty. Given its unusually high degree of dependence on administrative structures and imperial patronage, this can hardly be a matter for wonder.

Salt, Waterways, and Regional Decline

Salt merchants began to desert Yangzhou toward the end of the Qianlong era, well before the cataclysmic wars of the nineteenth century. In 1795, investigations into a case of peculation led to the finding that significant numbers of merchants were living in their native places and employing brokers, friends, or relatives to pay taxes in Yangzhou on their behalf.[9] The use of brokers was not new: Li Dou noted of the Shanxi merchants Wang Lütai and Yu Jimei that "Wang used Chai Yiqin, and Yu used Chai Binchen, both of whom were intimately acquainted with the salt regulations."[10] The mere mention of this fact suggests that the practice was relatively rare to Li's knowledge, but it must have become increasingly prevalent during the period in which he was writing. This was not necessarily unfavorable to the functioning of the salt monopoly, but it was certainly disadvantageous to Yangzhou. If merchants were not living in Yangzhou, then neither were they spending their money there. Merchants based in Huizhou, or perhaps across the Yangzi in Zhenjiang (a new source of salt merchants) or even Hankou, were spending their profits elsewhere.

Lin Sumen's catalogue of consumer items makes it clear that there was still plenty of money for discretionary spending in Yangzhou around the turn of the nineteenth century. Lin also commented, however, on the increasing constraints. "For weddings and funerals," he observed, "people used to spend a lot on wine for guests and invitations [to a feast]. In recent days, many [guests] simply receive a card and do not attend any feast."[11] This might be attributed to altered priorities, but relatively speaking, people in Yangzhou appear to have felt a decline in spending power. As Lin elsewhere observed: "These days everyone knows poverty, and no one cares to exaggerate his position. Whenever a boy and a girl get married, the two families talk of frugality and of the boy going to live with his in-laws."[12]

Conspicuous consumption is sometimes advanced as the reason for the decline of salt merchant wealth,[13] but the logic of this argument is difficult to defend. In the seventeenth and eighteenth centuries, many a Huizhou family outspent itself, but there was always an upwardly mobile family to take its place. Part of the reason for the diminishing presence of the Huizhou merchants in

Yangzhou might lie in changes in Huizhou itself during the eighteenth century. Harriet Zurndorfer suggests that an increase in land values due to the growing value of rice as an export commodity helped keep Huizhou men at home. She further speculates that the strong lineage system that had helped sustain the successful ventures of Huizhou men abroad over a period of two or more centuries was weakened by that same large-scale emigration.[14] The relationship between circumstances in Huizhou and the weakening of the salt monopoly deserves further investigation.

Whatever the situation in Huizhou, it is clear that in the late eighteenth century the Lianghuai salt monopoly was beginning to encounter problems that reduced its attractiveness to merchants. Quotas were becoming more difficult to clear, and smugglers were acting with ever greater impunity. In 1795, the Huaibei sector, like the Huainan, experienced the phenomenon of the disappearing merchant. Only twelve of an original twenty transport merchants were still active; the remainder had retired because of difficulties in disposing of their salt quotas.[15] Again in 1795, the Qianlong emperor sought to stem the decline in Huainan salt sales by creating a no-salt zone along the boundaries of the distribution area. In Zhejiang, Fujian, and Liangguang, all salt shops on the border of the Lianghuai zone were to be moved 30 *li* away, to prevent infiltration of salt into neighboring zones.[16] Just such a move had been urged sixty years before, debated at great length, and rejected on the grounds of hardship to the "little people" (*xiaomin*) and the likely inefficacy of the measure.[17] Its implementation in 1795 signaled that the balance between the official and illicit sales was tilting in favor of the latter, and the border areas, where smuggling had been more or less accepted as the price for protecting the core supply regions, were no longer acting as a buffer between salt zones.

One immediate cause of slow sales and loss of profits in 1795 was the Miao Rebellion, which, among other effects, sent Hunanese boatmen scurrying home to defend their villages and left the salt carriers from Yizheng queued at Wuchang with full cargoes of salt. Another cause, especially apparent in this year, was the *fenglun* (seal and sequence) system, introduced in 1792 to guarantee the timely disposal of cargoes. Under this system, the salt ships were sealed as they arrived and their cargoes subsequently released in the order the ships had arrived. This impeded the rapid disposal of salt by small carriers, which handled 30–40 percent of the salt transshipped at this port.[18]

There were also longer-term pressures on the salt merchants, which again must have particularly affected small or recently established businesses. Thomas Metzger, citing Saeki Tomi, has pointed to an increase in *baoxiao* from the 1770s on. Between 1738 and 1771, roughly the first half of the Qianlong reign, merchants repaid the imperial grace to the tune of nearly 10 million taels. Over a similar period between 1773 and 1804, the amount increased to nearly 27 million taels.[19]

The pressure on merchant capital increased in the early years of the eighteenth century, when the salt tax, the salt transport treasury, and private merchant resources were raided to cover military expenses incurred by the White Lotus Rebellion and to cope with crises in the waterway system. In 1803, the grand secretaries and ministers suggested raising the price of salt to finance waterworks, but at the time the profits from the salt monopoly were being diverted to cover the costs of military campaigns to suppress rebellions in the provinces and help rehabilitate war-damaged areas.[20] Obviously, funds being used for defense could not also be used for water control. The limited ability of salt merchants to return the imperial grace became apparent. In 1805 Governor-General Tiebao (1752–1824) and Director-General of River Conservancy Xu Duan (d. 1812) were clearly perplexed about how best to raise the funds needed for repairs to the hydraulic infrastructure following the floods of that year. The choice was between demanding merchant contributions—certainly the most common source of such funds during the eighteenth century—and imposing a levy on the land. In the end, a sum of 135,000 taels was extracted from the merchants, but the greater portion, a total of 90,000 taels, was contributed by just two merchants. The rest of the head merchants could jointly supply only half this amount.[21]

The court appeared to be aware that the merchants were under pressure and was solicitous of their interests. In 1813, the Jiajing emperor ordered that a massive overhaul of the canals and dikes in Xiahe be financed largely through a surtax levied on the land tax. In addition to the 321,000 taels allotted from the land tax, the salt merchants were to provide 88,000 taels, slightly more than a quarter of the land tax proceeds, and a further sum of 18,000 taels was to be disbursed from river conservancy funds. Provincial and river officials had other ideas about the breakdown of finances and in a memorial upgraded the merchant contribution considerably.[22] The merchant dollar, however, was not going as far as it used to. In 1814 it was estimated that the cost of materials had doubled over the preceding seven years.[23]

Inflation added greatly to pressures on the salt trade, and in the Daoguang era (1821–50) it was regularly blamed for difficulties in disposal of Lianghuai salt and the increased competitiveness of smuggled salt.[24] Salt was sold in the marketplace for copper cash, but the salt taxes were paid in silver. According to Wei Yuan (1794–1856), the exchange rate rose from around 1,000 cash per silver tael during the Qianlong era to 1,500–1,600 in the early Daoguang reign, the rise being especially steep in the years 1827–33.[25] The price of salt increased accordingly, and the backlog of the unsold salt mounted.

Problems in the salt monopoly, hinted at in the retirement of salt merchants during the 1790s, reached a crisis point in the 1820s. Zhang Liansheng has linked the problem of inflation to the opium trade, thus attributing the problems of the salt monopoly to the nefarious influence of the West.[26] It is not clear when opium began to appear in Yangzhou, but the means of transporting it were easy enough; barges from Chaozhou in Guangdong, an important source of opium in Shanghai at a later date, regularly delivered charcoal to Yangzhou in the early nineteenth century.[27] By the 1840s, when Zhou Sheng was living in Yangzhou, its use was widespread there.[28]

The problem with Zhang's hypothesis is that serious difficulties in the disposal of the salt quotas emerged around the turn of the century, and inflation was not marked until the 1820s. Nonetheless, it is true that in the 1820s, when the opium traders were flourishing, the salt merchants were visibly experiencing difficulties. According to Li Cheng, a Yangzhou scholar writing at this time:

Substantial merchants at present do not add up to half the number of former times, and every day increasing numbers of bankrupt merchants announce their retirement. All the wealthy merchants and great entrepreneurs of the empire regard Lianghuai with aversion, as a place exhausted of its wealth and without a future. So it is that great difficulties exist with a few dozen merchants managing over a million *yin* of salt.[29]

The merchants themselves, as Li also noted, contributed to the problem.[30] In 1822 Grand Councilor Cao Zhenyong (1755–1835), himself the grandson of a Huizhou salt merchant active in Yangzhou, addressed a range of problems in the salt trade. Among these he counted the corruption of Huang Zhiyun, the most prominent merchant of the time. According to Cao's memorial, Huang had monopolized the post of principal merchant for years on end, exploiting and abusing his position to a point that no other head merchant was willing to come forward to replace him. He was accused by fellow merchants of adding 200,000 taels to the quota of 700,000 set for joint merchant undertakings, of imposing excess levies, and of coercing loans. It was a matter of public outrage. Accordingly, the Grand Council requested (apparently in vain) imperial approval to abrogate the post of principal merchant and to return to a rotating schedule of head merchants.[31]

Corruption in the salt sector was not new. The Lu Jianzeng case of 1768 rivals the Huang Zhiyun case of 1822 in the scale of the misappropriation of funds. The difference is that the Lu incident occurred at a time of flourishing salt sales, while Huang's misdeeds were carried out at a time of crisis in the trade. In the summer of 1822, 413 boatloads of salt that had arrived at Wuchang over the preceding six months were still waiting to be relieved of their cargo, which amounted to nearly half of the total Huguang quota.[32]

Huang Zhiyun appears to have survived the scandal of 1822, and in any case his financial practices were merely emblematic of widespread malfeasance by merchants. Bao Shichen commented around this time on the mismanagement of the Yangzhou orphanage. This sprawling institution, which contained some 1,400 foundlings and their wet-nurses around 1830, was sustained by salt monopoly funds. According to Bao, the funds were managed in a highly irregular manner, going in part to the gratuitous support of an excessive number of directors and an endless stream of doctors. He recommended that the number of doctors entitled to provide attention be limited to seven or eight and that a strict eye should be kept on the provision and cost of medicines. In his opinion, the funds that supported the Yangzhou and Yizheng academies, the poorhouse, the old people's home, the widow relief association, and the lifesaving boats also needed auditing.[33]

Bao Shichen had the ear of Tao Zhu, who assumed responsibility for reforming the Lianghuai salt monopoly after becoming governor-general of Liangjiang in 1830. Tao found the head merchants to be engaged in extremely corrupt practices in the management of discretionary funds. For example, they claimed tens of thousands of taels of silver for refurbishing the salt offices when only a few thousand taels had been spent and a thousand taels of silver for visiting cards that cost only a few cash. Moreover, salt treasury funds that were supposed to finance performances by the Spring Stage and Virtuous Fame opera companies went to pay for private banquets for the merchants, and tens of thousands of taels were spent monthly on the support of the "sons and grandsons" of impoverished merchants.[34]

With respect to these stipends, Tao Zhu echoed Li Cheng's comments:

The Huai merchants originally numbered several hundreds of families. Recently they have exhausted their resources, and there now remain only several tens. Moreover, many borrow in order to manage the salt transport; not all [rely on] their own capital. Further, there are those who are merchants in name but have not the capacity to take a shipment [of salt] and are actually not engaged in trading salt.[35]

This portrait is worth considering from the vantage point of Hankou. William Rowe has found that local records on the Lianghuai salt merchants active in Hankou "consistently identify the Jiaqing and Daoguang periods as the golden age of this group's prosperity."[36] Hankou was the point of sale for salt destined for Hubei and Hunan and accounted for the greater part of the Huainan salt quota. It seems impossible that the Hankou salt merchants could have thrived without deep complicity in salt smuggling. If Hankou merchants were prospering while Yangzhou merchants were floundering, then Hankou was doing well at the expense of Yangzhou, the great market town flourishing while the administrative center declined, and the illegal trade in salt booming while the official trade continued to struggle. Indeed, Rowe notes that the Huguang governor-general found in the 1830s that "a great percentage of the [salt] trade [in Hankou] went unreported."[37]

Tao Zhu's investigations shed light on contradictions between different accounts of the prosperity of Yangzhou in the first half of the nineteenth century. Apparently, a few powerful merchants continued to live in great luxury while weaker families went under. But since many people in Yangzhou depended on the mismanagement of salt funds, it is no surprise that Tao Zhu's proposed reforms met with opposition in the city. His opponents stuck up two placards in the city, one depicting a peach (tao, a pun on Tao's name), the other a beautiful girl called Miss Tao, to which irreverent verses were added by witty critics.[38] Zhou Sheng, as noted in Chapter 9, described the consequences of the reforms as visible in the ever increasing number of prostitutes from "lazy and extravagant" families, who when suddenly deprived of their resources, "had nothing to which they could turn a hand."[39] In the same period, perhaps for the same reason, young women from good families were also being sent out to work as governesses, much like the daughters of the impoverished gentility in England.[40]

In fact, the changes to the Huainan sector of the trade were relatively conservative. The post of principal merchant was abolished, the price of salt at Hankou (critical to reducing the backlog of salt) was lowered, excess fees were ended, the amount of capital outlay needed to obtain monopoly rights was reduced, and salt was packaged in lots of 500 catties, which relieved merchants of packing costs.[41] More radical changes occurred in the Huaibei sector, where the monopoly was completely restructured. The hereditary monopoly enjoyed by wealthy merchants over the right to transport large quantities of salt was replaced by a "ticket system," in which individuals could annually secure a license for the sale of as little as ten lots of salt at a time. These reforms, Metzger notes, brought small entrepreneurs within the ambit of the monopoly. Many of these were "previously smugglers," whose limited capital had precluded them from buying into the shipment system.[42]

In Huainan the ticket system was not introduced until 1849. In this year, a great fire at the Wuchang dock destroyed 400 salt boats with cargoes worth five million taels. Salt merchants retired en masse, and the way was opened for the extension of the ticket system to the Huainan sector.[43] The Taiping Rebellion interrupted its implementation, and it was not until the 1860s that the Huainan trade was restructured. It is notable, however, in contrast to the situation in Huaibei, that the imperial government was forced to give substantial merchants a central role in the management of the Huainan trade. Toward the end of the Qing, Huainan merchants were still a formidable force in regional affairs in Jiangbei.[44]

As the salt trade declined in the early nineteenth century, the rural areas of Jiangbei suffered. The perennial difficulties of maintaining the inland waterway system in the Jiangbei basin intensified in the first half of the nineteenth century. In 1796, the collapse of sections of the Yellow River dikes resulted in floods that caused several million taels worth of damage to the Shandong section of the Grand Canal. This has been identified as the starting point of the collapse of the waterway system, which culminated in its complete breakdown in the 1820s.[45] Temporary or total remission of taxes in full or in part and provision of disaster relief in some part or another of Jiangbei was required every year between 1797 and 1809.[46] Floods occurred in 1804, 1805, and 1806. In 1807, the embankments protecting farmland in Xiahe needed repair, but the damage was considered too extensive for the owners of the fields to sustain the cost. Governor-General Tiebao requested that 17,759 taels be issued for the task, to be repaid over a period of six years.[47]

Associated damage to the dikes of Hongze Lake proved irreversible for a combination of engineering and fiscal reasons. The hydraulic system was caught in a spiral of increasing inefficiency, bad weather, and fiscal problems. Repairs to the Gaojia Dike were proposed in 1808, at an estimated cost of 1.5 million taels. Modifications reduced the projected outlay to just over a million taels, but as the memorials and edicts were passing to and fro, further floods descended on Xiahe. Three years later, only stopgap measures had been undertaken, and the cost of full restoration was now reckoned at 2 million taels. By 1812, only one of the five dams on the Gaojia Dike was still operational.[48] These dams were the key to preventing the dike from rupturing and the lakewaters from spilling down the Grand Canal into Xiahe.

The Daoguang reign is known in the hydraulic history of China as the time when the inland waterway system collapsed. Randall Dodgen, however, has pointed out that the years 1821 to 1841 were a period of remarkable quiescence along the Yellow River.[49] This record of successful water control was partly achieved by draining the waters of Hongze Lake into Xiahe, a solution that created a more or less ongoing disaster in Jiangbei. In 1828, Pan Zengshou, son of former Jiangsu local magistrate and salt controller Pan Shi'en (1770–1854), eloquently summed up the situation: "The waters of the Yangzi and Huai spill out for miles around, while the crows and frogs feed on the carcasses."[50] In 1832, troops had to be called in to control the thousands of peasants who had rallied to prevent the opening of the dikes along the Grand Canal at Gaoyou to release the waters into Xiahe.[51]

The Ruined City

In 1850, a year after the death of Ruan Yuan, the Taiping Rebellion erupted. Three years later Yangzhou was attacked and occupied by Taiping forces, the first of three incursions by the rebels.[52] Yizheng licentiate Ruan Zuo, probably a clan nephew of Ruan Yuan's, "led a group of men to resist the bandits" and was killed.[53] The frightened people labeled their doors with the word "zun," "submit," but whole families of local gentry perished, and some commoner families as well.[54] Stanislaus Clavelin, a Jesuit with the Jiangnan Mission who stayed with a Christian family on the outskirts of the city in 1854, heard that hundreds of women and children put on boats bound for the Taiping capital of Nanjing had perished in the Yangzi when imperial forces attacked the rebel flotilla. He described the spectacle in Yangzhou as frightful: poor food, extreme crowding, malodorous air, polluted water, unburied corpses creating conditions of pestilence that had reached epidemic proportions. "The angel of extermination," he wrote, "hovers over this recently rich, voluptuous city, striking repeatedly, never pausing to sheath his sword."[55]

Not only the city but the countryside was ravaged. The second and third incursions brought the rebels to Beihu, where a heavy toll was exacted. One member of Jiao Xun's clan was killed in 1858, and two others died in 1860, in the course of attacks at Gongdao Bridge.[56] The names of a new cohort of martyrs fill the pages of the local gazetteers, among them thousands of women who died protecting their virtue. The sacking of the Four Treasuries library in the Fountain of Letters Pavilion stripped Yangzhou of the most prominent sign of its relationship with Beijing, the Qing dynasty, and the imperial past.[57]

Given the nexus of the imperial administration, the empire as a whole, and the two institutions historically underpinning Yangzhou's fortunes—the Grand Canal and the salt monopoly—it is difficult not to see the Taiping Rebellion as exacerbating the loss of household income noticed by Lin Sumen, the downturn in employment observed by Zhou Sheng, and the ruin of the city's gardens described by both Qian Yong and Ruan Yuan. The Taiping Rebellion delivered a shattering blow to a city that had no resilience. In 1866, well after the cessation of local hostilities, it was still "in a half-ruined state, a sad relic of the past"[58]—in striking contrast to Hankou, which also suffered during the rebellion but which in 1859 presented to the eyes of Lord Elgin "a scene of commercial activity such as only Shanghai or Canton can equal."[59] According to Rowe, Hankou's fortunes as reflected in land values were not fully restored to their prewar levels until 1892, but the same passage of time failed to ensure a comparable recovery for Yangzhou.[60] In 1894, Dominic Gandar, sometime Catholic pastor in Yangzhou, virtually allegorized this past when he described the city as home simply to retired officials. He called attention to the still neglected state of the city, of which the dilapidated Wenchang Tower was emblematic (see Fig. 25):

Look you well to this tower situated in the middle of Yangzhou. It is a symbol: originally it was the glory of the city, animating the local poets and exciting the curiosity of all who passed by. Today, a thing of shame and sadness, it is no more than a skeleton. The Changmao [Taiping rebels] smashed all its adornments, and in thirty years not a penny has been spent on its repair.[61]

Fig. 25 The Wenchang Tower, which straddled the Old City Canal before the latter was filled in to create a road in 1952. This was among the buildings badly damaged during the Taiping Rebellion (photograph by the author, 1980).

Not only Yangzhou but the entire Jiangbei region was suffering. In 1853, the year of the Taipings' first attack on Yangzhou, the Yellow River broke its banks and resumed its old course north of the Shandong peninsula. Disorder in the region prevented repairs to the dikes, for apart from the Taipings advancing from the south, in 1862 the Nian rebels poured down from north of the Huai, advancing from Funing in the northwest to the Grand Canal and occupying Baoying while the Taipings held Shaobo further south.[62] The heart of Jiangbei began to resemble a feudal state of siege. The rebels were gone by 1864, but Yangzhou was thereafter periodically swamped with refugees, fleeing now a rumor, now a famine. Flooding decreased, only to be replaced by drought. According to the magistrate of Yancheng, writing in 1895:

In the three hundred years from early in the Wanli period [1573–1620] to the Xianfeng reign [1851–61] of the present dynasty, on the whole floods have been many and droughts few. . . . In Xianfeng 5 [1855], the Yellow River moved north, and the waters of the Huai, although they have not returned to the ancient route of Shen and Yu, flow peacefully down to the Yangzi. From when Qingshuitan [on the

Grand Canal] was sundered in Tongzhi 5 [1866], they have not caused disaster. These past thirty years, the great worry for Xiahe has lain not in floods but in drought.[63]

All of Jiangsu north of the Yangzi suffered from the neglect described by Kenneth Pomeranz for the Huang-Yun area of Shandong.[64] In 1906 it was for this reason briefly made a separate province, known as Jianghuai.[65]

The decline of Yangzhou might be attributed to the decline of the Grand Canal, a consequence in turn of the shift in the course of the Yellow River, but the Grand Canal continued to operate as far north as Huai'an. In the latter half of the nineteenth century, it was an efficient transport route for boats bearing foreign imports, and it continued to be used beyond Huai'an for grain tribute vessels from districts north of the Yangzi.[66] Although both Yangzhou and Huai'an suffered from the loss of through trade from the south, the main problem was that the import of goods could not be balanced with local products. In a place where "women had no employment," Yangzhou had nothing to fall back on. Xue Fubao (1840–81), whose family fled Wuxi to find a temporary haven north of the river in Baoying during the Taiping Rebellion, found occasion to report on economic conditions in Jiangbei in the second half of the nineteenth century. "The present poverty of Jiangbei," he wrote, "lies not only in the unproductive land but also in people's failure to exert themselves to the full." The situation was different in Jiangnan, his home region:

Why is this? The men [in Jiangnan] toil hard outside, and the women raise silkworms and do weaving at home. In a family of five, each lives by his or her own exertions, not relying on someone else. . . . In Jiangbei only in the Tongzhou-Haimen area do people know how to weave, and there the land is saline, the harvests poor, and the people survive with difficulty. In the Huai'an-Yangzhou area, people have not heard of the advantages of sericulture, have not witnessed the labor of weaving. Women and children just enjoy themselves all day, relying on a single person for their existence. As for the townspeople, they sit around at leisure, living off their rents, the greater number doing their "weaving and tilling" by attending to their profits, using whatever surplus they might have to double their returns. The more anxious the peasant, the heavier the interest. In a lean year, the peasants just turn around and head off elsewhere, without uttering a word to anyone.[67]

The peasants to whom Xue Fubao referred were crossing the Yangzi River to eke out a living on the streets of Shanghai, where they were employed to collect garbage and nightsoil and pull rickshaws.[68] They were not alone. Artists were deserting Yangzhou for the more profitable milieu of Shanghai, and so were scholars. In 1866, the Longmen Academy in Shanghai, established as a bulwark of Confucian learning to stem the tide of foreign influence, attracted as its director the Neo-Confucian scholar Liu Xizai, from the Yangzhou county of Xinghua. Barry Keenan captures the tensions between old and new, past and present, in describing Liu's return to Xinghua after fourteen years in Shanghai. In a steamboat from the Jiangnan Arsenal, he was carried "away from the coast and past Jiangnan strongholds of Confucian scholarship earlier in the Qing: Suzhou, Changzhou and Yangzhou, to find refuge and final rest in the more undisturbed hinterland of northern Jiangsu."[69]

The View from Shanghai

Shanghai, with its foreign settlements, great buildings, paved roads, running water, electricity, motor cars, and trolley buses, provided the general context for a new, twentieth-century apprehension of Yangzhou. Yu Dafu, a native of Zhejiang but long resident in Shanghai, set off in search of historical Jiangnan in 1928. In this year, the Nationalist Revolution was brought to fulfillment, and a new state was established, centered in a new national capital, Nanjing. The reunification of the country seemed imminent. All the same, the Nationalist Revolution was marked by a murderous internecine purge of Communist supporters, dashing any hopes for an easy creation of one country, one people. There is a certain poignancy in the timing of Yu Dafu's journey in search of the past.

His travels were to culminate in a visit to Yangzhou, the city of gardens and dreams, capable as no other place of "causing a man to lose his soul." He crossed the river full of romantic expectations instilled in him by the paeans of generations of poets and the historical writings of one and a half millennia. Alas for his hopes: even as he embarked on the road leading to Yangzhou, he found the scenery "flat and desolate, without a place capable of inspiring the slightest nostalgia." A day spent visiting the city's famous sites confirmed this melancholy introduction. Temples and pavilions were decayed and gardens untended. The trees, flowers, and rock arrangements that had made the city's gardens famous were represented only by a few sad relics.[70]

In 1935, Yu wrote a memoir of this visit in a letter addressed to Lin Yutang, editor of the literary magazine *This Human World* (*Renjianshi*). He concluded with a piece of advice:

If you dare not travel to Hangzhou, I urge you even more not to travel to Yangzhou. Lord Ouyang's Pingshan Hall, the Red Bridge of Wang Yuanting (Shizhen), Defense Minister Shi [Kefa] in *The Peach Blossom Fan*, Lin Ruhai in *Dream of Red Mansions*, together with the villas of the salt merchants and the minions of the local gentry, are better imagined in your dreams in Shanghai.[71]

Yu Dafu attributed Yangzhou's sorry state to the absence of a rail connection.[72] This was not a plausible explanation for the city's decline, but it certainly increased the effective distance between the lands south and north of the river. Emily Hahn commented on the remoteness of Yangzhou when reflecting on her visit there in the 1930s: "It is a very old, moss-grown city, dank and green and untouched and cut off from the bustle of Chinkiang [Zhenjiang] by a fiercely flowing river. The day we went across this river to see it a storm blew up and cut us off for three days, as no boats would cross."[73]

At that time, Yangzhou was an isolated backwater, enjoying prominence only by virtue of the contrast it offered to a hinterland that supplied the rural population with "food insufficient to fill them, clothes inadequate to cover them, dwellings of narrow and mean aspect."[74] When in 1936 Li Changfu described it as the premier city of Jiangbei, he was not saying much: the rest of Jiangbei, with the exception of Nantong (Qing Tongzhou), was impoverished, unlike the great cities and market towns that characterized the landscape and economy of the southern part of the province and encapsulated new ideas of what society, the economy, and cities should be.[75] Nantong and Haimen boasted a modernized cotton industry but were firmly oriented toward

Shanghai. Exports from the rest of Jiangbei consisted almost entirely of primary products, such as Yangzhou's pickled vegetables. Industry in this formerly large, wealthy city focused on the production of cosmetics and toothbrushes, marketed within Jiangbei.[76]

The context for the publication of Yu Dafu's letter was a furious public debate over a book called *Chatting at Leisure About Yangzhou*, written by onetime radical turned KMT party official Yi Junzuo, an old acquaintance of Yu Dafu's.[77] Yi had spent a few months in the city in the first half of 1932, during the Japanese occupation of Shanghai. He was at that time director of the Department of Publications and Censorship in the Jiangsu Education Commission. The entire staff of the department had evacuated to Yangzhou from Zhenjiang in consequence of the crisis, and his impressions of the city were thus formed at a time of great national anxiety, as well as colored, no doubt, by the inconvenience of life in a thoroughly unmodern city.

The book's portrayal of the city is remarkable mostly for its depiction of what the city lacked. There were no middle-class residences. The great mansions of the salt merchants were falling into disrepair, and ordinary houses were dreadfully antiquated little single-story dwellings. There were no department stores. There were no cars, only rickshaws. Public sanitation was conspicuous for its absence, and the locals often performed their ablutions in public. Drainage was poor, the bridges dilapidated, and the streets full of rubble.

The decayed city served Yi as a metaphor for the decadent society he observed, which he vividly depicted in a description of the rhythms of daily life: the teahouse in the morning, the bathhouse in the afternoon, the theater in the evening. Yangzhou men frittered away their days and nights in these unproductive activities, while the nation was in crisis. Lacking energy, initiative, in brief any spirit of struggle, they were themselves responsible for their poverty. Superstition, a preoccupation with eating and drinking, idleness, had combined to reduce the city to its present state. "Yangzhou," he wrote, "is like a great family in decline."[78]

The people of Yangzhou were incensed at this portrait of themselves, particularly at the gratuitous allegations in the book that most collaborators and prostitutes came from Yangzhou.[79] A local committee sued both author and publisher in the Jiangsu provincial court and succeeded in having the book suppressed.[80] It was a Pyrrhic victory. The plethora of articles on Yangzhou that accompanied publication of the case, although in the main critical of Yi Junzuo, confirmed that the Yangzhou dream was over. This was nowhere more obvious than in the debate over Yi's representation of Yangzhou women as prostitutes; the celebration of courtesan culture had loomed large in the dream but it had had its day.

In many respects, Yangzhou had little left but what had been bequeathed by the Qing. Yi Junzuo, not inappropriately, portrayed a society in a time warp. Following him through the streets of Yangzhou, past the decaying great mansions, the walled dwellings with their shrines to the Earth God, the bathhouses and teahouses, the reader recognizes an older Yangzhou: a city surrounded by walls, with a stream of people flowing past the customs station; boats small and large crowding the canals, life spilling out of the doorways into the narrow streets, and out of the city itself onto the lake, where visitors were taken to gaze upon famous historical sites.

The rhythm of life in this city was slow to alter. In this it resembled Minas Velhas, in Brazil, "an old-time town, with poor revenues and a mediocre population"[81] that intrigued Fernand Braudel because it had so long sustained social practices developed when the city was rich and flourishing. Yi Junzuo described just such a town, in another time and another place. His eye was jaundiced, but his account captured some defining characteristics of Yangzhou society and did not fail to evoke the gracious city of the eighteenth century that Li Dou had described with loving detail in an earlier, grander era.

ɔઠ

Appendixes

County	Town	Distance*	Location†
Jiangdu 江都	Guazhou 瓜洲	40 *li* S	On the Yangzi
	Yangzi 揚子	15 *li* S	Sancha Canal junction
	Wantou 灣頭	15 *li* NE	GC-STC junction
	Xiannümiao 仙女廟	30 *li* NE	STC
	Zhanggang 張綱	30 *li* E	Land route
	Wanshou 萬壽	60 *li* E	On Baita Canal
	Sima 嘶馬	70 *li* E	Jiangdu-Taixing border
Ganquan 甘泉	Shaobo 邵伯	40 *li* N	GC
	Huangjue 黃珏	40 *li* N [W]	Beihu
	Gongdaoqiao 公道橋	40 *li* N [W]	Beihu
	Chenjiaji 陳家集	60 *li* W	Near Anhui border
	Shangguan 上官	60 *li* W	Near Chenjiaji
Yizheng 儀徵	Xincheng 新城	15 *li* E	GC [Sancha Canal]
	Pushuwan 撲樹灣	30 *li* E	GC [Sancha Canal]
	Shirentou 石人頭	40 *li* E	GC [Sancha Canal]
	Doushanpu 陡山鋪	20 *li* W	
	Hejiagang 河家港	20 *li* E [S]	On Yangzi
	Baisha 白沙	S	Shoal formation, Yangzi River
Taizhou 泰州	Ningxiang 寧鄉	60 *li* E [NW]	Near Ganquan border[a]
	Hai'an 海安	120 *li* E	STC junction
	Xixi 西溪	120 *li* NE	Salt Canal
	Gangkou 港口	28 *li* N	Land or water route
	Qintong 秦潼	60 *li* NE	On Salt Canal
	Jiangyan 姜堰	45 *li* E	On STC, on way to Xixi
	Baimi 白米	65 *li* E	On STC
	Qutang 曲塘	75 *li* E	On STC
Gaoyou 高郵	Sanduo 參垛	40 *li* E	On a junction of STC
	Zhangjiagou 張家溝	40 *li* N	GC
	Cheluo 車邏	15 *li* S	GC
	Bei'a 北阿	80 *li* NW	Anhui border
	Jieshou 界首	60 *li* N	GC, Baoying border
	Linze 臨澤	90 *li* NE	Baoying border
	Shibao 時堡	120 *li* NE	Baoying border
	Yongan 永安	60 *li* SE	Ganquan border
	Fancha 樊汊	60 *li* SE	Ganquan border

Appendix A, *cont.*

County	Town	Distance*	Location†
Baoying 寶應	Huailou 槐樓	20 *li* S	GC
	Waxun 瓦甸	30 *li* S	GC
	Fanshui 范水	40 *li* S	GC
	Gangqiao 扛橋	50 *li* S	GC
	Huangpu 黃浦	20 *li* N	GC, Shanyang border
	Lucun 蘆村	40 *li* S	GC, near Fanshui
	Licheng 黎城	90 *li* [S] W	Gaoyou border
	Yongfeng 永豐		Near Licheng
	Sheyang 射陽	40 *li* E	Yancheng border?
Xinghua 興化	Tangzi 唐子	60 *li* E	
	Anfeng 安豐	70 *li* NE	
	Chang'an 長安	35 *li* N	
	Furong 芙蓉	35 *li* NE	
	Huzijiao 瓠子角	25 *li* S	
	Hekou 河口	45 *li* W	Gaoyou border

*Distances and orientation from county-level capital according to *YZYZFZ*, 6.1b–9b, modified by reference to *JSQSYT* and *JQYZFZ* 16.

† GC = Grand Canal; STC = Salt Transport Canal.

*a*The location is given as 60 *li* east of Taizhou in both the Yongzheng and Jiaqing editions of the prefectural gazetteer, but is marked in a northwest location in *JSQSYT*, 1.36a, which also indicates that a sub–assistant magistrate was stationed here, which may not have been the case in 1733.

Appendix B
Ranked Officials of the Lianghuai Salt Administration
in the Eighteenth Century

Office	Rank	Seat	Stipend in silver taels
Liangjiang governor-general as director-general of the salt monopoly (1731) (*Liangjiang zongdu zongli yanwu* 兩江總督總理鹽務)	[1b]	Nanjing 南京	3,000 (1734–74)
Salt censor (*xunyan yushi* 巡鹽御使)	[*]	Yangzhou (Old City) 揚州舊城	3,000 (1770) 5,000 (1794)
Salt controller (*yanyunshi* 鹽運使)	3b	Yangzhou (New City) 揚州新城	6,000 (1734) 4,000 (1737) 2,000 (1780)
Secretary (*jinglisi* 經歷司)	7b	Yangzhou (New City) 揚州新城	600 (1734)
Archivist (*zhishi* 知事)	8b	Yangzhou (New City) 揚州新城	400 (1734)
Treasury keeper (*guangyingku dashi* 廣盈庫大使)	8a (1728)	Yangzhou (New City) 揚州新城	700 (1734)
Baita Canal salt inspector (*Baita he xunjiansi* 白塔河巡檢司)	9b	Yangzhou (New City) 揚州新城 (formerly at Baita Canal 白塔河)	400
Subprefect for inspection of Huainan salt (1732) (*jianche tongzhi* 監掣同知)	6a 5a (1760)	Yizheng 儀徵	2,400 (1736)
Huainan salt examiner (*piyansuo dashi* 批鹽所大使)	8a	Yizheng 儀徵	700
Taiba weight inspection officer (*Taiba jiancheguan* 泰壩監掣官)	8a (1728)	Taizhou 泰州	700
Subprefect for inspection of Huaibei salt (1760) (*jianche tongzhi* 監掣同知)	5a	Huai'an 淮安	2,400

Appendix B, cont.

Office	Rank	Seat	Stipend in silver taels
Huaibei salt examiner (*piyansuo dashi* 批鹽所大使)	8a	Huai'an 淮安	400
Wusha Canal salt inspector (*Wushahe xunjiansi* 烏沙河巡檢司)	9b	Huai'an 淮安	210
Yongfengba weight inspection officer (*Yongfengba jiancheguan* 永豐壩監掣官)	8a (1728)	Qinghe county 清河縣	700?
Taizhou sub–assistant salt controller (*Taizhou yanyunsi yunpan* 泰州鹽運司運判)	6b	Dongtai yard 東泰場	2,700
Fu'an salt receiver (*Fu'an yankesi dashi* 富安鹽課司大使)	8a (1728)	Fu'an yard 富安場	400
Anfeng salt receiver (*Anfeng yankesi dashi* 安豐鹽課司大使)	8a	Anfeng 安豐場	500
Liangduo salt receiver (*Liangduo yankesi dashi* 梁垛鹽課司大使)	8a	Liangduo 梁垛場	400
Dongtai salt receiver (*Dongtai yankesi dashi* 東泰鹽課司大使)	8a	Dongtai 東泰場	500
Heduo salt receiver (*Heduo yankesi dashi* 何垛鹽課司大使)	8a	Heduo 何垛場	400
Dingxi salt receiver (*Dingxi yankesi dashi* 丁谿鹽課司大使)	8a	Dingxi 丁谿場	500 (1768)
Caoyan salt receiver (*Caoyan yankesi dashi* 草堰鹽課司大使)	8a	Caoyan 草堰場	500 (1737)
Liuzhuang salt receiver (*Liuzhuang yankesi dashi* 劉莊鹽課司大使)	8a	Liuzhuang 劉莊場	400
Wuyou salt receiver (*Wuyou yankesi dashi* 伍祐鹽課司大使)	8a	Wuyou 伍祐場	400
Xinxing salt receiver (*Xinxing yankesi dashi* 新興鹽課司大使)	8a	Xinxing 新興場	400
Miaowan salt receiver (*Miaowan yankesi dashi* 廟灣鹽課司大使)	8a	Miaowan 廟灣場	400
Tongzhou sub–assistant salt controller (*Tongzhou yanyunsi yunpan* 通州鹽運司運判)	6b	Shigang yard 石港場	2,700
Fengli salt receiver (*Fengli yankesi dashi* 豐利鹽課司大使)	8a (1728)	Fengli 豐利場	400

Appendix B, cont.

Office	Rank	Seat	Stipend in silver taels
Juegang salt receiver (*Juegang yankesi dashi* 掘港鹽課司大使)	8a	Juegang 掘港場	400
Shigang salt receiver (*Shigang yankesi dashi* 石港鹽課司大使)	8a	Shigang 石港場	500 (1737)
Jinsha salt receiver (*Jinsha yankesi dashi* 金沙鹽課司大使)	8a	Jinsha 金沙場	500 (1768)
Lüsi salt receiver (*Lüsi yankesi dashi* 呂四鹽課司大使)	8a	Lüsi 呂四場	400
Yuxi salt receiver (*Yuxi yankesi dashi* 餘西鹽課司大使)	8a	Yuxi 餘西場	500 (1737)
Yudong salt receiver (*Yudong yankesi dashi* 餘東鹽課司大使)	8a	Yudong 餘東場	400
Jiaoxie salt receiver (*Jiaoxie yankesi dashi* 角斜鹽課司大使)	8a	Jiaoxie 角斜場	400
Bingcha salt receiver (*Bingcha yankesi dashi* 栟茶鹽課司大使)	8a	Bingcha 栟茶場	400
Haizhou (formerly Huai'an) sub–assistant salt controller (*Haizhouyanyunsi yunpan* 海州鹽運司運判)	6b	Banpu yard 板浦場	2,700
Banpu salt receiver (*Banpu yankesi dashi* 板浦鹽課司大使)	8a (1728)	Banpu 板浦場	500
Zhongzheng salt receiver (*Zhongzheng yankesi dashi* 中正鹽課司大使)	8a	Zhongzheng 中正場	500
Linxing salt receiver (*Linxing yankesi dashi* 臨興鹽課司大使)	8a	Linxing 臨興場	400

NOTES: Yard officials are grouped according to the sector to which they belonged (Taizhou, Tongzhou, Haizhou). Ranks in square brackets signify a rank that did not derive from the office held within the salt administration. The salt censor continued to hold the rank of his previous appointment. Dates in parentheses indicate the creation of a new office or rank, or a change in stipend. Details for the administration of the salt yards represent the status quo in 1768.
SOURCE: *GXLHYFZ, juan* 129.

Appendix C
Natural Disasters in Gaoyou Department, 1645–85

Year	Nature of disaster	Consequences
1647	floods	
1649	great floods*	famine
1653	drought	famine
1659	floods	fields inundated
1663	drought, floods	
1665	floods*	great famine
1668	great floods	countless drowned
1669	floods*	people drowned, fields inundated
1670	floods*	fields inundated
1671	floods*	fields inundated
1672	floods*	famine
1673	floods*	harvest ruined
1674	floods, drought, floods	
1676	floods,* drought, floods	Shanghe and Xiahe inundated
1677	floods	
1678	floods, drought	crops withered
1679	drought, locusts, floods	famine
1680	floods*	crops and dwellings ruined
1683	floods	Xiahe inundated
1685	floods*	countless drowned; Shanghe and Xiahe both inundated

*Floods arising from the opening or breach of the dike system on the Grand Canal and Hongze Lake.
SOURCE: *QLGYZZ, juan* 12.

Appendix D
Select Items of Expenditure on Water Control by Salt Merchants in Jiangbei,
1727–1806

Year	Project	Cost in silver taels	Source of funds
1727–28	Overhaul of water control system in Xiahe	270,850+	Merchant contributions, through STT
1734–	Annual replacement of ropes and boards for sluice gates at Yizheng Grand Canal	70 p.a.	STT
1735	Repairs to sluice gates in Baiju salt yard	900	
1738	Overhaul of water control system in Xiahe	300,000	Merchant contributions over 5 years; STT
1739	Dredging of canals in Xinxing and Miaowan yards	27,830	[From contributions of 1738]; STT
1741	Partial dredging of Sancha Canal	600	STT, repaid by merchants over 8 years
Proposed 1745; approved 1754	Extensive dredging in Jiangbei	487,861	STT
1753	Repairs to Lanchao sluice gate, Yizheng	1,506	STT
1753	Dredging of Sancha Canal and construction of a dam	28,270	STT
1755	Dredging of Salt Yards Canal	122,000+	STT; repaid by merchants over 2 years
1755	Dredging of Sancha canal and construction of a dam	11,712	STT
1756	Repairs to sluice gates at Mangdao Canal	5,016	Salt tax
Proposed 1756; approved 1766	Repairs to sluice gates on Fangong Dyke	34,657	STT
1761	Dredging and excavation of canals on coast	87,241	STT

Appendix D, cont.

Year	Project	Cost in silver taels	Source of funds
Proposed 1765; approved 1766	Dredging of Salt Yards Canal and minor canals in Anfeng and other salt yards	20,000 16,160	Recouped from merchants' illicit profits
1766	Sluice gate established at Guangfu Bridge, Jiangdu	669	STT
1767	Dredging of Sancha Canal	22,618	STT
1772	Dredging of Salt Transport Canal; establishment of a sluice gate	65,853 1,684	Salt tax, repaid by merchants over 2 years
1776	Partial dredging of Sancha Canal	9,862	STT
1777	Repairs to Xiangshui sluice gate	2,392	STT
1790	Dredging of Salt Transport Canal and salt yards' canals	290,000	STT; repaid by merchants over 5 years
1799	Repairs to Tongji and Luosi sluice gates, Yizheng Grand Canal	8,390	STT
1806	Dredging of canals and erection of a rush sluice in Xiahe	60,000 30,000 45,000	Individual salt merchant Individual salt merchant Head merchants' contribution

NOTES: This is not an exhaustive list; cost of some items is not given. Estimate costs may differ from actual costs. STT, Salt transport treasury.
SOURCES: *GXLHYFZ, juan 65; XXSJJ; JQYZFZ.*

Appendix E
Private Gardens Visited by the Qianlong Emperor

Garden	1762	1765	1780	1784	Owner	Native place/ancestry
Jiufengyuan 九峰園	X	X	X	X	Wang Changxin 汪長馨	Shexian 歙縣
Yihongyuan 倚虹園	X	X	X	X	Hong Zhengzhi 洪徵治	Shexian 歙縣
Jingxiang-yuan 淨香園	X	X	X	X	Jiang Chun 江春	Shexian 歙縣
Kangshan caoting 康山草堂			X	X	Jiang Chun 江春	Shexian 歙縣
Quyuan 趣園	X	X	X	X	Huang Lüxian 黃履暹	Shexian 歙縣
Shugang chaoxu 蜀岡朝旭	X				Zhang Xuzeng 張緒增	Shanxi
Shuizhuju 水竹居		X		X	Xu Shiye 徐士業	Shexian 歙縣
Xiaoyuan huarui 篠園花瑞				X	Wang Tingzhang 汪廷璋	Shexian 歙縣
Xiaoxiang-xue 小香雪		X			Wang Lide 汪立德	Shexian 歙縣

SOURCES: *GLMST*; *NXSD*; *JQYZFZ*; *YZHFL, JQLHYFZ*.

Appendix F

Individualist Painters Associated with Yangzhou in the Eighteenth Century

Name	*zi*	*hao**	Native place
Wang Shishen 汪士慎 (1686–1759)	Jinren 近人	Chaolin 巢林 Xidong waishi 溪東外史	Shexian 歙縣, Huizhou
Li Shan 李鱓 (1686–1760)	Zongyang 宗揚	Futang 復堂 Aodaoren 懊道人	Xinghua 興化, Yangzhou
Jin Nong 金農 (1687–1764)	Shoumen 壽門	Dongxin 冬心 Jiliu shanmin 稽留山民	Renhe 仁和, Zhejiang
Huang Shen 黃慎 (1687–1770)	Gongshou 恭壽	Yingpiaozi 癭瓢子 Donghai buyi 東海布衣	Ninghua 寧化, Fujian
Gao Xiang 高翔 (1688–1752)	Fenggang 鳳岡	Xitang 西唐	Yangzhou
Zheng Xie 鄭燮 (1693–1765)	Kerou 克柔	Banqiao 板橋	Xinghua 興化, Yangzhou
Li Fangying 李方膺 (1695–1755)	Qiuzhong 虯仲	Qingjiang 晴江 Qiuchi 秋池	Tongzhou 通州
Luo Ping 羅聘 (1733–99)	Dunfu 遯夫	Yufeng 雨峰 Huazhi siseng 花之寺僧	Shexian 歙縣, Huizhou
Gao Fenghan 高鳳翰 (1683–1748)	Xiyuan 西園	Nancun 南村 Nanfu 南阜	Jiaozhou 膠州, Shandong
Hua Yan 華嵒 (1682–1756)	Qiuyue 秋岳	Xin luoshanren 新羅山人	Shanghang 上杭, Fujian

*Some of these painters, including Zheng Banqiao, were known pricipally or commonly by their *hao*, but a number also had a number of different *hao*. Only the principal *hao* are listed.

Appendix G
Scholars Associated with the Yangzhou School

Name	Native-place ancestry	Family background	Degree/office
Wang Zhong* 汪中 (1745–94)	Jiangdu (Shexian)	Scholars	*bagong*
Wang Xixun 汪喜荀 (1786–1847)	Jiangdu	Scholars	*jr* 1807
Wang Niansun* 王念孫 (1744–1832)	Gaoyou	Scholar-gentry, landowners	*js* 1775
Wang Yinzhi 王引之 (1766–1834)	Gaoyou	Scholar-gentry, landowners	*js* 1799
Jiao Xun 焦循 (1763–1820)	Ganquan	Military, scholar-officials, landowners	*jr* 1801
Ruan Yuan 阮元 (1764–1849)	Yizheng	Military, scholar-officials, landowners	*js* 1789
Ling Shu 淩曙 (1775–1829)	Yizheng	Humble	
Liu Wenqi 劉文淇 (1789–1856)	Yizheng	Medicine	*bagong*, 1819
Liu Baonan 劉寶楠 (1791–1855)	Baoying	Scholars, landowners	*js* 1840
Jiang Fan 江藩 (1761–1831)	Ganquan (Shexian)	Landowners	
Ling Tingkan 淩廷堪 (1757–1809)	Shexian (b. Haizhou)	Merchants	*js* 1793
Huang Chengji 黃承吉 (1771–1824)	Jiangdu (Shexian)	Merchants?	
Liu Taigong 劉台拱 (1751–1805)	Baoying	Scholars, landowners	*jr* 1771
Zhu Bin 朱彬 (1753–1834)	Baoying	Cousin to Liu Taigong	
Ren Dachun 任大椿 (1738–89)	Xinghua	Gentry	*js* 1769

ℭℬ

Reference Matter

CB

Notes

Part I Introduction

1. Zhu Ziqing, "Yangzhou de xiari," p. 38.

Chapter 1

1. On the veracity of Polo's claims, see Wood, *Did Marco Polo Go to China?*

2. Burke, *Venice and Amsterdam,* p. 111. Burke refers to Yangzhou in a comparative essay on consumption; see his "Res et verba: Conspicuous Consumption in the Early Modern World," p. 151.

3. Li Dou, *Yangzhou huafang lu.*

4. Zheng Xie, "Si jia."

5. See Watson, *Chuang Tzu: Basic Writings*, p. 45.

6. See Wei Minghua, "Xi Yangzhou meng."

7. For an alternative translation, see Payne, "Happy Regret," in *The White Pony,* p. 252.

8. Jiaodong Zhou Sheng, *Yangzhou meng.*

9. Yu Dafu, "Yangzhou jiu meng ji Yutang."

10. Feng Zikai, "Yangzhou meng." I thank Geremie Barmé for this reference.

11. Guinness, *In the Far East,* pp. 32–43.

12. Yu Dafu, "Yangzhou jiu meng ji Yutang."

13. Honig, *Creating Chinese Ethnicity*; Finnane, "The Origins of Prejudice."

14. See Yi Junzuo, *Xianhua Yangzhou;* and also Chapter 12 of this book. On the controversy surrounding Yi's interpretation of Yangzhou culture, see Finnane, "A Place in the Nation."

15. Fan served as director of a school of journalism in a communist base area north of Yangzhou during the anti-Japanese war (*BDRC* 2: 10–11; see also Hung, "Paper Bullets").

16. Fan Changjiang, "You Yangzhou."

17. Wang Xiuchu, "Yangzhou shiri ji." For translations, see Aucourt, "Journal d'un bourgeois de Yang-tcheou (1645)"; Backhouse and Bland, "The Sack of Yang Chou-fu"; Mao, "A Memoir of Ten Days' Massacre in Yangchow"; and Struve, *Voices from the Ming-Qing Cataclysm,* pp. 32–48 (partial translation only).

18. Ho Ping-ti, "The Salt Merchants of Yang-chou."

19. Ho Ping-ti, *The Ladder of Success in Imperial China,* pp. 249, 254. Among many possible citations of Ho's work, see Elman, *From Philosophy to Philology,* pp. 93–95; Martin Huang, *Literati and Self-Re/Presentation,* p. 31; and Mackerras, *The Rise of the Peking Opera,* p. 77.

20. Ho Ping-ti, *Zhongguo huiguan shilun.*

21. Zurndorfer, "From Local History to Cultural History," pp. 387–96.

22. See also Chapter 10 of this book.

23. Johnson, *Shanghai.*

24. Hay, *Shitao,* p. 25.

25. Naquin, *Peking,* p. 704.

26. Wang Zhenzhong, *Ming Qing Huishang yu Huai Yang shehui bianqian.*

27. Rowe, *Hankow.* The focus of this study is, however, the nineteenth century.

28. Shiba, "Ningpo and Its Hinterland," p. 403.

29. Ibid.

30. See Mann, "The Ningpo *Pang* and Financial Power at Shanghai."

31. Honig, *Creating Chinese Ethnicity.*

32. Hsü, *A Bushel of Pearls.*

33. Hay, *Shitao,* p. 19.

34. Ibid., p. 18.

35. Prominent examples include Yuan Jiang (ca. 1671–ca. 1746), his kinsman Yuan Yao (fl. 1739–1778), Wang Yun (1652–after 1735), and Yu Zhiding (1647–after 1709), active mostly in the late Kangxi to Yongzheng periods. These painters have received some critical attention; see Cahill, "Yuan Jiang and His School"; Mu Yiqin, "An Introduction to Painting in Yangzhou in the Qing Dynasty," pp. 22–25;

Barnhart, *Peach Blossom Spring*, pp. 104-18; and Hay, "Shitao's Late Work," pp. 235-37, 242-52. A host of others are known by name; for brief biographical details, see *YZHFL, juan* 2, pp. 36-52 (since numerous editions of this oft-cited work are available, the *juan* number is provided for ease of reference); and Wang Yun, *Yangzhou huayuan lu*. For a discussion of Gong Xian's statement, see Hsü, *A Bushel of Pearls*, p. 214.

36. Meyer-Fong, "Making a Place for Meaning in Early Qing Yangzhou." See also idem, *Building Culture in Early Qing Yangzhou*. The latter work appeared too recently to be considered in the writing of the present book.

37. *YZHFL*, preface, p. 5. See Wang Yinggeng, *Pingshantang lansheng zhi*; Zhao Zhibi, *Pingshangtang tuzhi*; Wang Zhong, *Guanling tongdian*; and Cheng Mengxing, *Pingshantang xiaozhi*. A copy of the last item is held in the Library of Congress. I am grateful to Tobie Meyer-Fong for the reference.

38. See also Chapter 11 of this book.

39. On the confusion between Yangzhou and Yizheng, see Chapter 3. According to the 1881 local gazetteer, "the greater half of people with Yizheng registration live in the prefectural city" (*GXGQXZ, fanli* [introduction], 4b).

40. Hanshang mengren, *Fengyue meng*, p. 30.

41. See Chapter 11.

42. Hay, *Shitao*, p. 32.

43. Ibid., pp. 4-11.

44. Four of Scott's sketches are reproduced in Wang Hong, *Lao Yangzhou*, pp. 77-78, 137. Of these, one has no attribution, two are described as being by a "foreign artist," and one as by an "Australian artist." Scott, an American who lived and taught for decades in the Philippines and is well known as a historian of that country, served briefly as a missionary in Yangzhou after World War II. He presented his collection of 21 sketches to the Yangzhou Municipal Committee in 1991, and they were then deposited in the Yangzhou Museum. I had the pleasure of meeting Professor Scott at a conference in Hong Kong that same year, while he was en route to Yangzhou, and he kindly provided me with photographs of his drawings.

45. I think that Hay (*Shitao*, p. 6) must be right in his speculative identification of the pagoda and gate. The Alexander watercolor, *Suburbs of a Chinese City*, is discussed by Susan Legouix (*Image of China*, p. 69), who notes that the "scene of this picture has yet to be identified." But in the published engravings of Alexander's illustrations of the Macartney embassy's journey, the engraving correspond-

ing to the watercolor is placed between a picture of Baoying Lake, north of Yangzhou (see Fig. 10 in the present book), and one of Jin-shan (Gold Mountain), to the south, which appears to reflect the order in which the sketches were made by Alexander, as the embassy passed down the Grand Canal. The representation of the pagoda is visually clear enough. It even has the right number of stories, which is not the case in Shitao's painting. See Alexander, *Costumes et vues de la Chine*, vol. 2, facing p. 33. The order of engravings in *Costumes et vues* is the same as in the folio of engravings by A. T. Pouncy that accompany the two volumes of Staunton, *An Authentic Account of an Embassy from the King of Great Britain to the Emperor of China.*

46. YZHFL, *juan* 7, p. 158.

47. Rowe, *Hankow*, pp. 3–9.

48. Burke, *Venice and Amsterdam*, p. 139.

Chapter 2

1. Barrow, *Travels in China*, p. 516.

2. From Xie Tao (464–99), "Song of the Men of Chin-ling," in Waley, *Chinese Poems*, p. 110.

3. Legge, *The Chinese Classis*, vol. 3, *The Shoo King*, p. 107.

4. Ibid., p. 112.

5. Legge, *The Chinese Classics*, vol. 5, *The Ch'un Ts'ew with the Tso Chuen*, pp. 818–19.

6. Zhu Xie, *Zhongguo yunhe shiliao xuanji*, pp. 3–5.

7. The site of Hancheng was identified with that of Guangling (Yangzhou) by Yue Shi (903–1007) (*Taiping huanyu ji*, 124.4a), but the Qing scholar Liu Wenqi (*Yangzhou shuidao ji*, 1.3b), citing conflicting evidence, concluded that this could not be known. Archeological finds confirm Yue Shi's statement; see Nanjing bowuyuan, "Yang-zhou gucheng 1978 nian diaocha fazhuo jianbao," p. 35.

8. *Hou Hanshu*, 6: 21.2561.

9. Ji Zhongqing, "Yangzhou gucheng zhi bianqian chutan," pp. 48–49.

10. Zheng Zhaojing, *Zhongguo shuili shi*, pp. 133, 144; Liu Wenqi, *Yangzhou shuidao ji*, 1.3a–b.

11. At this time the current of the canal was from south to north, following the natural lie of the land. It was reversed when the Yellow River began shifting to a southern course in the twelfth century (Bao Shichen, "Xiahe shuili shuo," 1.30a–b).

12. Bao Zhao, "Wucheng fu."

13. Wright, *The Sui Dynasty*, pp. 158–61.

14. Meng Shi, "Shi Tao's 'Brilliant Autumn in Huaiyang,'" pp. 146–48. Jonathan Hay (*Shitao,* pp. 79–80) translates the painting's title as "Desolate Autumn in Huaiyang" and discusses it at length, providing a full translation of the 86-line poem that occupies the top quarter of the work. The painting is reproduced in color in Meng's article and in black and white in Hay's book.

15. *Suishu,* 2: 26.873–74.

16. *Jiu Tangshu,* 40.1572. There is a saying dating from the fifth century: "If girding yourself with cash is your aim, / Fly to Yangzhou on the back of a crane." This is sometimes quoted with reference to the city of that time, but the reference was to the province of Yangzhou, not to Guangling (Zhu Fugui and Xu Fengyi, *Yangzhou shihua,* p. 3).

17. Xu Yuanchong et al., *Tangshi sanbaishou xinyi,* p. 94; Payne, *The White Pony,* p. 170.

18. Schafer, *The Golden Peaches of Samarkand,* pp. 17–18.

19. Quan Hansheng, *Tang Song diguo yu yunhe,* pp. 17, 32–35.

20. Li Tingxian, *Tangdai Yangzhou shikao,* p. 373.

21. Ibid., p. 378.

22. *Jiu Tangshu,* 40.1572; Twitchett, *Financial Administration Under the T'ang Dynasty,* pp. 52–95.

23. Quan Hansheng, "Tang Song shidai Yangzhou," p. 154. On soldiers, see *Ennin's Dairy,* p. 38. On migrants, see *Jiu Tangshu,* 146.3963.

24. *Jiu Tangshu,* 82.4715–16. For a good summary of the struggle for the city in the period 887–92, see Zhu Fugui and Xu Fengyi, *Yangzhou shihua,* pp. 82–89. The early phase of this struggle, involving the demise of the disgraced Tang commander Gao Pian (d. 887) at the hands of former Huang Chao rebel Bi Shiduo (d. 888) and the subsequent blockade of the city by Gao's former subordinate Yang Xingmi (d. 905), provided the context for a much-discussed seventeenth-century story by Langxian (fl. 1627); see Widmer, "Tragedy or Travesty? Perspectives on Langxian's 'The Siege of Yangzhou'"; Carlitz, "Style and Suffering in Two Stories by 'Langxian'"; Wu, "Her Hide for Barter"; and Finnane, "Langxian's 'Siege of Yangzhou.'"

25. Ji Zhongqing, "Yangzhou gucheng zhi bianqian chutan," p. 53.

26. Translated by Hsiung Ting in Payne, ed., *The White Pony,* p. 258. Romanization altered.

27. Zhu Fugui and Xu Fengyi, *Yangzhou shihua,* p. 47.

28. Quan Hansheng, "Tang Song shidai Yangzhou," pp. 170–74. On the building of the city wall in the Later Zhou and the improb-

ability of further wall building under the Northern Song, see Ji Zhongqing, "Yangzhou gucheng zhi bianqian chutan," p. 53.

29. Jiao Xun and Jiang Fan, *Jiaqing Yangzhoufu tujing*, 7.33a–34a. On the controversy surrounding Li's defense of Yangzhou, see Davis, *Wind Against the Mountain*, pp. 122–23.

30. Chang, "The Morphology of Walled Capitals," p. 75.

31. Hsiao, *The Military Establishment of the Yuan Dynasty*, p. 56.

32. Yule, *The Book of Ser Marco Polo*, p. 154.

33. *JQYZFZ*, 15.26b.

34. Barrow, *Travels in China*, p. 516.

35. *JQLHYFZ*, 44.2a–b. See also Chapter 3.

36. Zhu Fugui and Xu Fengyi, *Yangzhou shihua*, p. 1.

37. See, e.g., Yao Wentian, *Guangling shilüe*, preface.

38. *JQYZFZ*, 5.43b. Jiangdu county was divided in 1732, from which time Yangzhou had two home counties, Jiangdu and Ganquan.

39. Yao Wentian, *Guangling shilüe*, preface.

40. See Lo, "The Controversy over Grain Conveyance."

41. Worthy, "Regional Control in the Southern Song Salt Administration," pp. 109–10; *Song shi*, 477.13841.

42. *Yuan shi*, 21.467.

43. Achilles Fang, trans., *The Chronicle of the Three Kingdoms*, 1: 187.

44. Shiba, "Urbanization and the Development of Markets in the Lower Yangtze Valley," pp. 15–20.

45. Quan Hansheng, "Tang Song shidai Yangzhou," pp. 170–74. Quan attributes the reduced circumstances of Yangzhou under the Northern Song to the rise of Zhenzhou, but Zhenzhou's failure to maintain its ascendancy points to bureaucratic rather than geo-economic factors as responsible for this reversal of positions.

46. *Yuan shi*, 15.320, 131.3187.

47. Ho Ping-ti, *The Ladder of Success in Imperial China*, pp. 248–49; Fei Xiaotong, "Xiao chengzhen," p. 77; Honig, "The Politics of Prejudice," p. 253.

48. Kennelly, *L. Richard's Comprehensive Geography of China*, pp. 156–57 (romanization altered).

49. *QLJNTZ*, juan 69–71.

50. Kinkley, *The Odyssey of Shen Congwen*, p. 9.

51. *QLJNTZ*, 4.14b, 106.2a–b; Dan Shumo and Wen Pengling, " 'Kangxi liunian Jiangsu jian sheng' shuo queqie wuwu."

52. Leung, *The Shanghai Taotai*, p. 12.

53. On the use of Jiang-Huai, see *QLGYZZ*, 11A.16a, for a Tang reference, and 11B.32a, for a Qing reference.

54. Xie Zhaozhe, *Wu za zu*, p. 92.

55. In the context of an environmental study, Li Bozhang ("Changes in Climate, Land, and Human Efforts," p. 450) uses "Jiangnan" to denote the Lake Tai drainage basin, incorporating the prefectures of Suzhou, Taicang, Songjiang, Changzhou, Zhenjiang, Jiangning, Hangzhou, Jiaxing, and Huzhou. Honig ("The Politics of Prejudice," p. 245) defines it as "the Ningbo-Shaoxing region of Zhejiang and the Wuxi/Changzhou area of Jiangsu," notably excluding western Jiangsu, which is included in Li's formulation.

56. Finnane, "Bureaucracy and Responsibility," pp. 167–68.

57. See Li Changfu, *Jiangsusheng dizhi*, p. 130.

58. Tong Jun's 1937 survey of the historical gardens of Jiangnan, *Jiangnan yuanlin zhi*, includes Yangzhou among the Jiangnan cities.

59. Huang Junzai, "Jinhu langmo," 1.7b.

60. *JJWYZ*, 8.30b.

61. Ibid.

62. The status quo in 1606 is described in *KXYZFZ*, 2.12b–13a. At this time, Luhe county, west of Yizheng, and Chongming, the island county near Tongzhou, both belonged to Yangzhou, along with the various counties of what would later become the independent department of Tongzhou.

63. *WLYZFZ*, 1.5a–b.

64. For the sake of simplicity, the name Yizheng rather than Yizhen is used throughout the book.

65. *WLYZFZ*, 20.5b.

66. Public Record Office, China: F.O. 228/747, 28/6, 1884, p. 117.

67. *XFXHXZ*, 3.1a.

68. Taizhou and Gaoyou were both classed as departments (*zhou*) rather than counties (*xian*), a legacy of the Ming dynasty when each had neighboring counties under its jurisdiction. Xinghua and Baoying were under Gaoyou; Rugao under Taizhou (*JQYZFZ*, 5.43b).

69. Yang Dequan, "Qingdai qianqi Lianghuai yanshang ziliao chuji," p. 45.

70. *JQYZFZ*, 5.15a–16b.

71. *QLJNTZ*, 20.23b; *Qingchao tongdian*, 92.27135.

72. *QLJNTZ*, 6.12b.

73. *GXFNXZ*, 2.24a.

74. *GXJJXZ*, 5.14a.

75. Zhou Zhenhe and You Rujie, *Fangyan yu Zhongguo wenhua*, p. 89.

76. *HGTZ*, 11.2a–b, 6A.25a.

77. For a complete guide to the route of the Grand Canal in simple tabular form, see Gandar, *Le Canal impérial*, pp. 66–75. Gandar relied

for his information on merchant route books, on which see Wilkinson, "Chinese Merchant Manuals and Route Books."

78. Route books from the late imperial period bear this out. In the eighteenth-century *Routes of the Empire* (Chen Zhoushi, *Tianxia lucheng*), the first chapter is devoted to the route between the southeastern province of Fujian and the imperial capital, Beijing. To the mind of the author, the route fell into three sections: Fuzhou to Hangzhou, Hangzhou to Suzhou, Suzhou to Beijing. The distances from Yangzhou to Hangzhou, Suzhou, and Beijing, respectively, are separately listed, denoting a recognition of the Yangzi's division of north and south, but the importance of cities such as Zhenjiang, Yangzhou, and Huai'an, all situated at critical junctures in communication networks, is subsumed in that of the link between Jiangnan and Beijing.

79. *WXTK*, 27.5088, 5091. Yangzhou was outranked by the southern customs stations of Suzhou, Hangzhou, Wuhu, and Jiujiang, and also, less predictably, by Huai'an and Fengyang in the north, but its revenue gains in significance when the number and distribution of customs barriers under the various customs stations is considered. The size of the Huai'an customs quota, for instance, is less impressive when account is taken of the fact that it was based on revenue derived from 29 ports, reaching from Xuzhou, 480 *li* northwest of Huai'an, to Funing, 160 *li* to the east (*HGTZ, juan* 5). Of these 29 ports, four were inspection points only, no revenue being received. On the expansion of the Huai'an customs, see Takino, "Shindai Gaiankan," pp. 117–22. Yangzhou's port system, by contrast, was the smallest of any of the high-yielding customs stations except for that at Beijing.

80. Shiba, "Ningpo and Its Hinterland," p. 408.

81. For a detailed comparison of markets towns around Yangzhou in the early seventeenth century, see *JQYZFZ*, 60.3b–4a.

82. Qingjiangpu was made capital of Qinghe county in 1760 and thereafter was known simply as Qingjiang until 1914, when it became Huaiyin, the name by which it is presently known (Dan Shumo, *Zhonghua renmin gongheguo diming cidian: Jiangsu sheng*, p. 270).

83. Ibid., 16.4a; *MGJDXZ*, 6.5a.

84. *JQYZXZ*, 3.17a–18a, 16.12b–13b.

85. *DGTZZ*, 8.18a.

86. *QLJNTZ*, 26.21a.

87. Hohenberg and Lees, *The Making of Urban Europe*, pp. 66–70.

88. *YZHFL, juan* 10, p. 230.

89. Jiaodong Zhou Sheng, *Yangzhou meng*, p. 1. Nothing is known of the author of this work apart from the details of his activities in the pleasure quarters of Yangzhou in the 1840s, supplied in his apparently autobiographical account. The first two characters of the nom de plume are a term for Zhenjiang and, no doubt, an indication of the author's native place. The second two characters I have treated in references as a surname and personal name.

90. Ibid.

91. Wang Zhenzhong, *Ming Qing Huishang yu Huai Yang shehui bianqian*, p. 38.

Chapter 3

1. Lust, *Chinese Popular Prints*, p. 309.

2. Ibn Khurradādhbeh, cited in Lewis, *The Arabs in History*, p. 90.

3. For a discussion of Arab and Persian merchants in Tang Yangzhou, which takes Du Fu's verse as its point of departure, see Zhu Fuwei and Xu Fengyi, *Yangzhou shihua*, pp. 77–81.

4. Quan Hansheng, "Tang Song shidai Yangzhou," pp. 158–61; Jiang Hua, "Yangzhou gang yu Bosi wenhua de jiaoliu."

5. Yule, *Cathay and the Way Thither*, 1: 256.

6. Zhu Fugui, "Yisilanjiao yu Yangzhou," pp. 2–3.

7. Liu Binru and Chen Dazuo, "Yangzhou 'HuiHui tang' he Yuandai Alabowen de mubei," p. 49.

8. Rouleau, "The Yangchow Latin Tombstone as a Landmark of Medieval Christianity in China"; Rudolph, "A Second Fourteenth-Century Italian Tombstone in Yangchou." The gravestones can be viewed in the Yangzhou Museum. The surname of this family is given in these sources as Vigloni on the basis of a reading of the Latin engraving as VILONIS. I follow Robert S. Lopes ("Nouveaux documents sur les marchands italiens en Chine a l'époque mongole"), who reads the surname as YLLIONIS and attaches it to one Domenico Ilioni, identified in a Genoese statute of 1348. I owe this reference to the late William Henry Scott (see note 44 to Chapter 1, p. 335).

9. Zhu Jiang, "Jewish Traces in Yangzhou."

10. Mention has already been made of Ho Ping-ti, "The Salt Merchants of Yang-chou." The classic work on the salt monopoly during the Qing is Saeki Tomi's 1956 *Shindai ensei no kenkyū*. On the Lianghuai salt monopoly, an enormous amount of research has been published in the form of articles, as is clear from the footnotes in the present chapter and also Chapter 6. The most detailed book-length study is Xu Hong, *Qingdai Lianghuai yanshang de yanjiu*. For an excel-

lent brief description of how the monopoly functioned, see Metzger, "Organizational Capabilities of the Ch'ing State in the Field of Commerce."

11. Yule, *Cathay and the Way Thither*, 2: 210–11.

12. For Northern Song and early Yuan figures, see Yoshida, *Salt Production Techniques in Ancient China*, p. 28. On comparative revenues for the late Ming, see *Ming jingshi wenbian, juan* 474, cited in Wang Fangzhong, "Qingdai qianqi de yanfa," p. 4n1. The figures given are: Lianghuai, 680,000 taels; Changlu, 180,000; Shandong, 80,000; Liangzhe, 150,000; Fujian, 20,000; Guangdong, 20,000; Yunnan, 38,000; Henan, 120,000; Sichuan and Shaanxi, unfixed.

13. Ho Ping-ti, "The Salt Merchants of Yangchou," p. 153.

14. Brook, *The Confusions of Pleasure*, p. 49.

15. Marmé, "Heaven on Earth."

16. Tian Qiuye and Zhou Weiliang, *Zhonghua yanye shi*, p. 266.

17. I have never come across a detailed gloss of this term, but for the relevant quote, see Xue Zongzheng, "Mingdai yanshang de lishi yanbian," p. 27.

18. *JJSXTZ*, 39.21b. The exchange rate varied over time; see Tian Qiuye and Zhou Weiliang, *Zhonghua yanye shi*, p. 267.

19. Fujii, "Mingdai yanshang de yi kaocha," p. 258.

20. Ibid., pp. 254–55.

21. Li Ke, "Mingdai kaizhongzhi xia shang zao gouxiao guanxi tuojie wenti zai tan," p. 35. Xu Hong (*Qingdai Lianghuai yanshang de yanjiu*, pp. 87–95) examines in detail the evidence for the starting point for commutation of the grain supply system, reviewing the theories of Zeng Yangfeng, Zuo Shuzhen, and Fujii. He concludes that the trend toward commutation was established during the Chenghua reign (1465–87), beginning in 1468 in the Changlu salt zone and spreading to other sectors of the salt monopoly, being well established in Lianghuai by 1486. This is consistent with Fujii's conclusion in "Mingdai yanshang de yi kaocha," p. 252.

22. Xu Hong, *Qingdai Lianghuai yanshang de yanjiu*, p. 94.

23. *WLYZFZ*, 11.7a.

24. For this reason, the frontier merchants are sometime portrayed as simply securing the rights to trade salt and then selling these rights to the inland merchants (Xue Zongzheng, "Mingdai yanshang de lishi yanbian," p. 31). Xu Hong, however, cites evidence that early in the period of commutation, the frontier merchants continued to buy salt at the salt yards and then sold it at the export ports of Yizheng (then known as Yizhen) or Huai'an (for Huaibei salt). He cites

the doubling of the size of the *yin*, which caused long delays at checkpoints, as the most significant reason for their withdrawal from the actual trading process in the 1540s (Xu Hong, *Qingdai Lianghuai yanshang de yanjiu*, pp. 94–95). This is consistent with Fujii Hiroshi's analysis, on which Xu draws.

25. Xu Hong, *Qingdai Lianghuai yanshang de yanjiu*, pp. 94–95.

26. Ibid., p. 93; Fujii, "Mingdai yanshang de yi kaocha," p. 288.

27. Hu Puan, *Zhonghua quanguo fengsu zhi*, 7.11.

28. *JJSXTZ*, 40.2b.

29. Hu Puan, *Zhonghua quanguo fengsu zhi*, 7.11.

30. Zhang Zhengming, *Jinshang xingshuai shi*, p. 7.

31. Cressey, *China's Geographic Foundations*, pp. 45–46.

32. Bell, "The Great Central Asian Trade Route from Peking to Kashgaria," p. 60.

33. Zhang Zhengming, *Jinshang xingshuai shi*, p. 19.

34. See tabulation in ibid., pp. 98–103.

35. Fujii, "Mingdai yanshang de yi kaocha," p. 298.

36. Xie Zhaozhe, *Wu za zu*, cited in Ho Ping-ti, "The Salt Merchants of Yangchou," p. 143.

37. Liu Yuxi, "Wan bu Yangzi you nan wang sha wei."

38. Cressey, *China's Geographic Foundations*, p. 197.

39. On He's parentage, see *WLYZFZ*, *juan* 18; cited in Fujii, "Mingdai yanshang de yi kaocha," p. 286.

40. On this phenomenon in Suzhou, see Marmé, "Heaven on Earth," p. 36.

41. *JJWYZ*, 9.7a.

42. Fujii, "Mingdai yanshang de yi kaocha," p. 293.

43. Jiao Xun and Jiang Fan, *Jiajing Yangzhoufu tujing*, 8.28b–29b.

44. *JQLHYFZ*: 44.1b. On the pirate raids, see So, *Japanese Pirates in Ming China During the 16th Century*. So does not refer to the attack on Yangzhou, nor is the city marked as one of the places besieged in Jiangnan. Marmé ("Heaven on Earth," p. 36) notes that in Suzhou, too, the sojourner residential area outside the city wall was the chief target of the pirate attacks on the city.

45. *JQLHYFZ*, 44.2a–b.

46. *JJWYZ*, 8.30a.

47. Ibid., 37.3a.

48. Zurndorfer, *Change and Continuity*, p. 52.

49. Brook, *The Confusions of Pleasure*, p. 128.

50. Cited in Wei Minghua, "Kao 'Yangzhou luantan,' " p. 194. On Clapper Opera (*bangzi qiang*), see Mackerras, *The Rise of the Peking Opera*, pp. 8–9.

51. Wei Minghua, "Kao 'Yangzhou luantan.'" See also Chapter 11.

52. Translated by Wolfram Eberhard in "What Is Beautiful in a Chinese Woman?" p. 287. On Li, see *ECCP*, pp. 495–97.

53. Cited in Bao-Hua Hsieh, "The Acquisition of Concubines in China," p. 147.

54. Zhao Zhibi, *Pingshantang tuzhi*, 8.12a.

55. *JJWYZ*, 37.4a–6b, 8a–10a; *LQYZXZ*, 14.37a–38b.

56. Xu Hong, *Qingdai Lianghuai yanshang de yanjiu*, p. 95.

57. From the Wanli edition of the Shexian county gazetteer, cited in Fujii, "Mingdai yanshang de yi kaocha," p. 294.

58. See Cahill, *Shadows of Mt. Huang*.

59. See Zurndorfer, "The *Hsin-an ta-tsu chih* and the Development of Chinese Gentry Society," pp. 158–61; idem, "Local Lineages and Local Development," pp. 20–24; Shiba, "Urbanization and the Development of Markets," pp. 13–48.

60. Jiqi is rendered Jixi on contemporary maps, although with the same characters. The second character in the name has alternative readings, and Jiqi is still used on English-language websites run from that county.

61. Li Longqian, "Mingdai yan de kaizhongfa yu yanshang ziben de fazhan," p. 502.

62. Fujii, "Mingdai yanshang de yi kaocha," p. 289.

63. Zurndorfer, *Change and Continuity*, p. 132.

64. See the bibliography in Liu Sen, *Huizhou shehui jingji shi*, appendix 1, pp. 510–17.

65. *JJHZFZ*, 2.38a.

66. Zurndorfer, *Change and Continuity*, pp. 132, 145.

67. Kuhn, *Soulstealers*, p. 40. On Wang, see *ECCP*, pp. 834–35.

68. The routes, with distances between stopping stations, are listed in a Ming dynasty route book; see Dan Yizi, *Shishang yaolan*, 1.61a–b, 2.3b–4a. The compiler was a native of Xin'an, i.e., of Huizhou.

69. Fujii, "Mingdai yanshang de yi kaocha," p. 289.

70. Zurndorfer, *Change and Continuity*, p. 69.

71. Ibid., p. 50; see also the excellent chap. 3 in ibid., "Commercial Wealth and Rural Pauperism in late Ming and Early Ch'ing China: Hui-chou Prefecture in Transition."

72. Brook, *The Confusions of Pleasure*, p. 4.

73. Xu Hong, *Qingdai Lianghuai yanshang de yanjiu*, p. 96.

74. Dan Yizi, *Shishang yaolan*, 1.61a–b.

75. Quan Hansheng, "Tang Song shidai de Yangzhou," pp. 170–74.

76. See, e.g., *JQLHYFZ*, 43.26b, 43.28b, 45.8a, 45.12b, 46.8a, 46.9b, 46.10a. I have elsewhere remarked on the number of changes in registration to Yizheng, which I mistakenly speculated to have been due to the greater availability of land in Yizheng county; see Finnane, "Yangzhou: A Central Place in the Qing Empire," p. 139.

77. *KXHZFZ*, 2.66a; adapted from Zurndorfer, *Change and Continuity*, p. 137.

78. Zurndorfer, *Change and Continuity*, p. 114.

79. On the demimonde, see Chapter 9.

80. Fujii, "Mingdai yanshang de yi kaocha," pp. 297–98.

81. Zurndorfer, *Change and Continuity*, p. 97.

82. Metzger, "Organizational Capabilities of the Ch'ing State in the Field of Commerce," pp. 23–24.

83. Zurndorfer, *Change and Continuity*, p. 143.

84. *YZHFL, juan* 8, p. 169.

85. Ibid., p. 171.

86. Yuanxun's birthdate is indicated in his colophon on the painting shown in Fig. 6. His death is well documented, as we will see in the following chapter. Yuanhua was active in Yangzhou in the early Qing and is mentioned in association with his support for the Yangzhou orphanage, established in 1655 (*JQLHYFZ*, 56.4b). Xiaru lived to 64 years (*sui*) of age and died in 1673, giving a probable birth year of 1610 (Zheng Qingyou, *Yangzhou Xiuyuan zhi, shixi* [genealogical chapter], 2a, 3.1a).

87. Yuanxun, a metropolitan graduate of 1643, is described in the graduate lists only as registered in Yangzhou prefecture (*Yangzhoufu ji*). In the gazetteers of Yangzhou's home county, Jiangdu, he is described as "person of Jiangdu" (*Jiangduren*) but not as "registered in Jiangdu" (*Jiangdu ji*). The prefectural gazetteer states that his grandfather changed the household's registration to Jiangdu but elsewhere defines him as "registered in Yangzhou, person of Shexian." Yuanxun is nonetheless claimed as a native son in the Qing county gazetteer of Yizheng and the lists of graduates show all his nephews as registered in this country. See *JSBL*, 2: 1346; *YZYZFZ*, 39.23a–b, 24b; and *JQYZXZ*, 33.5b–7a.

88. The surname Zheng with corresponding generational personal names appears more than once in membership lists of the Restoration Society for both Yizheng and Yangzhou's home county, Jiangdu. In Yizheng, these were Zheng Yuanxi (*js* 1631), who bore the same

generational name, "Yuan," as Yuanxun; and Zheng Zhilun, who bore the same generational name as Zheng Zhiyan, father of the four brothers. In Jiangdu county, members included Zheng Yuanxun and also one Zheng Yuanbi. See *Fushe xingshi* (Members of the Resoration Society), an anonymous and undated manuscript. The manuscript is unpaginated, and names are listed by district. Zheng Yuanxi appears to have been an elder cousin of Zheng Yuanxun, and perhaps the son of Zheng Zhilun. Zheng Yuanxun is described in one source as the "younger brother" of Yuanxi, a conventional way of referring to cousins on the paternal side (*QLLHYFZ*, 34.21).

89. *JQYZXZ*, 33.5a. This edition of the Yizheng gazetteer was edited by the noted Yangzhou scholar Liu Wenqi, who cites all the relevant passages from the preceding prefectural and county gazetteers.

90. Ibid.

91. The Zhuxi Society was named after the Zhuxi Pavilion, west of Yangzhou city, immortalized in a poem by Du Mu (*JQYZFZ*, 31.27a). On the membership of the Restoration Society, see *Fushe xingshi*. On its origins and activities, see Atwell, "From Education to Politics." For a list of the composite local literary societies, see Xie Guozhen, *Ming Qing zhi ji dangshe yundong kao*, p. 165, although the list does not include any from Yangzhou.

92. *BHXZ*, 3.8a.

93. On earlier gardens, see Zhu Jiang, *Yangzhou yuanlin pinshang lu*, p. 53.

94. See Wu Zhaozhao, "Ji Cheng yu Yingyuan xingzao"; and Cao Xun, "Ji Cheng yanjiu," pp. 12–13.

95. *YZHFL, juan* 8, p.171.

96. Ji Cheng, *The Craft of Gardens*.

97. Li Dou points out that references to the location of the garden in local gazetteers vary, but concludes on the basis of the remains extant in his day that it was located south of the city. Southwest would be more precise, according to Wu Zhaozhao, "Ji Cheng yu Yingyuan xingzao," pp. 170–71.

98. Zhu Jiang, *Yangzhou yuanlin pinshang lu*, p. 75.

99. Ibid., p. 171.

100. *YZHFL, juan* 8, pp. 168–69.

101. Zheng Qingyou, *Yangzhou Xiuyuan zhi*. For changes in garden architecture in the late eighteenth century, see Chapter 8.

102. Zheng Qingyu, *Yangzhou Xiuyuan zhi*, compiler's preface [*zi xu*], n.p.

103. Chen Congzhou, *Shuo yuan,* p. 26.

104. Zheng Yuanxun, "Yingyuan zizhi," 31.51b.

105. *JQYZXZ,* 39.23a.

106. Ibid., 31.55a.

107. Fang Xiangying, *Xiuyuan ji,* 31.56a.

108. *YZHFL, juan* 8, p. 172. See also Borota, "The Painted Barques of Yangzhou," p. 63; Hsü, *A Bushel of Pearls,* pp. 29–31; and Chapter 8.

109. Cahill, "Introduction," p. 11; and idem, "The Older Anhui Masters," p. 68.

110. Wang Shiqing, "Dong Qichang de jiaoyou," p. 481.

111. *JQYZFZ,* 31.52a.

112. Pang, "Late Ming Painting Theory," p. 28; *YZHFL, juan* 8, p. 168. On Chen, see *ECCP,* pp. 83–84, 789.

113. Hsü, *Bushel of Pearls,* pp. 58–59.

114. Métailé, "Some Hints on 'Scholar Gardens' and Plants in Traditional China," p. 248.

115. Artificial mountains emerged as a feature of the Suzhou garden in the mid-Ming and by the late sixteenth century were "one of the visually dominant features of the fashionable garden elsewhere in China." In the eyes of critics of this fashion, they were especially associated with the deplorable habits of the *nouveaux riches,* the vulgar practices of "foppish great merchants" (Clunas, *Fruitful Sites,* pp. 74–75).

116. Ji Cheng, *The Craft of Gardens,* p. 29.

Chapter 4

1. Crossley, *The Manchus,* p. 41.

2. Mo Dongyin, *Manzushi luncong,* pp. 60–61.

3. Crossley, *The Manchus,* pp. 52–53, 212; see also *ECCP,* pp. 594–99.

4. *YZSRJ,* pp. 240–41.

5. On the structural narrative, see Maier, "Consigning the Twentieth Century to History."

6. *YZSRJ.*

7. *JQYZFZ,* 44.49a.

8. Ibid., 33.5b.

9. Parsons, *The Peasant Rebellions of the Late Ming Dynasty,* pp. 2–6.

10. *JQYZFZ,* 44.50a.

11. Ibid., 44.49a, 50b–51a.

12. Ibid., 69.28a. On Zhang, see *ECCP,* pp. 37–38.

13. *JQYZFZ,* 70.22b.

14. See Chapter 1, note 17, p. 334.

15. *YZSRJ*, pp. 232, 240. In the second case, involving a rape, the native-place identity of the woman is not stated, but the surname Jiao (the same as that of the eminent scholar Jiao Xun) is a common local surname.

16. Ibid., p. 232. On Shan-Shaan and Huizhou lineages in Yangzhou, see *JQJDXZ*, 12.21b–22a.

17. *YZSRJ*, p. 231. Lynn Struve's translation (*Voices from the Ming-Qing Cataclysm*, p. 47) suggests that Wang might have been a scholar. This is a plausible but not precise interpretation of a passage concerning Wang's ambivalence of whether to identify himself to a Manchu commander as a man of high social position.

18. See also Chapter 8.

19. *YZSRJ*, p. 235.

20. Ibid., p. 241.

21. Ibid., p. 230.

22. Cited in Mote, "The Transformation of Nanking," p. 104, p. 689n4.

23. Wakeman, *The Great Enterprise*, 1: 573.

24. *JQYZFZ*, 49.41a.

25. There are graphic descriptions from the anti-Japanese war in the twentieth century of large cities being virtually emptied of their populations in advance of the battle; see Tong and Li, *The Memoirs of Li Tsung-jen*, p. 334; and George Johnston's description of Guilin in 1944 in Kinnane, *George Johnston*, pp. 52–53.

26. See Wakeman, *The Great Enterprise*, 1: 362; *ECCP*, pp. 37–38, 491–93.

27. *ECCP*, pp. 410–11.

28. Dai Mingshi, "Yangzhou chengshou jilüe," p. 51.

29. Ibid.

30. *YZSRJ*, p. 229.

31. Ibid., p. 231.

32. Ibid., p. 128.

33. Wakeman, *The Great Enterprise*, 1: 547.

34. Struve, *Voices from the Ming-Qing Cataclysm*, p. 31.

35. *YZSRJ*, p. 229; Mao, "A Memoir of Ten Days' Massacre in Yangchow," p. 516.

36. Trans. Mao, "A Memoir of Ten Days' Massacre in Yangchow."

37. In Jiading, south of the Yangzi, 20,000 people are said to have died in a twelve-hour bloodbath after the city was occupied (Wakeman, *The Great Enterprise*, 1: 659).

38. Ibid., 1: 547, 563–64n158.

39. Ibid., 1: 547, citing Maurice Collis, *The Great Within*, p. 63, which in turn relies on Backhouse and Bland, *Annals and Memoirs of the Court of Peking*, p. 187.

40. *YZSRJ*, pp. 241–42.

41. *JQYZFZ*, 45.1b.

42. Ibid., 49.39b.

43. Trans. Mao, "A Memoir of Ten Days' Massacre in Yangchow," p. 522.

44. Wakeman, *The Great Enterprise* 1: 562n155.

45. *QLJDXZ*, 29.25b.

46. Ibid., 29.26a.

47. Xu Fengyi and Zhu Wugui, *Yangzhou fengwu zhi*, p. 85.

48. Quan Zuwang, "Meihualing ji," p. 211. Cf. Wakeman, *The Great Enterprise*, 1: 564–66. On Quan, see *ECCP*, pp. 203–5; and Mote, "The Intellectual Climate in Eighteenth-Century China," pp. 47–51. Mote examines this version of Shi Kefa's death in the context of Quan's disillusion with the Qing and growing interest in the history of Ming loyalists.

49. Quan Zuwang, "Meihualing ji," p. 211.

50. Wei Minghua, "Shuo Yangzhou shiri."

51. Yi Junzuo, *Xianhua Yangzhou*, p. 22.

52. *ECCP*, pp. 701–2.

53. Compare Quan, "Meihualing ji," p. 211; and Dai Mingshi, "Yangzhou chengshou jilüe," p. 51.

54. See Wilhelm, "The Po-hsüeh Hung-ru Examination of 1679"; and Struve, "Ambivalence and Action."

55. Yangzhou wenwu yanjiushi, Shi Kefa jinian guan, *Liangjie guazhong Shi Kefa*, p. 2.

56. For these and other literary tributes to Shi Kefa, see ibid.

57. See *Shi Kefa pingjia wenti huibian*.

58. Zhu Ziqing, "Shuo Yangzhou," p. 36.

59. Wakeman, *The Great Enterprise*, 1: 351n108; *DMB*, 1: 216.

60. On Wu Sheng, see *DMB*, 2: 1494–95; on Yuan Jixian, see *ECCP*, pp. 948–49, which does not, however, mention Yuan's appointment to Yangzhou.

61. *YZHFL*, juan 8, p. 171.

62. On the Beihu families, see Chapter 5. Wang was the son-in-law of Li Zhi (*js* 1577), one of a number of censors demoted and removed from office in 1585 for their attacks on the powerful grand secretary Shen Shixing (1535-1614), part of the chain of political controversies that led to the formation of the Donglin party (*Mingshi*, 20: 124.6141–45; *QSG*, 45: 500.13827; *DMB*, 1: 331, 2: 1188). Li Zhi's grandson, Li

Zongkong, was a prominent figure in early Qing Yangzhou; see index.

63. According to Ruan Yuan (*Guangling shishi*, 3.5a-b), in the early Qing the Zong family lived in Yiling, a market town in Jiangdu east of Yangzhou.

64. YZYZFZ, 21.23a.

65. Ibid., 29.7b.

66. JQYZFZ, 49.40a.

67. Ibid., 49.40b.

68. YZYZFZ, 29.57b.

69. JQYZXZ, 33.6a.

70. Dai Mingshi, "Yangzhou chengshou jilüe," p. 45.

71. JQYZXZ, 33.6a. See also Luo Zhencheng, "Shi Kefa biezhuan," p. 22.

72. JQYZFZ, 49.33b-37a.

73. BHXZ, 3.8a.

74. YZHFL, *juan* 8, pp. 167-72.

75. JQYZFZ, 49.33b-37a.

76. Ibid., 49.41a.

77. Ibid., 49.40b.

78. QSG, 45: 287.13826-27.

79. YZYZFZ, 29.57b; DGTZZ, 24.6a.

80. See note 63 to this chapter.

81. Mori Noriko, "Yanchang de Taizhou xuepai," p. 62.

82. Shen Defu, *Wanli yehuo bian*, pp. 482-83. On Shen Defu, see DMB, 2: 1190-91.

83. Ibid., p. 483.

84. XFXHXZ, 8/6.6b; JQLHYFZ, 45.6a-b.

85. DMB, 1: 562-66.

86. For Zong's biography, see QSG, 44: 484.13333. For a potted biography in English, see Strassberg, *The World of K'ung Shang-jen*, p. 371n61. Strassberg's notes constitute a veritable biographical encyclopaedia of literati active in Yangzhou in the late seventeenth century; see ibid., pp. 371-72.

87. Meyer-Fong, "Site and Sentiment," pp. 125-27. Li Dou (*YZHFL*, *juan* 10, p. 217) incorrectly describes Zong Guan as Yuanding's father, instead of as his cousin.

88. QLJNTZ, 107.25a.

89. XFXHXZ, 8/6.6a-7b.

90. JQYZFZ, 53.14a.

91. YZHFL, *juan* 8, p. 170.

92. Ibid., p. 171.

93. Shi Dewei, *Weiyang xunjie jilüe*, pp. 4b–5a.

94. Wakeman, *The Great Enterprise*, 1: 569–79.

95. Mao, "A Memoir of Ten Days' Massacre in Yangchow," p. 537.

Chapter 5

1. *BHXZ*, 3.8a.

2. Wang Zhangtao, *Ruan Yuan zhuan*, p. 5.

3. *BHXZ*, 3.29b.

4. Ibid., 3.1a.

5. Ibid., 3.9b.

6. Struve, "Ambivalence and Action," p. 327.

7. Ruan Yuan, *Guangling shishi*, 4.6b.

8. *ECCP*, pp. 240, 265; *QSG*, 45: 501.13859. On Cheng Sui, see Jang, "Cheng Sui." Sun Mo's dates are given in Li Huan, *Guochao qixian leizheng*, 79.40b.

9. On Shitao's Yangzhou years, see Hay, *Shitao*; and idem, "Shitao's Late Work."

10. *Yinyi* are listed for earlier dynasties, including the Ming; see *JQYZFZ*, 1a–7b.

11. For an important genre of Wu Jiaji's poetry, that dealing with female fidelity, filiality, and chastity, see Chaves, "Moral Action in the Poetry of Wu Chia-chi." Chaves appends to this study a number of translations of Wu's poems, together with thematically comparable works by some of his circle, including Sun Zhiwei and Chen Weisong. For a meticulously annotated edition of Wu's work, together with biographical materials written by his associates, see Yang Jiqing, *Wu Jiaji jianjiao*. An unpunctuated Qing dynasty edition is available in reprint; see Wu Jiaji, *Louxuan shiji*.

12. *JQYZFZ*, 53.14a.

13. For a discussion of the documentation of the "twice-serving officials," see Crossley, *A Translucent Mirror*, pp. 290–96.

14. *JQYZFZ*, 53.40a, 53.46a.

15. Yang Jiqing, *Wu Jiaji jianjiao*, p. 18. The same entry lists other loyalist associates of Wu Jiaji.

16. Strassberg, *The World of K'ung Shang-ren*, pp. 136–37.

17. *QSG*, 45: 501.13857.

18. Zong Yuanding, "Maihua laoren zhuan," p. 60; Ruan Yuan, *Guangling shishi*, 3.5a–b.

19. *BHXZ*, 3.16b; *JQYZFZ*, 53.17b.

20. Ibid., 3.15a.

21. Wang Zhangtao, *Ruan Yuan zhuan*, pp. 5–7.

22. *BHXZ*, 6.10b. Jiao noted among earlier generations of his family one who was a military *juren* of 1615; another who topped the *juren* examination list in 1687 (see *JQYZFZ*, 42.12b); and a third who won the same degree in 1726 (*BHXZ*, 6.3b, 5b).

23. Winston Hsieh, "Triads, Salt Smugglers and Local Uprisings," pp. 155–57.

24. *BHXZ*, 3.11a.

25. *JQYZFZ*, 53.16b.

26. Yang Jiqing, *Wu Jiaji jianjiao*, p. 312.

27. Chaves, "Moral Action in the Poetry of Wu Chia-chi," p. 403.

28. Jang, "Cheng Sui," p. 111. See also Nelson, "Rocks Beside a River."

29. Ayling and Mackintosh, *A Collection of Chinese Lyrics*, p. 165.

30. Jiaodong Zhou Sheng, *Yangzhou meng*, p. 5.

31. Du Fuweng, "Shitao yu Yangzhou," p. 5.

32. Hay, "Shitao's Late Work," 1: 21.

33. Ibid., p. 24.

34. William Ding Yee Wu, "Kung Hsien," p. 5. Gong remained poor, but this has been attributed to his otherworldliness (Kim, "Zhou Liang-kung and His *Tu-hua lü* Painters," p. 200).

35. Andrews, "Zha Shibiao," pp. 102–8.

36. Yang Jiqing, *Wu Jiaji jianjiao*, p. 15. On the attack on Sanyuan by Li Zicheng, against which Sun is said to have mobilized local resistance, see *BZJ*, 139.12b.

37. *JQLHYFZ*, 146.12a.

38. Yang Jiqing, *Wu Jiaji jianjiao*, p. 15.

39. *JQLHYFZ*, 146.12a.

40. Ibid., 44.5b.

41. Wang Sizhi and Jin Chengji, "Qingdai qianqi Lianghuai de shengshuai," p. 67.

42. *QLLHYFZ*, 34.17b.

43. See Oertling, "Patronage in Anhui During the Wanli Period," p. 171.

44. Contag, *Chinese Masters of the 17th Century*, pp. 13–14.

45. Hay, "Shitao's Late Work," 2: app. I, p. 661.

46. Wei Xi, "Shande jiwen lu," 27.27a. See Smith, "Social Hierarchy and Merchant Philanthropy," p. 433. Wei does not provide a time line for Min's activities but records that he died at 72; given Wei's death date of 1681, we must suppose that Min was born quite early in the seventeenth century.

47. Wei Xi, "Shande jiwen lu," 27.33a. This must have been between the years 1668 and 1674 (see Appendix C). The chronology of

Min's business activities, as noted above, is not clear. Either he made his money before the fall of the dynasty and preserved it through the period of warfare, or he made it very quickly after the change of dynasty.

48. *JQYZFZ*, 19.1a. On the institution of the "school-temple," see Feuchtwang, "School-Temple and City God," p. 583.

49. *JQYZFZ*, 14.3a–b.

50. Ibid., 19.1a.

51. *Ming Qing like jinshi timing beilu*, 3: 1407–22.

52. There were, for example, no candidates from the later highly successful prefecture of Jiaxing in Zhejiang, nor any from Fuzhou or Guangzhou.

53. On accommodation, see Beattie, "The Alternative to Resistance," pp. 241–76.

54. *KXLHYFZ*, 21.5b; *JQYZFZ*, 48.8a.

55. *ECCP*, pp. 809–10, 864–65. Wu Qi's family changed its registration to Jiangdu through the salt monopoly, as did Wang Ji's to Yizheng (*JQLHYFZ*, 46.8a, 10a).

56. *JQYZFZ*, 38.4a.

57. Ibid., 38.2a–b.

58. *QLJNTZ*, 105.14a–15b.

59. Wilhelm, "The *Po-hsüeh hong-ru* Examination of 1679," p. 65.

60. *ECCP*, pp. 173–74.

61. Kim, "Zhou Liang-kung and His *Tu-hua-lu* Painters," p. 194.

62. Ibid.

63. Zhou Lianggong, *Du hua lu*.

64. For a translation of this preface, see Kim, "Zhou Liang-kung and His *Tu-hua-lu* Painters," pp. 193–94.

65. Meyer-Fong, "Making a Place for Meaning in Early Qing Yangzhou."

66. *ECCP*, pp. 831–33.

67. Wang Shizhen, *Yuyang jinghualü jianzhu*, "Nianpu," 18b–19b.

68. Yang Jiqing, *Wu Jiaji jianjiao*, p. 489. The newly established orphanage was thrown into crisis at this time due to the flight of its supporters (see Smith, "Social Hierarchy and Merchant Philanthropy," p. 437).

69. Wang Shizhen, *Yuyang jinghualü jianzhu*, "Nianpu," 5b.

70. Ibid., 6b. See, e.g., on Taizhou, "Dongri deng Hailing Dexiang ge" (Mounting the Dexiang Pavilion in Hailing one winter's day); and on Gaoyou, "Qin you qu er shou" (Two verses of a refrain on the Qin post-station), in ibid., 2.24a, 25a.

71. Wang Shizhen, "Yechun jueju shier shou" (Seductive spring: twelve quatrains), in ibid., 3.4a, 3.4b–5a.

72. Ibid., 3.27a, 3.7a, 3.43b.

73. Ibid., 2.81. For a detailed discussion of Wang Shizhen's Red Bridge project, see Meyer-Fong, "Making a Place for Meaning in Early Qing Yangzhou."

74. *YZHFL, juan* 2, p. 212.

75. Yang Jiqing, *Wu Jiaji jianjiao,* p. 312.

76. Ibid., p. 505.

77. Ibid., pp. 487–90. The manuscript was published under the title *Louxuan ji* (Louxuan collection).

78. Ibid., pp. 80, 92, 126, 193–94, 212, 217, 362, 363.

79. Ibid., pp. 82–83; *JQYZFZ,* 28.17a; *GXGQXZ,* 8.6a–b.

80. Kong's life, particularly his activities in Yangzhou, is the subject of a careful study by Richard Strassberg, *The World of K'ung Shang-jen;* I draw on his work extensively for the few comments made here.

81. Chen Wannai, *Kong Shangren yanjiu,* pp. 24–25.

82. See Strassberg, *The World of K'ung Shan-jen,* pp. 135–87, 362–91nn. For a full list of Kong's poet and painter friends, with brief biographies, see Chen Wannai, *Kong Shangren yanjiu,* pp. 46–90.

83. Yang Jiqing, *Wu Jiaji jianjiao,* p. 514.

84. Strassberg, *The World of K'ung Shang-jen,* pp. 135, 187.

85. For translations of fourteen of these, see ibid., pp. 155–62.

86. Kong Shangren, *Kong Shangren shiwenji,* 1: 170.

87. Strassberg, *The World of K'ung Shang-ren,* p. 277.

88. Kong Shangren, *The Peach Blossom Fan,* p. 258.

89. Strassberg, *The World of K'ung Shang-jen,* p. 245.

90. Kong Shangren, *The Peach Blossom Fan,* p. 1.

91. Hay, *Shitao,* p. 99. For a discussion of the "long eighteenth century," see Mann, *Precious Records,* pp. 20, 236n1.

92. *KXYZFZ,* cited in Yang Jiqing, *Wu Jiaji jianjiao,* p. 143.

93. *KXLHYFZ,* 13.4b.

94. Zhu Xie, *Zhongguo yunhe shiliao xuanji,* p. 109; *JQYZFZ,* 10.3b; *XSJJ,* 65.963.

95. Yu Chenglong (1638–1700) complained about the regulated opening of dams north and south of Gaoyou, a key drainage mechanism in Jin Fu's scheme that helped protect the dike from too much pressure but had adverse effects on Xiahe (*JQYZFZ,* 10.20b).

96. See Chapter 7.

97. Owen, "Salvaging Poetry," p. 105. On Wei Xi, see *ECCP,* pp. 847–48.

98. *JQYZFZ*, 15.8b.

99. Owen, "Salvaging Poetry," p. 105.

100. Strassberg, *The World of K'ung Shang-jen*, p. 144.

Part III Introduction

1. *YZHFL, juan* 18, p. 403.

Chapter 6

1. *YZHFL, juan* 9, pp. 197-98.

2. *JQYZFZ*, 31.47a; *QLJDXZ*, 8.21a-22b.

3. See Loewe, "Imperial Sovereignty."

4. The yamen now serves as the offices of Yangzhou Municipality.

5. *Jinshen quanshu*, 2.46a. On the rankings of posts in the field administration, see Skinner, "Cities and the Hierarchy of Local Systems," pp. 314-16.

6. *KXLHYFZ*, preface, 5a-6b.

7. *DMB*, 2: 1429-30.

8. *KXLHYFZ*, preface, 5a-6b.

9. See Smith, "Social Hierarchy and Merchant Philanthropy," p. 426.

10. Yang Lien-sheng, "Government Control of Urban Merchants in Traditional China," p. 190.

11. Eisenstadt, *The Political Systems of Empires*, pp. 365-66.

12. Yang Lien-sheng, "Government Control of Urban Merchants in Traditional China," p. 190.

13. *KXLHYFZ*, 7.6a-b.

14. Wang Zhenzhong, *Ming Qing Huishang yu Huai Yang shehui bianqian*, p. 19.

15. Ibid., pp. 12-18; Spence, *Ts'ao Yin and the K'ang-hsi Emperor*, pp. 124-51; Silas Wu, *Passage to Power*, p. 88; Hay, *Shitao, passim*.

16. *YZHFL, juan* 7, pp. 154-55; *GLMST*, p. 4.

17. Yang Dequan, "Qingdai qianqi Lianghuai yanshang ziliao chuji," p. 47. On the titles, which also had substantive offices attached to them, see Brunnert and Hagelstrom, *Present Day Political Organization of China*, pp. 19-20. Examples of titles holders are Shexian merchants Wang Tingzhang and Hong Zhengzhi, each of whom played host to the emperor in his garden (*GLMST*, pp. 17-18, 21-22).

18. From 2,760,000 taels in 1682 to 5,740,000 taels in 1766; see comparative tables in Xiao Guoliang, "Lun Qingdai gangyan zhidu," p. 66.

19. See Metzger, "T'ao Chu's Reform of the Huai-pei Salt Monopoly," pp. 2-4.

20. Torbet, *The Ch'ing Imperial Household Department*, pp. 106-10.

21. Wang Zhenzhong, *Ming Qing Huishang yu Huai Yang shehui bianqian*, p. 4.

22. Liu Jun, "Daoguangchao Lianghuai feiyin gaipiao shimo," p. 126.

23. Yang Dequan, "Qingdai qianqi Lianghuai yanshang ziliao chuji," p. 45.

24. Xu Hong, *Qingdai Lianghuai yanshang de yanjiu*, pp. 103-4.

25. Li Cheng, *Huaizuo beiyao*; cited in Wang Fangzhong, "Qingdai qianqi de yanfa," p. 32.

26. *KXLHYFZ*, 13.8a.

27. Wang Sizhi and Jin Chengji, "Qingdai qianqi Lianghuai yanshang de shengshuai," p. 51. Ho Ping-ti ("The Salt Merchants of Yangchou," p. 134) erred in correlating the number of head merchants with the number of salt yards. There were 30 salt yards early in the dynasty (matching the number of head merchants in the Yongzheng period) but apparently only 24 head merchants before this time, and the yards themselves were reduced in number through a series of amalgamations beginning in 1678; there were only 23 by the end of the Qianlong reign (Xu Hong, *Qingdai Lianghuai yanshang de yanjiu*, p. 8).

28. *JQLHYFZ*, 44.9a-b; see also Ho Ping-ti, "The Salt Merchants of Yangchou," pp. 138-39.

29. *JQLHYFZ*, 46.14b-5a. See also Chapter 8.

30. *JQLHYFZ*, 42.13a-b.

31. Cheng Mengxing, *Pingshantang xiaozhi*; Wang Yinggeng, *Pingshantang lansheng zhi*.

32. Wang Zhenzhong (*Ming Qing Huishang yu Huai Yang shehui bianqian*, p. 34) dates this development to no later than 1768, noting that Ruan Yuan referred to the merchant Huang Yuande by this title in the context of the Qianlong emperor's visit to Yangzhou that year.

33. Ruan Yuan (*Guangling shishi*, 7.2b) notes that the lapse of time since the Kangxi emperor's last visit had left local authorities in confusion as to the proper procedures to be followed for the impending first visit of the Qianlong emperor.

34. *JQLHYFZ*, 44.15a.

35. Ho Ping-ti, "The Salt Merchants of Yangchou," pp. 160-61.

36. Ibid., p. 35. Wang lists these merchants by their business names: Jiang Guangda for Jiang Chun, Bao Youheng for Bao Shufang, and Huang Yingtai for Huang Zhiyun. Huang Yuande may have

been the business name of Huang Lüxian, owner of a garden visited by the Qianlong emperor (see Chapter 8). Another business name listed is Hong Zhenyuan. This is probably Hong Zhengzhi, owner of three gardens in Yangzhou. Both these merchants held the imperial title of director of the imperial gardens and hunting grounds; see *YZHFL, juan* 12, pp. 275–76, *juan* 10, p. 225; *GLMST*, 17–18, 20–21, 36–37. On Bao Zhidao and his son, see Liu Sen, "Huishang Bao Zhidao ji qi jiashi kaoshu." On Huang Zhiyun and the Ge garden, see Luo Weiwen, "Shanguan ji jing yi zhu, linglong qishi you cun." For more on this garden, see Chapter 8. On Huang's role as principal merchant, see Chapter 12.

37. See Appendix E.

38. Wang Zhenzhong, *Ming Qing Huishang yu Huai Yang shehui bianqian*, p. 37. On *baoxiao*, see Metzger, "T'ao Chu's Reform of the Huai-pei Salt Monopoly," p. 2; and Chapter 12 of this book.

39. Tian Qiuye and Zhou Weiliang, *Zhonghua yanye shi*, p. 305.

40. Data from *JQLHYFZ*; cited in Metzger, "Organizational Capabilities of the Ch'ing State in the Field of Commerce," pp. 18–19.

41. Wang Fangzhong, "Qingdai qianqi de yanfa," p. 15.

42. Cahill, *The Painter's Practice*, p. 58.

43. *GXLHYFZ*, 151.2b, 4b–5a.

44. Fang Yujin, "Daoguang chunian Chu an yanchuan fenglun sanmai shiliao," p. 40.

45. *JQYZFZ*, 18.8b.

46. The fullest study of this case is Takino Shōjirō, "Shindai Kenryū nenkan ni okeru kanryō to enshō." See also *ECCP*, pp. 541–42; and Saeki Tomi, *Shindai ensei no kenkyū*, pp. 228–29.

47. Takino Shōjirō, "Shindai Kenryū nenkan ni okeru kanryō to enshō," p. 103.

48. Wu Jianyong, "Qing qianqi queguan ji qi guanli zhidu," p. 93.

49. Li Linqi, "Ming Qing Huizhou liangshang shilun," p. 75.

50. *GXLHYFZ*, cited in Wang Zhenzhong, "Ming Qing Yangzhou yanshang shequ wenhua ji qi yingxiang," p. 9.

51. There are no good population statistics available for the city, but on the basis of comparative data for Chinese urbanization ca. 1843, Skinner ("Regional Urbanization," p. 238. Fig. 1) has estimated its population for that time at around 175,000. The ranks of the salt merchants in Yangzhou had been severely depleted by this time, but they were probably replaced by refugees from the city's hinterland, which was in a state of crisis due to the collapse of the inland waterway system. When we take into account that the population as a

whole was still growing, 100,000 or so seems a reasonable estimate for the urban population toward the end of the Qianlong reign.

52. *YZHFL, juan* 1, p. 16.

53. Ibid., *juan* 13, p. 279.

54. See Wang Zhenzhong's excellent study, *Ming Qing Huishang yu Huai Yang shehui bianqian*.

55. *JQLHYFZ, juan* 5, *passim; GXLHYFZ, juan* 16-17, *passim.* The distribution of the yards is conveniently presented in tabular form in Saeki Tomi, *Shindai ensei no kenkyū*, pp. 16-17, but Saeki wrongly lists Juegang and Fengli yards as divided between Tongzhou and Rugao, with Shigang lying wholly within Tongzhou. According to *GXLHYFZ*, 16.15b, 17b, these yards lay in Rugao alone, and a comparison of the salt yard maps in this source with district-level maps shows this to be correct. Shigang, including Matang as of 1736, straddled the Tongzhou-Rugao border, although in its original form it was contained within the borders of Tongzhou (ibid., 16.4b-5a). As for the dates of amalgamation, Li Cheng errs in stating that the amalgamation of Yuxi and Yuzhong, Shigang and Matang, and Caoyan and Baiju took place in Yongzheng 10 (1732) (*HZBY*, 8.8b). This measure followed a recommendation made by Gao Bin in 1736 (*QLQSL*, 14.16a-b). See Appendix B for the list of salt yards and relevant officials.

56. See the map in the 1905 edition of the Lianghuai salt gazetteer, which is superimposed on a grid of units ten *li* by ten *li* (*GXLHYFZ*, 16.2b-3a, 17.5b-6a).

57. Xu Hong, *Qingdai Lianghuai yanshang de yanjiu*, p. 15. On Zhang Shicheng, see *DMB*, 1: 99-102.

58. Xu Hong, *Qingdai Lianghuai yanshang de yanjiu*, p. 16.

59. *KXLHYFZ*, 38.56a.

60. Ibid.

61. *JQLHYFZ*, 4.8b-24b.

62. The on-site cost of salt remained fairly steady throughout the eighteenth and early nineteenth centuries; see Wang Fangzhong, "Qingdai qianqi de yanfa," p. 14.

63. *JQLHYFZ*, 27.1a-2a.

64. Fang Zhuofen et al., "Hedong Lake Salt and Huainan Sea Salt," p. 354.

65. *JQLHYFZ*, 29.1a-b.

66. *JQDTXZ*, 15.8a.

67. *GXFNXZ*, 6.11b.

68. *JQLHYFZ*, 29.2b-6a.

69. Ho Ping-ti, "The Salt Merchants of Yangzhou," p. 132.

70. *QLQSL*, 41.9b, 25.21b.

71. *JQDTXZ*, 18.36b.

72. Tian Qiuye and Zhou Weiliang, *Zhonghua yanye shi*, p. 307.

73. Takino Shōjirō, "Shindai Kenryū nenkan ni okeru kanryō to en-shō," pp. 102–3. See *ECCP*, pp. 411–13, under Kao Pin (Gao Bin, 1683–1755) and Kao Chin (Gao Jin, 1707–79), for individual biographies.

74. *ECCP*, pp. 541–42.

75. *GXLHYFZ*, 129.11a; *HZBY*, 8.2a.

76. *QCTD*, 35.2213.

77. The total number of bureaus was formally nineteen (see Ho Ping-ti, "The Salt Merchants of Yangzhou," p. 131), but the population bureau (*hu fang*), which handled the business of the local or "consumer" salt ports (*shi an*), was composed of a north bureau (*beifang*), which dealt with Huaibei local salt, and a south bureau (*nanfang*), which handled Huainan (*HZBY*, 8.4b–7a).

78. *GXLHYFZ*, 129.13b.

79. *KXLHYFZ*, 2.8b–9a, 17b–18a; *JQLHYFZ*, 5.15b, 27b.

80. *GXLHYFZ*, 130.2a; *HZBY*, 8.8b.

81. The ten yards under the control of the Tongzhou subcontroller were contained with Tongzhou and the neighboring, later dependent district of Rugao, and those under the Taizhou subcontroller lay wholly or partly within the jurisdiction of what was then Taizhou. All these yards lay within Yangzhou prefecture. The Huai'an subcontroller had authority over two yards in the northern part of Taizhou, but the remainder, in Yanchang (including later Funing) and Haizhou, all lay within Huai'an prefecture (*KXLHYFZ*, *juan* 3, *passim*).

82. The Taizhou subcontroller stationed at Dongtai Yard now had responsibility for the twelve (later eleven) yards between Miaowan in Funing district and Fuan in Taizhou (later Dongtai). His domain stretched through four district-level jurisdictions in two different prefectures but did not extend to Jiaoxie and Bingcha, the southernmost yards in Taizhou (Dongtai), which, together with the remaining seven Huainan salt yards, were under the control of the Tongzhou subcontroller, stationed at Shigang (*GXLHYFZ*, *juan* 16–17, *passim*).

83. Ibid., 17.20a–22a.

84. Areas are given in Bureau of Foreign Trade, *China Industrial Handbooks: Kiangsu*, pp. 6–8.

85. *GXLHYFZ*, 130.12a–13a.

86. Ibid., 130.11a–12a.

87. Ibid., 12a–13a.

88. Wang Zhenzhong, *Ming Qing Huishang yu Huai Yang shehui bianqian*, p. 107.

89. *GXLHYFZ*, 4.1b–2a.

90. Xu Hong, *Qingdai Lianghuai yanshang de yanjiu*, pp. 127–81 *passim*.

91. *QLQSL*, 2.10a–b.

92. *WXTK*, 29.5115.

93. *JQYZFZ*, 21.4a–b.

94. *QLQSL*, 14.71a–b.

95. Ibid., 1060.3a.

96. The tax on salt at the consumer ports varied. For Jiangdu and Ganquan counties the base rate in the eighteenth century was around three-quarters of the rate for shipment salt; see *GXLHYFZ*, 93.3a–b.

97. *CYQS*, 83.9b.

98. *GXLHYFZ*, 59.11b–12a.

99. Ibid., 59.4a–b.

100. Ibid., 59.21b.

101. Saeki Tomi, *Shindai ensei no kenkyū*, p. 203.

102. *GXLHYFZ*, 59.12b–13a; *JQYZFZ*, 21.6a–b.

103. *GXLHYFZ*, 59.13a–b.

104. *QLJNTZ*, 81.26a; *GXLHYFZ*, 82.2b.

105. *JQYZFZ*, 21.3b.

106. *WXTK*, 29.5120.

107. The numbers rose from seven to ten patrol merchants and from twelve to twenty watchers; one merchant and two watchers were to be stationed at each city gate, and all were placed under the authority of a specially deputed official (*GXLHYFZ*, 59.36b–37a).

108. *JQYZFZ*, 21.3b–4a.

109. Metzger, "Organizational Capabilities of the Ch'ing State in the Field of Commerce," p. 38.

110. *QLQSL*, 21.20a.

111. Ibid., 1060.15b–16a.

112. Ibid., 1058.9a–b.

113. Ibid., 21.20a.

114. Ibid., 1056.2b–3a.

115. Ibid., 1056.23a.

116. Ibid., 1060.2b.

117. Ibid., 1063.9b–10a.

118. *HZBY*, 5.17a.

119. Ibid., 4.1b.

120. Ibid., 5.22a–b.

121. Fang Yujin, "Daoguang chunian Lianghuai siyan yanjiu," pp. 86–87.

122. Ibid., p. 86.

123. Rowe, *Hankow*, p. 234.

124. See Chapter 10.

125. *QLLHYFZ*, 34.13a–b; Metzger, "Organizational Capabilities of the Ch'ing State in the Field of Commerce," p. 24*n*.

126. Quan Hansheng, "Tang Song shidai Yangzhou," pp. 170–74.

127. It was impossible for salt to be carried to its various destinations in lots of 364 catties. In the mountainous regions of Jiangxi and Huguang, the transport routes were tortuous, "the mountains remote, the canals in the lesser counties shallow and narrow." For this reason, the *yin* had to be broken up into smaller packages. These were differentiated by weight according to destination, to prevent salt destined for one port from being illegally sold at another: large packages for Huguang, smaller for Jiangxi (*JQLHYFZ*, 4.38b).

128. See the table in Wang Zhenzhong, *Ming Qing Huishang yu Huai Yang shehui bianqian*, p. 144.

129. Huang Junzai, *Jinhu langmo*, 1.7b.

130. Wang Zhenzhong, *Ming Qing Huishang yu Huai Yang shehui bianqian*, p. 96.

131. *JQDTXZ*, 15.8a.

Chapter 7

1. *Relations de la mission de Nan-king confiée aux religieux de la Compagnie de Jésus*, vol. 2, *1874–75*, pp. 28–33.

2. Gandar, *Le Canal impérial*, pp. 3–4.

3. Ibid., p. 4.

4. *JQYZFZ*, 10.5a.

5. Ibid., 10.3b.

6. *XSJJ*, 135.1962–63. On Jin Fu, see *ECCP*, pp. 161–63.

7. Fei Xiaotong, "Xiao chengzhen," p. 77. The rather free English translation of this work does not include the sentence cited here; see idem, "Small Towns in Northern Jiangsu," p. 90.

8. Fei Xiaotong, "Xiao chengzhen," p. 77.

9. Lamouroux, "From the Yellow River to the Huai," p. 545; Zheng Zhaojing, *Zhongguo shuili shi*, pp. 26–29.

10. Zheng Zhaojing, *Zhongguo shuili shi*, pp. 136–39.

11. Zhu Xie, *Zhongguo yunhe shiliao xuanji*, p. 87.

12. Jin Fu, *Jin Wenxiang gong zhihe fanglüe*, pp. 331, 333.

13. See Vermeer, "P'an Chi-hsün's Solutions for the Yellow River Problems of the Later 16th Century."

14. *JQYZFZ*, 11.4b, 9b–10a.

15. For identification of the Yangzi outlets, see ibid., 11.9a.

16. Jin Fu's strategy was developed during an early phase of reconstruction in the 1680s but was abandoned amid great political controversy on grounds of its inefficacy; see *QDHCZ*, pp. 26–28, 33–34. The controversy is outlined in Strassberg, *The World of K'ung Shang-jen*, p. 208. Zhang took office as director-general of river conservancy early in 1700, after a long period of conflict within the river administration. He set in place the hydraulic system that governed control of the confluence of the Yellow River, Huai River, and Grand Canal for the next century. On the Gao Dike protecting Hongze Lake, he replaced six reduction dams (*jian ba*), which allowed a continuous outflow of water from Hongze Lake, with three overflow dams (*gun ba*), which allowed only a limited release of water in times of flood. Seven canals were cut to lead the waters from the lake north to the Yellow River and thence east to the sea. On the Grand Canal's eastern dike, which protected Xiahe, three reduction dams south of Gaoyou were replaced with overflow dams, and repairs and dredging were carried out at the Yangzi outlets to improve the efficiency of drainage into the Yangzi. See *XSJJ*, 68.991, 68.997, 139.2011, 140.2019, 143.2021, 152.2183; *KXQSL* 220.15a; and *HXNB*, 11.24b.

17. By early in the Qianlong reign, the Huai River was again being drained via Hongze Lake and the Grand Canal eastward through Xiahe, but Gao Bin, appointed to a second term as Jiangnan director-general of river conservancy in 1748, sought to revive Zhang Pengge's policy. Floods in 1753 led to an abandonment of this approach; see *JQYZFZ*, 22.43b–44a; *XXSJJ*, 88.1975–76. Ji Huang (1711–94), appointed assistant director-general in 1757, was a strong advocate of continuous drainage through Xiahe (*JQYZFZ*, 12.6b). Then in 1760, Director-General Bai Zhongshan and Liangjiang Governor-General Yinjishan declared their intention of reverting to Gao Bin's strategy (ibid., 25.9a).

18. Will, "State Intervention," p. 298.

19. *XSJJ*, 141.2035.

20. Hay, *Shitao*, pp. 78–79.

21. Will, "State Intervention," p. 321; Perdue, *Exhausting the Earth*, p. 192.

22. On the last two terms, see Will, "State Intervention," p. 337.

23. *TZYZFZ*, 1.20b, 27b; *XFXHXZ*, 2A.11a.

24. See Chapter 10.

25. Ibid.

26. *XXSJJ*, 8.181. See Finnane, "Bureaucracy and Responsibility," pp. 164–65.

27. Finnane, "Bureaucracy and Responsibility," pp. 178–87.

28. Officials listed in *XSJJ*, 167.2415–16; and *WXTK*, 85.5620–21; tabulated in Finnane, "Bureaucracy and Responsibility," pp. 174–77.

29. Ch'ang-tu Hu, "The Yellow River Administration in the Ch'ing Dynasty," p. 508.

30. *XSJJ*, 136.1971.

31. *XFXHXZ*, 2A.14b. See the case of Wei Yuan's intervention in river conservancy measures when he was local magistrate in Xinghua in 1849, mentioned in Leonard, *Wei Yuan and China's Rediscovery of the Maritime World*, p. 30.

32. *XXSJJ*, 95.2184, 98.2215; *JQYZFZ*, 12.33a.

33. *JQYZFZ*, 12.15a.

34. *XXSJJ*, 75.1746.

35. *JQYZFZ*, 14.35a–b.

36. See Finnane, "Bureaucracy and Responsibility," pp. 173–84.

37. *YZQSL*, 72.12a–b.

38. *XXSJJ*, 132.3010.

39. Ibid., 75.1745.

40. Ibid.

41. *JQYZFZ*, 12.3b–7a.

42. Finnane, "Bureaucracy and Responsibility," p. 173.

43. *QLQSL*, 22.27b.

44. *XXSJJ*, 87.1977.

45. *XSJJ*, 87.1977.

46. *QLQSL*, 22.27b.

47. *HDSL*, 901.20b–21a, 902.4b–6a.

48. Ibid., 904.17a; *XXSJJ*, 15.344.

49. *QCTD*, 33.2207.

50. *JQQSL*, 40.23a–b.

51. Polachek, "Literati Groups and Literati Politics," pp. 153–56.

52. See, e.g., in the Yongzheng reign, *XXSJJ*, 75.1751; in the Qianlong reign, ibid., 11.265; and in the Jiaqing reign, *JQQSL*, 44.3a.

53. *XXSJJ*, 81.1875.

54. Ibid., 80.1851.

55. Ibid., 75.1736.

56. Ibid., 80.1845.

57. Ibid., 77.1785; *JQYZFZ*, 11.23b.

58. *XXSJJ*, 81.1870, 83.1910.

59. *QLQSL*, 251.18a.

60. *XXSJJ*, 81.1870.

61. Ibid., 83.1909.

62. Ibid., 82.1888.

63. *QLQSL*, 251.176.

64. *QDHCZ*, pp. 79–80.

65. *HDSL*, 904.6a–b.

66. *GXLHYFZ*, 65.23b.

67. Ibid., 6.13a.

68. Ibid., 6.13b.

69. *JQDTXZ*, 10.4b.

70. *GXLHYFZ*, 65.6b.

71. *XXSJJ*, 7.172.

72. *HDSL*, 904.6a–b.

73. *XXSJJ*, 7.172; *GXLHYFZ*, 65.5a.

74. See Perdue, *Exhausting the Earth*, p. 213, table 21.

75. Ibid., p. 218.

76. Perdue, "Water Control in the Dongting Lake Region," p. 750.

77. *XXSJJ*, 88.1999.

78. *QLQSL*, 22.26b–28a.

79. *GXLHYFZ*, 130.17b.

80. *XSJJ*, 136.1970.

81. *QLTZZLZZ*, 3.19a.

82. *XXSJJ*, 88.1975–76.

83. Ibid., 88.1977.

84. Will, "State Management of Water Conservancy," p. 81.

85. Takino Shōjirō, "Shindai Kenryū nenkan ni okeru kanryō to enshō."

86. *JQYZFZ*, 11.9a.

87. Will, "State Intervention"; Perdue, "Water Control in the Dongting Lake Region."

88. Will, "State Intervention," pp. 346–47.

89. *JQGYZZ*, 12.16b–18a.

90. Ho Pingti, *Studies in the Population of China*, p. 173.

91. Li Bozhong, "Changes in Climate, Land, and Human Efforts," pp. 452–55.

92. See Chapter 12.

Chapter 8

1. No figures exist for the urban population in this period. See note 51 to Chapter 6, p. 357.

2. Skinner, "Introduction," pp. 533–36.

3. Yinong Xu, *The Chinese City in Space and Time*, pp. 75-77.

4. *JQYZFZ*, 13.1b-2a.

5. Ibid., 15.2a.

6. *JJWYZ*, 1.16a-b. The faded condition of the map in this six-teenth-century gazetteer does not lend itself to reproduction, nor is it even entirely legible. Normally one would expect to find a map of this sort reproduced in later editions of the gazetteer (as with the map of the Song city). Why this was not so in the case of the Ming map is probably answered by the fact that the Old City (*jiucheng*) of Qing times basically preserved the Ming form.

7. Wanli edition of the Jiangdu county gazetteer, cited in *JQYZFZ*, 61.3b.

8. *JQYZFZ*, 19.1a, 13.3b, 5.20b. On the significance of the school-temple, see Feuchtwang, "School-Temple and City God."

9. Du Halde, *The General History of China*, 1: 141. Since the French-language original of this work, *Description géographique, historique, chronologique et physique de l'empire de la Chine et de la Tartarie chinoise*, was published in the first year of the Qianlong reign, it must have been in preparation in the Yongzheng reign. The source of the claim concerning the "Tartar garrison" is unclear except to the extent that other portions of Du Halde's description of Yangzhou appear to have been adapted from Martino Martini's mid-seventeenth-century *Novus atlas sinensis*, p. 128. Although Du Halde himself spent many years in China, it seems likely that he relied on literary sources such as this, along with reports from fellow Jesuits, for much of his infor-mation. The presence of Jesuits in the Chinese provinces up until the 1720s should indicate good-quality information on notable places in China, but according to Jules Crochet, S.J. (1857-1932), who served in Yangzhou in the early Republican era, there was no priest in Yang-zhou after the death there of Père Gabiani in 1696, that is, up until the reopening of the Jesuit mission after the Opium War (see Crochet, "Historie d'une Chrétienté," p. 103). On the distribution of Banner garrisons in the provinces, see Elliott, *The Manchu Way*, pp. 105-16.

10. *Yangzhou yingzhi*, 9.1b-2b.

11. Ibid., 9.9b.

12. Yinong Xu, *The Chinese City in Space and Time*, p. 77.

13. *Yangzhou yingzhi*, 9.9b.

14. Zhang Dai, *Tao'an mengyi*, p. 39; trans. adapted from Strass-berg, *Inscribed Landscapes*, pp. 347-48.

15. *QLJNTZ*, 79.22a. See also Ray Huang, *Taxation and Governmen-tal Finance in Sixteenth-Century Ming China*, p. 226.

16. See note to this effect in *GXJDXZ*, *juan shou*, 1a.

17. Ibid., 8a.

18. *YZHFL, juan* 9, p. 188.

19. Hanshang mengren, *Fengyue meng*, p. 37. The references (*Xianliang jie, Beiliu xiang, Tianshou'an*) can be followed on Map 7.

20. The salt merchant Kang, known as Western Kang, must have been from Shanxi. He was as well known at the same time as An Qi (b. 1683), which provides a clue as to when his fortunes were great (*YZHFL, juan* 9, p. 194).

21. Ibid., pp. 189–90. "Professor Crow" was a pun on Wu Lun-yuan's surname, which incorporates the character for crow and is pronounced in the same way.

22. Ibid., pp. 188–95. The literal meaning of *waichengjiao* is "base beyond the walls," meaning the bottom or southern end of the ex-tramural settlement. The name clearly dates from the time before the New City was enclosed.

23. *YZHFL, juan* 9, p. 199.

24. *Yangzhou yingzhi*, 16.17b. Satins Street was also known as Many Sons Street (Duozijie), apparently a case of dialect confusion; the Mandarin *duan* (satin) sounds very like *duo* (many) in the local patois (Ye Sen and Zhu Feng, "Yangzhou jiexiang zatan," p. 145).

25. *YZHFL, juan* 9, pp. 185-86. According to local lore, Kingfisher Blossom was originally the name of New Victory Street (Xinsheng jie) further north, but the latter had been renamed by the Manchus when they were stationed there during the occupation of the city in 1645. Later the name was transferred to the nearby street that Li Dou mentions, and indeed where he lived (Du Fuweng, "Shitao yu Yang-zhou," p. 5).

26. Yang Jiqing, *Wu Jiaji jianjiao*, p. 373.

27. Among them were the Kangshan garden, built by Jiang Chun, and the gardens of the four Huang brothers: Sheng, Lümao, Lüxian, and Lühao (*YZHFL, juan* 12, pp. 275–76).

28. *GXJDXZ*, 12b.8b–9a.

29. *YZHFL, juan* 8, p. 170, *juan* 15, p. 332. On the Zheng family, see also Chapters 3 and 4. The family of Wang Jiaoru, engaged in the salt trade for a number of generations, was one of the richest in Yang-zhou, with extensive property holdings shared among its many members. Wang Tingzhang, Jiaoru's son, purchased Cheng Meng-xing's Bamboo Garden (ibid., *juan* 15, pp. 332–33).

30. Jin Nong, Gao Fenghan (1683–ca. 1748), Gao Xiang, Wang Shi-shen, and Luo Ping lived in this quarter for some time (Bian Xiaoxuan, "Yangzhou baguai zhi yi de Gao Xiang," p. 233; Hay, *Shitao*, p. 8).

31. Lin Xiuwei, *Yangzhou huapai*, p. 90.

32. Xue Yongnian and Xue Feng, *Yangzhou baguai*, pp. 100, 103; Zheng Xie, *Zheng Banqiao quanji*, "Banqiao tihua," 7a. For the location of the Temple of the Bamboo Forest, see *JQYZFZ*, 28.26a.

33. *YZHFL, juan* 9, pp. 196-98.

34. *Yangzhou yingzhi*, 9.3b-4b, 5a-b.

35. Ibid., 9.8b.

36. Ibid., 16.17a-b.

37. Ibid., 9.8a-9a. In present-day Yangzhou, a small area west of Shi Kefa Road and south of the Yangzhou municipal offices (formerly the salt controller's yamen) still bears the name *jiaochang*.

38. Hanshang mengren, *Fengyue meng*, p. 14.

39. Tsang, "Portraits of Hua Yan," p. 71. In this article, Tsang quotes Hua Yan's young friend and former neighbor as saying that he was Hua's neighbor when the latter was living with Yuan in the north of the city, but Yuan later moved "to the east side of the city" (i.e., near the eastern wall). It seems reasonable to infer a north-central or northwestern location for the former residence.

40. *JQYZFZ*, 45.23b.

41. *YZHFL, juan* 18, pp. 404-7.

42. Ibid., *juan* 4, p. 61.

43. These became a byword for the city. In the oft-quoted words of Shandong scholar-official Liu Daguan, "Hangzhou abounds in mountains and lakes, Suzhou in markets, Yangzhou in gardens and pavilions; each stands on its own merits and cannot be compared with the others" (ibid., *juan* 6, p. 144).

44. Chen Congzhou, "Yangzhou yuanlin de yishu tese." For an introduction in English to extant gardens in Yangzhou, see Chen Lifang and Yu Sanglin, *The Garden Art of China*, pp. 180-81; and Johnston, *Scholar Gardens of China*, pp. 190-206.

45. Zhu Jiang, *Yangzhou yuanlin pinshang lu*, p. 83.

46. Chen Congzhou, *Yuanlin tancong*, p. 55. For identification of the garden, see *GLMST*, 35-36; and Zhu Jiang, *Yangzhou yuanlin pinshang lu*, pp. 154-55.

47. *JQYZFZ*, 48.50a.

48. The best contemporary survey of Yangzhou's gardens is Zhu Jiang, *Yangzhou yuanlin pinshang lu*. Good documentation of garden ownership can be found in *YZHFL*; *GLMST*; and Zhao Zhibi, *Pingshantang tuzhi*.

49. *YZHFL, juan* 12, pp. 275-76.

50. Wang Yinggeng, *Pingshantang lansheng zhi*.

51. Ibid.

52. Zhang Dai, *Tao'an mengyi*, p. 54.

53. Hay, "Shitao's Late Work," 1: 232.

54. Ibid., 1: 261.

55. Zong Yuanding, "Maihua laoren zhuan," p. 60.

56. YZHFL, juan 6, p. 136.

57. Ibid., juan 12, p. 275.

58. Ibid., juan 15, p. 325.

59. Ruan Yuan, Guangling shishi, 6.14b. On gardens around Red Bridge in the early Qing, see Wang Yinggeng, Pingshantang lansheng zhi, 3.18a.

60. YZHFL, juan 13, pp. 287-88.

61. Ibid., juan 4, p. 96, juan 13, p. 300.

62. A rather poor black-and-white reproduction of this work was published in Chen Congzhou, Yuanlin tancong, front matter, n.p.

63. YZHFL, juan 13, pp. 300-308.

64. Ibid., juan 15, p. 329. This garden is said to have been designed by Shitao although it is difficult to see how this could have been the case if Shitao died in 1707 (YZHFL, juan 2, p. 38). Yu was a first-degree graduate of Jiangdu county, but it is probable that his family was from Huizhou, perhaps the village of Yu'an, the home place of another prominent man in Yangzhou (ibid., juan 13, p. 299). His associates and his place of residence make it very likely that he was a salt merchant.

65. Ibid., juan 15, p. 327.

66. Ibid., juan 8, p. 172.

67. Ibid. I have used Ginger Hsü's (A Bushel of Pearls, p. 17) translation of Xiao linglong shanguan.

68. Zheng Qingyou, Xiuyuan zhi. According to Zhu Jiang (Yangzhou yuanlin pinshang lu, p. 107), a work by this name was edited by Zheng Xiaru's son, Zheng Weiguang, and subsequently proscribed. He may have been relying for this information on Li Dou, who wrote the same, but Li had not seen the book, because he did not know how many juan it included (YZHFL, juan 8, p. 172). The extant work is by a later descendant of Zheng Xiaru, but the work appears to be rare (a copy is held in the Oriental Collection of the British Museum Library).

69. Cited in Mu Yiqin, "An Introduction to Painting in Yangzhou in the Qing Dynasty," p. 21; trans. slightly adapted.

70. Zhao Zhibi, Pingshangtang tuzhi, 9. 21b-22a.

71. Qingchao yeshi daguan, 11.6.

72. Finnane, "Yangzhou," p. 137.

73. YZHFL, juan 15, pp. 334-35.

74. Ibid., p. 337.

75. *JQLHYFZ, juan shou* 4, 47a.

76. *YZHFL, juan* 7, pp. 161–64.

77. Ibid., *juan* 12, p. 269.

78. Naquin, *Peking*, p. 309.

79. Ibid., *YZHFL, juan* 4, p. 100.

80. These were Zhang Sike, originally of Lintong in Shaanxi, and Lu Zhonghui, from a Shexian family (*YZHFL, juan* 4, p. 86).

81. Hsü, *A Bushel of Pearls*, pp. 27–41.

82. *YZHFL, juan* 8, p. 172. For a full translation of the passage, see Borota, "*The Painted Barques of Yangzhou*," p. 63.

83. Hsü translates his account in the present tense, but the past is probably more appropriate. As noted in the Introduction to the present book, Li Dou composed his work between 1764 and 1795, beginning after the death of Ma Yueguan in 1755 and probably not long before that of Ma Yuelu, who died sometime after 1766.

84. See Hsü, *A Bushel of Pearls*, pp. 31–32. Cf. Bian Xiaoxuan, "Yangzhou baguai zhi yi de Gao Xiang," p. 239; and Lin Xiuwei, *Yangzhou huapai*, p. 120.

85. Zhang Geng, *Guochao huazheng lu*.

86. Qiu Liangren, "Yangzhou er Ma," p. 124. A poor reproduction in this periodical of Zhang Geng's lately rediscovered work shows the artist's intention was to illustrate the garden rather than to reflect on it, and he may have intended simply to provide a plan of the garden.

87. Ibid.

88. *YZHFL, juan* 4, p. 84. Works included paintings from the Song, Yuan, and Ming dynasties as well as some by contemporary artists active in Yangzhou, such as local professional painter Yu Zhiding, the individualist Gao Fenghan (1683–ca. 1748), and the Hangzhou artist Chen Zhuan (ca.1670–ca. 1740), with all of whom they must have been acquainted.

89. Hsü, *A Bushel of Pearls*, chap. 2.

90. The artist Chen Zhuan was a guest at Cheng Mengxing's Bamboo Garden for ten years before being invited to stay with head merchant Jiang Chun, in Jiang's garden residence in the city. On Chen Zhuan, see *ECCP*, pp. 85–86. Fang Chao-ying, probably relying on Hang Shijun's (1696–1773) biography of Chen, states simply that Chen was the guest of a rich salt merchant. Li Dou (*YZHFL, juan* 15, p. 328) takes Hang to task for mentioning that Chen was the guest of Xiang and Jiang Chun but failing to mention Cheng Mengxing. Likewise the Huizhou painter Wang Shishen (1685–1759) lived for a number of years at the Ma brothers' Little Translucent Mountain

Lodge, and his fellow painter Li Shan (1686–1760) stayed for a while at the Eastern Garden of the Shanxi merchant He Junzhao (Xue Yongnian and Xue Feng, *Yangzhou baguai*, p. 28).

91. Scott, "Yangzhou and Its Eight Eccentrics," p. 62.

92. Zheng Xie, *Zheng Banqiao quanji*, "Banqiao tihua," 7a.

93. Xue Yongnian and Xue Feng, *Yangzhou baguai*, p. 26.

94. Wang, a native of Gaoyou, served at court for seventeen years and was much admired there for his technical skills; see Wang Yun, *Yangzhou huayuan lu*, 1.6b–7a. For a partial reproduction of this painting, see *Zhongguo meishu quanji*, vol. 10, *Qingdai huihua*, entry 102, p. 103; and commentary in the appendix, "Tuban shuoming," p. 36.

95. Hsü, *A Bushel of Pearls*, p. 53.

96. Hsü's pioneering study tellingly focuses on a group of artists who were all dead by 1765. The famous "professional" painters of Yangzhou (to utilize a problematic categorization) came slightly earlier. They included Yu Zhiding, Yuan Yao, and Wang Yun. See note 35 to Chapter 1, p. 334.

97. On opera, or musical drama (discussed further in Chapter 11), see Mackerras, *The Rise of the Peking Opera*, pp. 49–80.

98. Trans. adapted from Chou and Brown, *The Elegant Brush*, p. 137.

99. *YZHFL, juan* 11, p. 251.

100. Ibid., *juan* 4, p. 77.

101. Ji Cheng, *Craft of Gardens*, p. 55.

102. *YZHFL, juan* 18, p. 404.

103. Ibid., pp. 404–7.

104. Ibid., *juan* 11, pp. 234–41.

105. Ibid., *juan* 9, pp. 186–88, 193–96.

106. Ibid., *juan* 6, p. 132.

107. Ibid., *juan* 13, p. 279.

108. Ibid., p. 284.

109. Respectively, Hanyuan, Liubu, Guo Hanzhang guan, Sushi xiaoyin, Liushang (ibid., *juan* 11, p. 254).

110. Wu Woyao, *Ershi nian mudu zhi guai xiazhuang*, cited in Zhu Jiang, *Yangzhou yuanlin pinshang lu*, p. 99.

111. Wei Peh-t'i, "Juan Yuan," p. 2.

112. *YZHFL, juan* 4, p. 101.

113. On public use of late Ming gardens, see Clunas, *Fruitful Sites*, pp. 94–97.

114. *YZHFL, juan* 11, pp. 241, 250.

115. Zhu Jiang, *Yangzhou pinshang lu*.

116. *YZHFL, juan* 9, p. 196.

117. Tsang, "The Relationships of Hua Yan and Some Leading Yangzhou Painters," p. 19.

118. *YZHFL, juan* 1, p. 25.

119. Ibid., *juan* 4, p. 78.

120. Ibid.

121. Ibid., *juan* 11, p. 254.

122. The most famous restaurants were two at Great East Gate, two by Little East Gate, one at the Parade Ground, one at Sanzhu Lane, and one near the salt merchant quarter at Quekou Gate (ibid., *juan* 1, pp. 23-24).

123. Yi Junzuo, *Xianhua Yangzhou*, p. 9.

124. There was one bathhouse at Ridge Street, near the courtesan quarter, and another at Guangchu Gate in the north wall, not far from where the Ma brothers had lived. In the Old City, there were two, located at Kaiming Bridge and Taiping Bridge, respectively, the latter convenient for officials working in the local government yamens and also for scholars working at the Anding Academy. Four more were variously at Quekou Gate, Xu Ning Gate, North Canalside, and Eastern Pass (Dongguan); see *YZHFL, juan* 1, p. 25.

125. Ibid., *juan* 2, pp. 36-52. For a discussion of the term "eight eccentrics" and its problems, see Giacalone, *The Eight Eccentrics of Yangzhou*, pp. 28-29; and Chou and Brown, *The Elegant Brush*, pp. 7-8. Various artists apart from those listed in Appendix F make an occasional appearance in lists of the "eight eccentrics," including Gao Qipei (1672-1734), Bian Shoumin (1684-1752), and Min Zhen (1730-ca. 1788), but these artists' links with Yangzhou are quite tenuous (Giacalone, *The Eight Eccentrics of Yangzhou*; Bai Jian and Ding Zhian, "Bian Shoumin san ti").

126. This school developed under the monk Shizhuang (d. 1732), who had studied painting with the Huizhou painter Zha Shibiao. One of Shizhuang's disciples, Zhutang, had a grandson, Ganting, who also became a painter-monk. All these men were natives of the Nanjing district of Shangyuan, but Ganting himself had a number of pupils, all of whom were Yangzhou natives. Three pedagogical generations after Shizhuang brings us, surely, to the middle of the eighteenth century and a cohort of Yangzhou painters who would be selling works through the latter part of the Qianlong reign (*YZHFL, juan* 2, p. 35).

127. *Catalogue of the Exhibition of Individualists and Eccentrics*, p. 53.

128. *YZHFL, juan* 2, pp. 47, 37, 45-46, respectively.

129. Wang Yun, *Yangzhou huayuan lu*, 2.13b.

130. *YZHFL, juan* 13, p. 327.

131. Ibid., p. 299.

132. Ibid., *juan* 2, p. 38.

133. Ibid., *juan* 15, p. 329.

134. Luo Weiwen, "Shanguan ji jing yi zhu, linglong qishi you cun." The first colophon was by Ma Yueguan, as discussed earlier in this chapter. The second was by Bao Shichen (1775-1855), who records its sale to one Wang Xuejiang, a salt merchant, and of the scroll itself to Wu Shanzun, director of the Anding Academy, who sought Bao's inscription. The colophon is undated, but Bao was in Yangzhou in 1801. A third colophon on this scroll dates from after the Taiping Rebellion and was written by the art historian Wang Yun (1816-after 1883), a descendant of Wang Xuejiang, who notes that not long after Bao had added his piece to the scroll, "the great stone was moved [from the garden], and the Mountain Hall was reduced to a pile of broken potsherds." As for the scroll itself, he had happened on it in the marketplace (Qiu Liangren, "Yangzhou er Ma," p. 124).

135. Saeki, "Yunshang de moluo he yanzheng de bihuai," p. 385.

136. Hanshang Mengren, *Fengyue meng*, p. 30.

Part IV Introduction

1. Hechter, *Internal Colonialism*.

Chapter 9

1. Zhu Zeqing, "Shuo Yangzhou," p. 35.

2. Zhu Ziqing, "Wo shi Yangzhouren."

3. Zhu Ziqing, "Shuo Yangzhou," p. 35.

4. Da Cruz, "Treatise in Which the Things of China Are Related at Great Length," p. 149; de Rada, "The Relation of Fr. Martin de Rada, O.E.S.A.," p. 282.

5. Pruitt, *A Daughter of Han*, p. 29.

6. Zhu Ziqing, "Ze'ou ji." For an English translation of this essay, see idem, "My Wife and Children."

7. Zhu Ziqing, "Shuo Yangzhou," p. 35. For the work to which Zhu refers, see Zhang Dai, *Tao'an mengyi*, pp. 56-58. Translations can be found in Lin Yutang, *Translations from the Chinese*, pp. 229-31; Pollard, "The Jades of Yangzhou"; Bao-Hua Hsieh, "The Acquisition of Concubines in China," pp. 147-50; and Wei Minghua and Antonia Finnane, "The Thin Horses of Yangzhou," pp. 56-57. This last reference is to my translation with introduction, annotations, and illustrations of an article of which the Chinese version eventually appeared in abbreviated form in Wei Minghua, *Yangzhou wenhua tanpian*,

pp. 150–67. The major omission in the published version is of an etymological discussion of the word "horse." An earlier and much briefer version of Wei's article appeared in *Dushu*, no. 3 (1983): 103–7.

8. Lin Yutang, *Translations from the Chinese*, pp. 160–62.

9. *WLYZFZ*, preface, 3a–b.

10. Xie Zhaozhe, *Wuzazu*, p. 196. On Xie, see *DMB*, 1: 546–50.

11. Zhang Dai, *Tao'an mengyi*, p. 40.

12. Wang Shixing, *Guangzhi yi*, p. 29. On Wang Shixing, see *DMB*, 2: 1405–6.

13. One of the few thin horses known by name was born in 1595 and died in 1612 (Ko, *Teachers of the Inner Chambers*, p. 91).

14. Xie Zhaozhe (*Wuzazu*, p. 186) mentioned thin horses, as did the historian Shen Defu, poet and playwright Kong Shangren, poet and social commentator Jin Zhi (1663–1740), and Zhang Dai; see Wei Minghua and Finnane, "The Thin Horses of Yangzhou." On Shen Defu, see *DMB*, 2: 1190–91. Jin Zhi was from a prominent family in the Nanjing area, and his forbears had served in the Ming bureaucracy. His father was a metropolitan graduate of 1653 (Jin Zhi, *Bu xia dai bian*, p. 1).

15. Cf. the rather free translation by Chen Shih-hsiang and Harold Acton in Kong Shangren, *The Peach Blossom Fan*, p. 190, in which there is no allusion either to Yangzhou or to the "thin horses," despite the annotation provided in the Chinese edition on which the translation is based. For the original, see Kong Shangren, *Taohua shan*, pp. 163–65.

16. Ko, *Teachers of the Inner Chambers*, p. 265.

17. The boldest statement on the growth of literacy in the Qing is to be found in Rawski, *Education and Popular Literacy in Ch'ing China*. For a discussion of the debate over Rawski's estimates, especially of female literacy, see Clara Wing-chung Ho, "Encouragement from the Opposite Gender," p. 308. See also Handlin [Smith], "Lü K'un's New Audience."

18. See the data presented in Mann, *Precious Records*, p. 231, based on Hu Wenkai, *Lidai funü zhuzuo kao*.

19. Shen Defu, *Wanli yehu bian*, p. 54.

20. Jin Zhi, *Bu xia dai bian*, pt. 5, p. 92.

21. Zhang Dai, *Tao'an mengyi*, pp. 56–57.

22. Shen Defu, *Wanli yehu bian*, p. 597.

23. Ibid.

24. Zhang Dai, *Tao'an mengyi*, p. 58.

25. Bao-Hua Hsieh, "The Acquisition of Concubines in China," pp. 162–63.

26. *KXYZFZ*, 7.3b; *YZHFL, juan* 9, p. 187. See also Wang Shizhen, *Xiangzu biji* (Notes from the first whiff of spring), 1702, cited in Wang Shunu, *Zhongguo changji shi*, p. 262.

27. Wang Shunu, *Zhongguo changji shi*, pp. 261–62.

28. Ibid.; Mann, *Precious Records*, pp. 126–27.

29. *YZHFL, juan* 9, p. 189.

30. Wu Jingzi, *Rulin waishi*, pp. 397–99; idem, *The Scholars*, pp. 509–11.

31. Wei Minghua and Finnane, "The Thin Horses of Yangzhou," p. 48.

32. Xu Ke, *Qingbai leichao*, vol. 38, *bai* 80: 38.

33. Ko, *Teachers of the Inner Chamber*, p. 256.

34. *YZHFL, juan* 2, p. 46.

35. Zhu Jiang, *Yangzhou yuanlin pinshang lu*, pp. 152–53.

36. *YZHFL, juan* 9, p. 190.

37. Ibid.

38. Ibid., p. 36.

39. Ibid., p. 35.

40. Pomeranz, *The Great Divergence*, p. 155.

41. Translation adapted from Hay, *Shitao*, p. 12. Hay translates *Yangzhou junyi* (Yangzhou prefecture and county/counties) as "the city and prefecture of Yangzhou," but it is unlikely that Li Gan was making a statement about social trends in the prefecture as a whole, since most of Yangzhou's dependencies were poor. Li Gan's comment is also cited in Wei Minghua, *Yangzhou shouma*, p. 143.

42. Translation slightly adapted from Hay, *Shitao*, p. 12.

43. Wei Minghua, *Yangzhou shouma*, p. 143.

44. *YZHFL, juan* 1, p. 28.

45. Wei Minghua, *Yangzhou shouma*, p. 153.

46. Ibid., p. 152.

47. Lin Sumen, *Hanjiang sanbai yin, juan* 6.

48. Ibid., 6.4b.

49. Rawski, *The Last Emperors*, p. 41; Elliott, *The Manchu Way*, p. 471n67.

50. Lin Sumen, *juan* 6.

51. *JQYZFZ*, 11b–12a.

52. Elvin, *The Pattern of the Chinese Past*, pp. 268–84.

53. Tanaka, "Rural Handicraft in Jiangnan," pp. 80–81.

54. I have used the translation from Lin Yutang, "Family Letters of a Chinese Poet," in idem, *Translations from the Chinese*, p. 492. Lin omits five of the sixteen letters originally published by Zheng, along with portions of two more. For a full translation of the letters into

French, with an introduction and detailed notes, see Diény, Les "Lettres Familiales" de Tcheng Pan-kiao.

55. QLQSL, 437.16b–17a.
56. GXHAFZ, 2.6b–7a.
57. JQYZFZ, 61.2b–3a.
58. JQDTXZ, 19.7b.
59. Zheng Changkan, Ming Qing nong cun shangpin jingji, p. 150.
60. Gui Chaowan, Huanyou jilüe, 5.3a. My thanks to Pierre-Etienne Will for this reference.
61. QLTZZLZZ, 1.33b.
62. Walker, Chinese Modernity and the Peasant Path, p. 56.
63. QLJDXZ, 29.4b, 77a, 98a.
64. Six out of 35 cases; ibid., 29.35b, 36a, 39b.
65. Wu Chengming, "Introduction: On Embryonic Capitalism," p. 9.
66. Ho Ping-ti, "The Salt Merchants of Yangchou," p. 144.
67. Jiaodong Zhou Sheng, Yangzhou meng, p. 39.
68. Zhao Hongen, Yuhua ji, 4. 6.35a–b. I thank Helen Dunstan for this reference.
69. JQYZXZ, 33.5b; TZYZFZ, 3.7b.
70. JQLHYFZ, 46.31b; MGSXZ, 9.39b.
71. GLMST, 30a.
72. Zhao Zhibi, Pingshantang tuzhi, 2.16a–17b. On Hu Yuan, see Sun Xianjun, "Gu jiaoyujia Hu Yuan."
73. QLJDXZ, 29.50b, 60b–62b; YZHFL, juan 16, pp. 366–67.
74. Mann, "Historical Change," p. 68.
75. QLJDXZ, 29.60b.
76. Ibid., 61a.
77. Wu Jingzi, Rulin waishi, pp. 397–99; idem, The Scholars, pp. 509–11.
78. Wu Jingzi, Rulin waishi, pp. 471–72; idem, The Scholars, pp. 600–602.
79. See Zurndorfer, Change and Continuity, pp. 50–51.
80. Suzuki, "Shindai Kishūfu no sōzoku to sonraku," p. 80.
81. Carlitz, "The Social Uses of Female Virtue in Late Ming Editions of Lienü Zhuan," pp. 139–40.
82. Cohen and Monnet, Impressions de Chine, pp. 103–4.
83. Qingchao yeshi daguan, juan 6, p. 46.
84. Jiaodong Zhou Sheng, Yangzhou meng, p. 4.
85. I am grateful to Wei Minghua for this point.
86. Marmé, "The Rise of Suzhou," p. 19.

87. Brook, *The Confusions of Pleasure*, 220–22; Ko, "Bondage in Time," p. 10.
88. Shen Fu, *Chapters of a Floating Life*, pp. 43–44.
89. *YZHFL, juan* 9, p. 189.
90. Ibid., p. 192.
91. Hershatter, *Dangerous Pleasures*, p. 54.
92. Silas Wu, *Passage to Power*, p. 102.
93. Jiaodong Zhou Sheng, *Yangzhou meng*, pp. 36, 41.
94. Hanshang mengren, *Fengyue meng*, pp. 12, 25, 27, 38, 39.
95. Ibid., p. 40.
96. Bray, *Technology and Gender*, p. 256.
97. Fei Xuan, *Yangzhou meng xiangci*, cited in Wei Minghua and Finnane, "The Thin Horses of Yangzhou," p. 63.

Chapter 10

1. *TZYZFZ*, 3.7b.
2. The premises of the Widow Support Society were located in the Hu Anding Shrine next to the Anding Academy in the Old City (*JQYZFZ*, 18.6b).
3. Liang Qizi, "Mingmo Qingchu minjian cishan huodong de xingqi."
4. Conflicting accounts of the founding of the orphanage are provided in the local and salt gazetteers. The account followed here is given in the Yongzheng and Jiaqing editions of the prefectural gazetteer. The Tongzhi edition and the Guangxu salt gazetteer provide a divergent account, citing a memoir by Fang Junyi (1815–89), who as Lianghuai salt controller undertook expansion of the new premises in 1870. (The original orphanage was destroyed during the Taiping Rebellion and re-established in Suzhou Song Street in 1869.) According to Fang's quite detailed narrative, three salt merchants were responsible for setting up the orphanage in 1655: two from Huizhou (Wu Ziliang and Fang Ruting) and the West merchant Yuan Hongxiu (*TZYZFZ*, 3.7a–b; *GXLHYFZ*, 152.13b). The Jiaqing edition of the salt gazetteer begins by mentioning these same three merchants but goes on to say that it was supported by Li Zongkong and Min Shizhang, along with Zheng Yuanhua, Cheng Yourong, Wu Bichang, and Xu Chengzong, all Huizhou men judging by their names, except for Li Zongkong (*JQLHYFZ*, 56.4b). Zheng Yuanhua was Zheng Yuanxun's brother. Xu Chengzong must have been a brother or cousin of Xu Chengxun, who gained the *jinshi* degree under Jiangdu registration in 1676, although he is described as a "She-

xian" man in the salt gazetteer (*JQLHYFZ*, 35.13b). Cheng Yourong, also of Shexian, is mentioned in the same source for his relief efforts during floods in the Kangxi period, probably those of which Li Zongkong wrote in 1672 (*JQLHYFZ*, 46.29a).

5. *JQYZFZ*, 18.5b.

6. E.g., Huang Lümao is described as *yiren* (*QLJDXZ*, 7.23b), and Ma Yueguan as *junren* (*JQYZFZ*, 19.10b).

7. See note 4 to this chapter.

8. Rowe, *Hankow*, pp. 228–29, 246.

9. Comparable circumstances in Kunming, where in the early eighteenth century sojourning merchants operated a poorhouse supported by rents from urban properties, suggest that a comparative study of the late imperial philanthropic movement with an eye to society rather than to the state might show a systematic relationship between the initiators of philanthropic institutions and other sectors of the local society; see Rowe, *Saving the World*, p. 370.

10. Chen Qubing, *Wu shi zhi*, p. 326.

11. These were identified by subtracting from the list of Qing *jinshi* graduates from Yangzhou to 1806 those surnames that also appear in the biographies section of the Huizhou gazetteer of Shexian (*MGSXZ*), along with the distinctive West surnames.

12. Furth, *A Flourishing Yin*, pp. 228–65.

13. Wang Zhenzhong, "Ming Qing Yangzhou yanshang shequ wenhua ji qi yingxiang," p. 3.

14. *YZHFL, juan* 13, p. 282.

15. Wu Jiaji, "Song Wang Zuoyan gui Xin'an." Wang Zuoyan, a teacher by profession, was the offspring of a Xiuning salt merchant family; see Yang Jiqing, *Wu Jiaji shi jianjiao*, p. 68.

16. Wang Zhengzhong, *Ming Qing huishang yu Huai Yang shehui bianqian*, pp. 134–35.

17. See references to shrine lands in ibid., pp. 70–71.

18. For Qing examinations up to 1770, 19 Jiangdu/Ganquan registrations and 21 Yizheng registrations are recorded for the first-level examination. For the *jinshi* exams, 7 Jiangdu/Ganquan and 8 Yizheng registrations are listed (*QLSXZ, juan* 8).

19. *MGSXZ, juan* 7–10.

20. Ho Pingti, *Zhongguo huiguan shilun*, p. 60.

21. *YZHFL, juan* 3, p. 58.

22. Lu Zuoxie, "Shilun Ming Qing shiqi huiguan de xingzhi he zuoyong."

23. Wang Zhenzhong, *Ming Qing Huishang yu Huai Yang shehui bianqian*, p. 166.

24. *GXJDXZ*, 12b.8b–9a; elsewhere I have discussed the question of native-place associations in Yangzhou and, at that time, was more tentative in demurring from Ho Ping-ti's view; see Finnane, "Yangzhou," p. 140.

25. Zhu Jiang, *Yangzhou yuanlin pinshang lu*, p. 109.

26. "Jianli huiguan beiji." The inscription is dated the fifth month of the tenth year of the Guangxu reign (1884).

27. The stele inscription recording the amounts is headed "Changyun geshang juankuan" (Amount levied from each yard and transport merchant).

28. Wang Fangzhong, "Qingdai qianqi de yanfa," p. 17.

29. Smith, "Social Hierarchy and Merchant Philanthropy."

30. Liang Qizi, *Shishan yu jiaohua*, chap. 4.

31. Elliott, *The Manchu Way*.

32. Mann, "Historical Change."

33. *KXLHYFZ*, 13.4b.

34. Will, *Bureaucracy and Famine*, p. 201.

35. *GXLHYFZ*, 152.1b, 3a.

36. Places and years for the founding of other Lianghuai salt charity granaries were: Taizhou, Tongzhou, Rugao, Yancheng, Haizhou, and Banpu *chang*, all in 1727; and Shigang *chang*, Dongtai *chang*, and Funing, in 1735 (ibid., 152.6b–7a).

37. Will, *Bureaucracy and Famine*, p. 201.

38. *QLTZZLZZ*, 4.21b–22a.

39. *GXLHYFZ*, 152.7b.

40. Ibid., 151.4b.

41. *JQYZFZ*, 18.5b, 8b.

42. Liang Qizi, "Mingmo Qingchu minjian cishan huodong de xingqi," p. 66.

43. *JQYZFZ*, 18.6b; *YZHFL*, *juan* 12, pp. 275–76, *juan* 9, p. 198. No date of establishment is provided for the *tongrentang*. On numbers of doctors, see Bao Shichen, *Anwu sizhong, Zhongqu yishao: shang*, 5b–6a.

44. *YZYZFZ*, 13.2a.

45. *JQYZFZ*, 30.41a–b.

46. *GXLHYFZ*, 137.27b–28a.

47. Rowe, *Saving the World*, p. 369. Discussing Liang Qizi's identification of Chen Hongmou as "*the* field official" active in the government co-optation of philanthropy, Rowe points out that Chen's involvement in this domain was periodic and depended on where he was serving. He goes on to detail Chen's particular involvement in the Yangzhou *pujitang*, but the *pujitang* he discusses was in fact in Guazhou, as Liang (*Shishan yu jiaohua*, p. 106) notes. See further dis-

cussion below on the differentiation between the prefectural city and other places with respect to the establishment and support of urban institutions. Rowe's source, the 1743 Jiangdu county gazetteer, did not list the Yangzhou city hospice.

48. *JQYZFZ*, 18.6a; *JQLHYFZ*, 56.1b.

49. Sheng Langxi, *Zhongguo shuyuan zhidu*, p. 132.

50. *GXLHYFZ*, 151.2b; *YZHFL, juan* 3, p. 60.

51. Cai Guihua, "Yangzhou Meihua shuyuan kao," pp. 374–76.

52. *GXLHYFZ*, 15.4b.

53. The students in the regular (*zheng*) category, one half of the total number, enjoyed stipends of 3 taels per month; the remainder, in the supplementary (*fu*) category, got 1 tael per month. Thus over 6,000 taels per annum were expended on stipends alone. As of 1794, residential students were paid an extra three fen per day, amounting to nearly 11 taels a year, and the top ten students in the regular category received another 18 taels a year. There were additional students, for whom no quota was set and no financial support provided (ibid., 151.2b, 4b–5a).

54. Ibid., 151.5b.

55. *JQYZFZ*, 19.15b–16a.

56. *MGJDXZ*, 8b.5a.

57. *YZHFL, juan* 16, p. 23.

58. *JQYZFZ*, 18.7b.

59. Ibid., 18.8b.

60. Liu Sen, "Huishang Bao Zhidao ji qi jiashi kaoshu."

61. *JQLHYFZ*, 56.7a.

62. *JQYZFZ*, 18.6a; *QLJDXZ*, 7.24a–25b.

63. *JQYZFZ*, 18.8b–9a.

64. *QLTZZLZZ*, 8.14b, 16b.

65. *GXLHYFZ*, 151.9a–b.

66. *JQDTXZ*,14.2b, 11b.

67. *JQRGXZ*, 9.54b.

68. *GXLHYFZ*, 151.9a–b.

69. *GXYCXZ*, 17.24b–25a.

70. Ho Ping-ti, *The Ladder of Success in Imperial China*, p. 249.

71. Metzger, "The Organizational Capabilities of the Ch'ing State in the Field of Commerce," p. 41.

72. *JQLHYFZ, juan* 48; *JQYZFZ, juan* 39.

73. Ho Ping-ti, *The Ladder of Success in Imperial China*, p. 249.

74. Ho Wai-kam, "The Literary Gathering at a Yangzhou Garden," p. 375. For other reproductions, see Chou and Brown, *The Elegant Brush* (1985), pp. 133–34; and *The Elegant Brush* (1986), pp. 109–10.

Note that these are not the same work. The 1985 publication is a detailed catalogue of the exhibition held at the Phoenix Art Museum; the 1986 version, produced for the same exhibition when it was held in Hong Kong, is a slighter volume without the detailed catalogue notes although it includes an introduction by Zhou Ruxi (i.e., Ju-hsi Chou). This painting has also been discussed at length by Frederick Mote ("The Intellectual Climate in Eighteenth-Century China"), who reads it as a statement of disaffection with the Qing.

75. Ho Wai-kam, "The Literary Gathering at a Yangzhou Garden," p. 375.

76. Liang Qizi, "Mingmo Qingchu minjian cishan huodong de xingqi," p. 64.

77. Ibid. Elsewhere Liang Qizi (*Shishan yu jiaohua*, p. 124*n*41) notes that the blurring of official and merchant categories was not limited to Yangzhou, which may well have been the case. For example, the position of the Huizhou salt merchants active in the Liangzhe sector must have been comparable. But the extent to which other places were comparable to Yangzhou in this respect remains to be documented. Liang's footnotes for this point indicate that present examples are almost fully limited to Anhui families and particularly those of Huizhou. The exception is an example offered by Susan Mann, *Local Merchants and the Chinese Bureaucracy*, pp. 89–90, but Mann's evidence is very tentative, as discussed further below.

78. Elman, *From Philosophy to Philology*, p. 94.

79. Martin Huang, *Literati and Self-Re/Presentation*, p. 31.

80. Mann, *Local Merchants and the Chinese Bureaucracy*, p. 89.

81. Rowe, *Hankow*, pp. 248–49. Although Rowe's analysis of Hankou society is at odds with the thesis I present here, his description of social interactions, particularly the activities of the Huizhou salt merchants, resonates with circumstances in Yangzhou; see especially pp. 98–106, 246–47.

82. Elman, *From Philosophy to Philology*, p. 95.

83. Hsü, *A Bushel of Pearls*, p. 41.

84. Zurndorfer, *Continuity and Change*, p. 132.

85. *ECCP*, pp. 559–60; Xue Yongnian and Xue Feng, *Yangzhou baguai*, pp. 11–12.

86. Wang Xiang, *Jiangsu suishi fengsu tan*, p. 76.

87. Of 104 local notables from Shexian described as active in Jiangbei during the Qing in *MGSXZ, juan* 3–7, there were fifteen surnamed Cheng, fifteen surnamed Wang (water radical), and fourteen surnamed Jiang (water radical).

88. The Cheng clan is one of Ho Ping-ti's case studies; see his "Salt Merchants of Yangzhou," pp. 158–59.

89. See also Chapter 3, pp. 66–67.

90. On Lu Zhonghui, see *YZHFL*, *juan* 4, p. 86; on Cheng Mengxing (*js* 1712) and Min Hua, see *MGSXZ*, 10.39b, 40a. Retired governor-general of Gansu Hu Qiheng, another participant in this gathering, is wrongly described as a native of Yangzhou in Chou and Brown, *The Elegant Brush*, p. 134. He was in fact a native of Wuling (now Changde) in Hunan (*YZHFL*, *juan* 4, p. 86).

91. *YZHFL*, *juan* 4, p. 88.

92. Chou and Brown, *The Elegant Brush*, p. 136. On the Zhangs of Lintong, see *YZHFL*, *juan* 15, pp. 337–38. Note that Zhang Shike (Shih-k'o) is given as Zhang Sike (Ssu-k'o) in both Ho Wai-kam, "The Literary Gathering at a Yangzhou Garden," p. 373; and Chou and Brown, *The Elegant Brush*, p. 136, although it is clear in the latter work that the figure in the painting has been identified correctly via his *zi* (used in Li E's colophon), through reference to Li Dou's entry in *YZHFL*, *juan* 4, p. 86. Although *si* is indeed a variant reading for *shi* (scholar), it is unclear why such a rendering might have been used in this case. For the sake of clarity, the conventional reading is used here, particularly because there was in Yangzhou a more famous Zhang Sike (different characters, *zi* Xinnan), also from Lintong but of a different family (*QLJDXZ*, 22.18a; *YZHFL*, *juan* 15, p. 338).

93. See, e.g., Qiu Wenbo (fl. mid-10th c.), *Wenhuitu* (Portrait of a literary gathering), Palace Museum, Taipei.

94. *YZHFL*, *juan* 4, p. 90. Wang Wenchong received his *jinshi* under Yizheng registration (*JQYZFZ*, 39.28a).

95. *YZHFL*, *juan* 13, pp. 300–308.

96. Of the garden owners mentioned here, Cheng Mengxing owned the Bamboo Garden (see Chapter 8); Wang Shu owned the South Garden, later redeveloped as the Garden of Nine Peaks (Jiufengyuan) by his grandson (*YZHFL*, *juan* 7, pp. 161–64). Zhang Shike and Lu Zhonghui shared ownership of a garden adjacent to the Temporary Retreat (ibid., *juan* 4, p. 86).

97. Ibid., *juan* 4, p. 83.

98. Luo Weiwen, "Shanguan ji jing yi zhu, linglong qishi you cun."

99. Ho Wai-Kam, "The Literary Gathering at a Yangzhou Garden," p. 375.

100. See my own early validation, with caveats, in "Yangzhou," p. 139, citing the reference to the "shenshang" Li Zongkong and Min Shizhang.

101. *JQYZFZ*, 18.5b; *JQLHYFZ*, 45.7a–b.

102. Li's family, originally from Datong in Shanxi, had long been established in Yangzhou. His great-grandfather and grandfather were both *jinshi* graduates (see note 62 to Chapter 4, p. 349; *JQLHYFZ*, 45.3b).

103. *JQYZFZ*, 18.6b.

104. Smith, "Social Hierarchy and Merchant Philanthropy," p. 422.

105. Mann, *Local Merchants and the Chinese Bureaucracy*, p. 232n17.

106. Ibid., p. 22.

107. *QLQSL*, 1076.5a–b.

108. For changes in Zhejiang and Shanxi, see ibid., 1069: 15a–b.

109. Kong Shangren, *Huhai ji*, p. 182.

110. For further discussion, see Chapter 11.

Chapter 11

1. Jiaodong Zhou Sheng, *Yangzhou meng*, p. 34. This is a reference to the classical cosmological correspondence between the divisions of heaven and earth. "Ox" is an abbreviation for "leading the ox" (*qianniu*), i.e., the herd boy. Wang Zhong made the same reference in his essay "Guangling dui" (Answering questions about Guangling). The reference is glossed in Wang Zhong, *Wang Zhong ji*, p. 425.

2. Zhao Hang, *Yangzhou xuepai xinlun*, p. 14. On Fang, see *ECCP*, pp. 238–40.

3. Zhang Shunhui, *Qing ruxue ji*, pp. 378–479.

4. Ruan Yuan, "Yangzhou huafanglu er ba," p. 7.

5. Zhang Shunhui, *Qingdai Yangzhouxue ji*, p. 166.

6. Liang Ch'i-ch'ao, *Intellectual Trends in the Ch'ing Period*, p. 75.

7. See Zhang Shunhui, *Qingdai Yangzhouxue ji*, pp. 9–10. Much of Zhang's pioneering study was incorporated into his later work, *Qing ruxue ji*, pt. VIII, "Yangzhouxue ji," pp. 378–479. See also Ōtani, "Yōshū, Jōshū gakujutsu kō," which draws in part on Zhang Shunhui's work.

8. *YZHFL*, *juan* 3, pp. 62–67. See also *ECCP* under individual names.

9. Zhang Shunhui, *Qingdai Yangzhouxue ji*, p. 9.

10. *YZHFL*, *juan* 10, p. 219. The Red Bridge (Hong qiao) was by this time written as "Rainbow Bridge" (also Hong qiao).

11. *YZHFL*, *juan* 10, p. 220.

12. Wu Zhefu, *Siku quanshu*, pp. 138–40.

13. Ibid., p. 264.

14. Guy, *The Emperor's Four Treasuries*, p. 22; *ECCP*, pp. 93–95.

15. *YZHFL, juan* 4, p. 99. On the numbers of books in each collection, see Guy, *The Emperor's Four Treasuries,* p. 110, fig. 2.

16. *JQLHYFZ,* 43.28a–b; *YZHFL, juan* 3, p. 64, *juan* 4, p. 99. Xie Rongsheng wrote a preface for Li Dou's *Painted Barques* and, according to Rowe (*Saving the World,* p. 28), was close to Ruan Yuan in later life.

17. Wu Zhefu, *Siku quanshu,* p. 41.

18. Ibid., p. 144.

19. Mote, "The Intellectual Climate in Eighteenth-century China," pp. 45–48; see also Fisher, "Loyalist Alternatives in the Early Ch'ing."

20. Du Jinghua, "Chai Shijin tousong citie an."

21. Probably only the elder of the Xu grandsons was beheaded. The younger brother's sentence was commuted to slavery in Heilongjiang. Tao Yi died in jail before he could be either beheaded or pardoned. Functionaries from the censorship bureau and the clerk from the Dongtai yamen were acquitted. The case is outlined in *ECCP,* pp. 320–21. For a detailed account of the case, see Zheng Yuan, *Qianlong huangdi quanzhuan,* pp. 631–36. This work is unsourced, but the generous quotations show it to be based on Qing documents. See also Yu Yuchu, "Xu Shukui 'Yi zhu Lou shi' an."

22. Zheng Yuan, *Qianlong huangdi quanzhuan,* pp. 631–36. Zhang Shucai and Du Jinghua, *Qingdai wenziyu an,* pp. 78–86.

23. Guy, *The Emperor's Four Treasuries,* pp. 174–77.

24. Wei Minghua, "Kao 'Yangzhou luantan,'" p. 207.

25. Ibid., p. 187.

26. Ren Zuyong, "*Yuewei caotang biji* so lu Ren Dachun yishi kao," pp. 118–19.

27. Wang Zhangtao, *Ruan Yuan zhuan,* p. 28.

28. *ECCP,* pp. 120–21. For a discussion of Ji Yun's notes on Ren Dachun, see Ren Zuyong, "*Yuewei caotang biji* so lu Ren Dachun yishi kao."

29. Wu Zhefu, *Siku quanshu,* p. 184. On Wang, see *ECCP,* pp. 805–7.

30. Wang Zhangtao, *Ruan Yuan zhuan,* pp. 97–98; *ECCP,* p. 400.

31. *YZHFL, juan* 5, p. 103.

32. *JQYZFZ,* 45.23b.

33. Li Huan, *Guochao qixian leizheng,* vol. 12, 247.29a.

34. Crossley, *A Translucent Mirror,* p. 290.

35. Wang Zhong, *Wang Zhong ji,* p. 161.

36. *Qingshi liezhuan,* vol. 9, 6.17a; see also *ECCP,* pp. 908–10, which does not mention Yan's connection with the salt trade.

37. *ECCP,* pp. 695–99.

38. Zhang Shunhui, *Qingdai Yangzhouxue ji*, pp. 165-66.

39. Ōtani, "Yōshū Jōshū gakujutsu kō," p. 330.

40. Zhang Shunhui, *Qingdai Yangzhouxue ji*, p. 12.

41. Ōtani, "Yōshū Jōshū gakujutsu kō."

42. Cited in full in Zhang Shunhui, *Qingdai Yangzhouxue ji*, p. 164.

43. Ibid.

44. Mori, "Yanchang de Taizhou xuepai," pp. 60-65.

45. De Bary, "Individualism and Humanitarianism," pp. 171, 173.

46. Wang Zhangtao, *Ruan Yuan zhuan*, pp. 8-10, esp. p. 8n1, for his relationship with the Jiang family. On the Jiang family itself, see Ho Ping-ti, "The Salt Merchants of Yang-chou," pp. 160-61. On Ling Tingkan, see *ECCP*, pp. 514-15. Ruan first met Ling in 1782, when he was nineteen by Chinese reckoning (Zhang Jian, *Leitang'an zhu dizi ji*, p. 6).

47. Zhang Jian, *Leitang'an zhu dizi ji*, p. 3.

48. Wang Zhangtao, *Ruan Yuan zhuan*, pp. 5-6.

49. Wei Peh-t'i, "Juan Ruan," p. 80.

50. Ruan Yuan, *Yanjingshi ji*, cited in *TZYZFZ*, 15.1a.

51. Zhou Cun, *Taipingjun zai Yangzhou*, p. 2.

52. *BHXZ*, 6.3b.

53. Ho Ping-ti, *The Ladder of Success in Imperial China*, p. 202; Elman, *Philosophy and Philology*, p. 95, citing Ho.

54. *YZHFL, juan* 3, p. 60.

55. On Ling's mother, see Ruan Yuan, "Ling mu Wang tairuren shou shi xu."

56. Ibid.

57. *ECCP*, p. 515.

58. Others were Huang Chengji (1771-1824), whose father moved from Shexian to Yangzhou, and Jiang Fan, whose family did in fact come from Anhui although not from Huizhou prefecture and who was properly localized by virtue of being established on the land in Ganquan county.

59. Shimada, *Pioneer of the Chinese Revolution*, pp. 91-92.

60. Ruan Yuan, *Dingxiangting bitan*, 3.54a.

61. Chen Qubing, *Wu shi zhi*, p. 328.

62. *ECCP*, pp. 528, 514, 815.

63. Wang Zhong, "Xianmu Zou ruren lingbiao," p. 240.

64. Wang Zhong, "Guangling dui."

65. *ECCP*, pp. 814-15.

66. Wang Zhong, "Xianmu Zou ruren lingbiao," p. 240.

67. *BHXZ*, 6.2a-b.

68. Wei Peh-t'i, "Juan Yuan," pp. 8-12.

69. See Zhang Shunhui, *Qingdai Yangzhouxue ji;* Zhao Hang, *Yangzhou xueji xinlun;* and various essays in Feng Erkang, *Yangzhou yanjiu.*

70. The Baoying Lius had a particularly significant position in the network of Yangzhou scholars. See *ECCP*, pp. 528-29, 530-31. Liu Taigong became acquainted with Wang Zhong around 1772 and with Wang Niansun on his trips to Beijing. Liu Baonan, who was Taigong's disciple as well as his cousin, became friends with the Yizheng/Yangzhou scholar Liu Wenqi when he was studying at the Anding Academy in Yangzhou sometime after 1806. He stayed with Wang Xixun, Wang Zhong's son, when he was in Beijing in 1822-23 and in 1833 served as assistant to Ruan Yuan's son Changsheng (1788?-1833) in Baoding, Zhili, where Changsheng was prefect. Liu Wenqi, for his part, was the nephew and student of Ling Shu, who served for a time as assistant to Ruan Yuan and tutor to his children. Liu Wenqi's son, Liu Yusong (1818-67) of Yizheng, and Liu Baonan's son, Liu Li of Baoying—each of whom helped further his father's work—were later colleagues in the Jiangnan Printing Office, established in Nanjing after the Taiping rebellion. See Zhang Shunhui, *Qingdai Yangzhouxue ji,* pp. 9-10; Zhao Hang, *Yangzhou xuepai xinlun, passim;* and *ECCP*, under individual names.

71. Wei Peh-t'i, "Juan Yuan," p. 38.

72. Zhang Shunhui, *Qingdai Yangzhouxue ji,* pp. 11-14. Gu Guangqi was from Suzhou but had well-established connections in Yangzhou, where his father had served as a physician and where he himself dwelt for long periods at a time in his middle years. His close acquaintance with Song texts appears to have been in part a consequence of his work on collating Song editions for Zhang Dunren (1754-1834), when the latter was prefect of Yangzhou in 1804-5 (*ECCP*, pp. 417-18).

73. Wang Zhong, *Guangling tongdian;* Ruan Yuan, *Guangling shishi;* idem, *Huaihai yingling ji;* Liu Baonan, *Shengchao xun Yang lu;* idem, *Baoying tujing;* Liu Wenqi, *Yangzhou shuidao ji;* Jiao Xun and Jiang Fan, *Jiaqing Yanghoufu tujing;* Jiao Xun, *Yangzhou zuzheng lu;* Jiao Xun, *Yangzhou Beihu xiaozhi.*

74. Ruan Xian, *Yangzhou Beihu xuzhi;* Ruan Heng, *Guangling mingsheng tu;* Wang Zhangtao, *Ruan Yuan zhuan,* p. 6; Lin Sumen, *Hanjiang sanbai yin.*

75. Hsü, *A Bushel of Pearls,* p. 227.

76. *JQJDXZ*, 12.21b-22a.

77. In the 1597 edition of the country gazetteer, the compiler made a lengthy comment on the number of sojourners in the city. Like his nineteenth-century successor, he noted that in some families differ-

ent members had different registrations. His comments were directed, however, at celebrating the literary and examination successes of Yangzhou's non-natives (WLJDXZ, 18.8b–9a).

78. Lin Sumen, *Hanjiang sanbai yin*, 8.11a–b.

79. Strassberg, *World of K'ung Shang-jen*, pp. 131, 360n38.

80. On the Zhang troupe, see YZHFL, *juan* 5, p. 119.

81. Mackerras, *The Rise of the Peking Opera*, pp. 116–18.

82. On theater in late seventeenth-century Yangzhou, see Strassberg, *The World of K'ung Shang-jen*, p. 128.

83. YZHFL, *juan* 5, p. 117.

84. Mackerras, *The Rise of the Peking Opera*, pp. 6–7. Strassberg (*World of K'ung Shang-jen*, p. 357n30) notes that the division between *ya* (elegant or high class) and *hua* ("flowery" or vulgar) was not yet in place in the late seventeenth century; see also Wei, "Kao 'Yangzhou luantan,'" p. 192.

85. YZHFL, *juan* 5, p. 125.

86. See Ho Ping-ti, "The Salt Merchants of Yangchou," pp. 160–61; and Wang Zhenzhong, *Ming Qing Huishang yu Huai Yang shehui bianqian*, pp. 35–36.

87. Hsü, *A Bushel of Pearls*, p. 62.

88. YZHFL, *juan* 5, p. 125.

89. Ibid.

90. Ibid.

91. Wei Minghua, "Kao 'Yangzhou luantan,'" p. 187.

92. YZHFL, *juan* 18, p. 406.

93. Ibid., *juan* 5, p. 125.

94. Zurndorfer, *Change and Continuity*, pp. 241–47.

95. Wei Minghua, "Kao 'Yangzhou luantan,'" p. 206.

96. Ibid., p. 216.

97. Moser, *The Chinese Mosaic*, p. 115.

98. The error is noted in Chou and Brown, *The Elegant Brush* (1985), p. 136.

99. Jiaodong Zhou Sheng, *Yangzhou meng*, p. 41.

100. Lin Sumen, *Hanjiang sanbai yin*, 5.8b–9a.

101. Zhao Xueli and Cao Yongquan, "Tan Yangzhou hunsu," p. 68.

102. Lin, *Hanjiang sanbai yin*, 9.11b.

103. Ibid., 9.3b.

104. Ibid., 3.18a.

105. Honig, "Pride and Prejudice: Subei People in Contemporary Shanghai," p. 146.

106. Lin Sumen, *Hanjiang sanbai yin*, 3.12b.

107. Ibid., 7.3b, 7.4b, 5b–6a, 6.7a, 8.7a–8a.

108. Ibid., 10.1b.

109. Ibid., 10.1a.

110. Roddy, *Literati Identity*, p. 82.

111. Ibid.

112. For a partial English translation of the novel, see Lu Ju-chen, *Flowers in the Mirror*.

113. Ono Kazuko, "*Kyōka en no ekai*," pp. 41–43.

114. Hanan, "*Fengyue Meng* and the Courtesan Novel," p. 349.

115. David Der-wei Wang (*Fin-de-Siècle Splendor*, p. 5) points to the rise of the urban landscape in fiction as a feature of late Qing literary modernity. *Dreams of Wind and Moon* must meet this criterion.

116. Wang Zhangtao, *Ruan Yuan zhuan*, p. 80.

Part V Introduction

1. Huang Puan, *Zhonghua quanguo fengsu zhi*, pt. II, 3.96.

Chapter 12

1. Qian Yong, *Lüyuan conghua*, 20.6b–7a.

2. *YZSZXJ*, 3.5a.

3. Ibid., 3.6a–b.

4. Liang Zhangju, "Wenxuan lou," in idem, *Guitian suoji*, 1.2b.

5. Hanshang mengren, *Fengyue meng*, pp. 10, 14, 28, 30, 37, 54.

6. Ibid., p. 14.

7. Gong Zizhen, *Gong Zizhen quanji*, p. 185.

8. Metzger, "The Organizational Capabilities of the Ch'ing State in the Field of Commerce," p. 41.

9. *QLQSL*, 1419.9a–10b.

10. *YZHFL, juan* 2, p. 54.

11. Lin Sumen, *Hanjiang sanbai yin*, 5.13b.

12. Ibid., 5.8a.

13. Hsü, *A Bushel of Pearls*, p. 15.

14. Zurndorfer, *Change and Continuity*, p. 165.

15. *GXLHYFZ*, 40.19b.

16. *YFTZ*, 11.16a.

17. Ibid., 11.5a–8b.

18. Fang Yujin, "Daoguang chunian Chu an yanchuan fenglun sanmai shiliao," p. 39. On the *fenglun* system, see Rowe, *Hankow*, p. 104.

19. Metzger, "T'ao Chu's Reform of the Huai-pei Salt Monopoly," p. 2.

20. Between 1801 and 1804, more than 2.5 million taels of silver from the Lianghuai salt tax, levies through the Salt Transport Treasury, and merchant contributions were used for rebel suppression and postwar rehabilitation in Hubei, Shaanxi, Henan, and Sichuan (*JQQSL*, 82.3b, 85.9b, 85.20a, 91.3a, 92.27a, 93.14b, 102.5a, 104.5b, 104.14a, 126.27b).

21. *TZYZFZ*, 1.9a. On Tiebao, see *ECCP*, pp. 717-18.

22. *JQQSL*, 120.12b.

23. *Qinding gongbu zeli*, 1875 ed., 39.11a-b.

24. Zhang Liansheng, "Qingdai Yangzhou yanshang de xingshuai yu yapian yuru," p. 23.

25. *Wei Yuan ji*, cited in ibid.

26. Zhang Liansheng, "Qingdai Yangzhou yanshang de xingshuai yu yapian yuru."

27. Lin Sumen, *Hanjiang sanbai yin*, 4.6b. On Chaozhou merchants in the opium trade in Shanghai, see Goodman, *Native Place, City, and Nation*, pp. 70-72.

28. Jiaodong Zhou Sheng, *Yangzhou meng*, p. 38.

29. *HZBY*, 3.1a.

30. Ibid., 4.1b.

31. Fang Yujin, "Daoguang chunian Chu an yanchuan fenglun sanmai shiliao," p. 45. As Wang Zhenzhong (*Ming Qing Huishang yu Huai Yang shehui bianqian*, p. 39) points out, the abrogation of the post of principal merchant is declared to have been one of the reforms introduced by Tao Zhu in the following decade.

32. I.e., 348,119 *yin* out of a total of 779,934. Fang Yujin, "Daoguang chunian Chu an yanchuan fenglun sanmai shiliao," p. 39.

33. Bao Shichen, *Anwu sizhong: Zhongqu yi shao, shang*, 5b-6a. On the orphanage's size and funding, see *GXLHYFZ*, 152.13b-14a.

34. Tian Qiuye and Zhou Weiliang, *Zhonghua yanye shi*, pp. 306-7.

35. Ibid., p. 305.

36. Rowe, *Hankow*, p. 104.

37. Ibid.

38. *Qingbai leishao, juan* 4, cited in Yang Dequan, "Qingdai qianqi Lianghuai yanshang ziliao chuji," p. 49.

39. Jiaodong Zhou Sheng, *Yangzhou meng*, p. 39.

40. Ibid.

41. Tian Qiuye and Zhou Weiliang, *Zhonghua yanye shi*, p. 308.

42. Metzger, "Tao Chu's Reform of the Huai-pei Salt Monopoly," p. 30.

43. Tian Qiuye and Zhou Weiliang, *Zhonghua yanye shi*, p. 315.

44. Chu, *Reformer in Modern China*, p. 139.

45. Polachek, "Literati Groups and Literati Politics," p. 94.

46. The range of relief measures provided makes these difficult to tabulate. The references are as follows: *JQQSL*, 14.11b, 21.14b (1797); 34.6b, 35.7a (1798); 51.4a, 52.12a–b, 53.11a (1797); 74.9b (1800); 87.29a (1801); 103.27a, 104.8b (1802); 107.2b (1803); 134.33b (1804); 139.3a, 150.25a–b, 150.38a (1805); 156.8a, 156.16a, 156.26a, 160.13a–b, 164.26a, 166.25b–26a (1806); 173.4a, 190.17a (1807); 192.9b, 201.37b, 204.17b (1808); 306.18b (1809).

47. *TZYZFZ*, 1.7a.

48. Ibid., 1.5a–13a, 25b.

49. Dodgen, "Hydraulic Evolution and Dynastic Decline," pp. 36–37.

50. Polachek, *The Inner Opium War*, p. 78. On Pan Shi'en, see *ECCP*, pp. 607–8.

51. *TZYZFZ*, 1.49a–51a.

52. Zhou Cun, *Taipingjun zai Yangzhou*.

53. *TZYZFZ*, 11.23b. Zuo bears the same radical as the single-character personal names of Ruan Yuan's sons, showing this to be a generational name in the clan. Only Ruan Yuan's adopted son failed to have a similar name. The Ruan family register would probably have clarified the relationship, but it was destroyed during the Cultural Revolution. See Wang Zhangtao, *Ruan Yuan zhuan*, p. 3n3.

54. *TZYZFZ*, *juan* 10–11, *passim*.

55. Brouillion, *Mémoire sur l'état actuel de la mission du Kiang-nan*, p. 442.

56. *TZYZFZ*, 11.20b, 22a.

57. Wu Zhefu, *Siku quanshu*, p. 159.

58. Harvey to Alcock, 29/10/1866, Public Record Office: China, F.O. 228/415.

59. *North China Herald*, 22/1/1859.

60. Rowe, *Hankow*, p. 78.

61. Gandar, *Le Canal impérial*, p. 62. Romanization altered.

62. *TZYZFZ*, 24.22a.

63. *GXYCXZ*, 3.29b.

64. Pomeranz, *The Making of a Hinterland*.

65. Hinton, *The Grain Tribute System of China*, p. 75.

66. Ibid., pp. 39-48; Carles, "The Grand Canal of China," pp. 106–7.

67. *Huangchao jingshiwen tongbian*, 40.10b–11a. On Xue Fubao, see *ECCP*, pp. 331–32.

68. Honig, *Creating Chinese Ethnicity*, pp. 58–91.

69. Keenan, *Imperial China's Last Classical Academies*, p. 59.

70. Yu Dafu, "Yangzhou jiu meng ji Yutang."

71. Ibid., p. 6.

72. Ibid., p. 3.

73. Hahn, *China to Me*, pp. 43–44.

74. Wang Muhan, *Jiangsu yankenqu tudi liyong wenti zhǐ yanjiu*, p. 22866.

75. Li Changfu, *Jiangsusheng dizhi*, p. 130.

76. Bureau of Foreign Trade, *China Industrial Handbooks*, p. 127.

77. Yi Junzuo, *Xianhua Yangzhou*. For a detailed discussion of the controversy, see Finnane, "A Place in the Nation."

78. Yi, *Xianhua Yangzhou*, pp. 2–21.

79. Ibid., p. 16.

80. *Xianhua Yangzhou* has consequently acquired the status of rare book. A copy is held in Princeton University Library. In Yangzhou, considerable sensitivities still attach to this book, which has been reprinted in recent years but for internal circulation only. That is, it has not been republished as such. The imprint under which it appears is *Yangzhou lishi wenhua fengsu neibu ziliao* (Internal document on the historical culture and customs of Yangzhou).

81. Braudel, "In Bahia, Brazil."

∽

Works Cited

Abbreviations: Gazetteers

DGGYZZ (*Daoguang Gaoyou zhouzhi*)   Gaoyou department gazetteer, 1845

DGTZZ (*Daoguang Taizhouzhi*)   Taizhou department gazetteer, 1827

GXFNXZ (*Guangxu Funing xianzhi*)   Funing county gazetteer, 1886

GXGQXZ (*Guangxu Ganquan xianzhi*)   Ganquan county gazetteer, 1885

GXHAFZ (*Guangxu Huai'an fuzhi*)   Huai'an prefectural gazetteer, 1884

GXJDXZ (*Guangxu Jiangdu xianzhi*)   Jiangdu county gazetteer, 1883

GXJJXZ (*Guangxu Jingjiang xianzhi*)   Jingjiang county gazetteer, 1879

GXLHYFZ (*Guangxu Lianghuai yanfazhi*)   Lianghuai salt gazetteer, 1905

GXYCXZ (*Guangxu Yancheng xianzhi*)   Yancheng county gazetteer, 1895

HGTZ (*Huaiguan tongzhi*)   Huai'an customs complete gazetteer, 1816

JJHZFZ (*Jiajing Huizhou fuzhi*)   Huizhou prefectural gazetteer, 1566

JJSXTZ (*Jiajing Shanxi tongzhi*)   Shanxi provincial gazetteer, 1564

JJWYZ (Jiajing Weiyang zhi) — Yangzhou prefectural gazetteer, 1542

JQDTXZ (Jiaqing Dongtai xianzhi) — Dongtai county gazetteer, 1817

JQGQXZ (Jiaqing Ganquan xianzhi) — Ganquan county gazetteer, 1810

JQGYZZ (Jiaqing Gaoyou zhouzhi) — Gaoyou department gazetteer, 1813

JQJDXZ (Jiaqing Jiangdu xianzhi) — Jiangdu county gazetteer, 1811

JQLHYFZ (Jiaqing Lianghuai yanfazhi) — Lianghuai salt gazetteer, 1806

JQRGXZ (Jiaqing Rugao xianzhi) — Rugao county gazetteer, 1808

JQYZFZ (Jiaqing Yangzhou fuzhi) — Yangzhou prefectural gazetteer, 1810

JQYZXZ (Jiaqing Yizheng xianzhi) — Yizheng county gazetteer, 1808

JSYHTZ (Jiangsu yanhai tuzhi) — Jiangsu coast illustrated gazetteer, 1889

KXYZFZ (Kangxi Yangzhou fuzhi) — Yangzhou prefectural gazetter, 1685

LQYZXZ (Longqing Yizhen xianzhi) — Yizhen county gazetteer, 1566

KXLHYFZ (Kangxi Lianghuai yanfazhi) — Lianghuai salt gazetteer, 1693

MGJDXZ (Minguo Jiangdu xianzhi) — Jiangdu county gazetteer, 1926

MGSXZ (Minguo Shexianzhi) — Shexian county gazetteer, 1937

QLGYZZ (Qianlong Gaoyou zhouzhi) — Gaoyou department gazetteer, 1783

QLJDXZ (Qianlong Jiangdu xianzhi) — Jiangdu county gazetteer, 1743

QLJNTZ (Qianlong Jiangnan tongzhi) — Complete Jiangnan gazetteer, 1737

QLLHYFZ (Qianlong Lianghuai yanfazhi) — Lianghuai salt gazetteer, 1748

QLTZZLZZ (Qianlong Tongzhou zhili zhouzhi) — Tongzhou independent department gazetteer, 1755

QLSXZ (Qianlong Shexian zhi) — Shexian county gazetteer, 1771

TZYZFZ (Tongzhi Yangzhou fuzhi) — Yangzhou prefectural gazetteer, 1874

WLJDXZ (*Wanli Jiangdu xinzhi*)	Jiangdu county gazetteer, 1597
WLTZZ (*Wanli Tongzhouzhi*)	Tongzhou departmental gazetteer, 1578
WLYZFZ (*Wanli Yangzhou fuzhi*)	Yangzhou prefectural gazetteer, 1601
XFXHXZ (*Xianfeng Xinghua xianzhi*)	Xinghua county gazetteer, 1852
YFTZ (*Yanfa tongzhi*)	Complete gazetteer of salt regulations, 1918
YZYZFZ (*Yongzheng Yangzhou fuzhi*)	Yangzhou prefectural gazetteer, 1733

Abbreviations: Other

BDRC	Boorman, ed., *Biographical Dictionary of Republican China*
BHXZ	Jiao Xun, *Yangzhou Beihu xiaozhi*
BZJ	Qian Yiji, *Beizhuanji*
CYQS	*Qinding hubu caoyun quanshu*
DMB	Goodrich, ed., *Dictionary of Ming Biography*
ECCP	Hummel, ed., *Eminent Chinese of the Qing Period*
GLMST	Ruan Heng, *Guangling mingshengtu*
HDSL	*Qinding daqing huidian shili*
HXNB	Wu Tongju, comp., *Huaixi nianbiao*
HZBY	Li Cheng, *Huaizuo beiyao*
JQQSL	*Daqing lichao shilu*, Jiaqing period
JSBL	*Ming Qing like jinshi timing beilu*
JSQSYT	Zhu Kebao, *Jiangsu quansheng yutu*
NXSD	Gao Jin, ed., *Nanxun shengdian*
QCTD	*Qingchao tongdian*
QDHCZ	Wang Huzhen and Wu Weizu, comps., *Qingdai hechen zhuan*
QLQSL	*Daqing lichao shilu*, Qianlong period
QSG	Zhao Erxun, *Qing shi gao*
WXTK	*Qingchao wenxian tongkao*
XSJJ	Fu Zehong, ed., *Xingshui jinjian*
XXSJJ	Li Shixu, ed., *Xuxingshui jinjian*
YZHFL	Li Dou, *Yangzhou huafang lu*
YZQSL	*Daqing lichao shilu*, Yongzheng period
YZSRJ	Wang Xiuchu, *Yangzhou shiriji*

Gazetteers Listed Alphabetically

Dongtai xianzhi 東泰縣志 — Dongtai county gazetteer, 1817

Funing xianzhi 阜寧縣志 — Funing county gazetteer, 1886

Ganquan xianzhi 甘泉縣志 — Ganquan county gazetteer, 1810, 1885

Gaoyou zhouzhi 高郵州志 — Gaoyou department gazetteer, 1783, 1813, 1845

Huai'an fuzhi 淮安府志 — Huai'an prefectural gazetteer, 1884

Huaiguan tongzhi 淮關通志 — Huai'an customs complete gazetteer, 1816

Huizhou fuzhi 徽州府志 — Huizhou prefectural gazetteer, 1566

Jiangdu xianzhi 江都縣志 — Jiangdu county gazetteer, 1597, 1743, 1811, 1883, 1926

Jiangnan tongzhi 江南通志 — Complete Jiangnan gazetteer, 1737

Jiangsu yanhai tuzhi 江蘇沿海圖志 — Jiangsu coast illustrated gazetteer, 1889

Jingjiang xianzhi 靖江縣志 — Jingjiang county gazetteer, 1879

Lianghuai yanfazhi 兩淮鹽法志 — Lianghuai salt gazetteer, 1693, 1748, 1806, 1905

Rugao xianzhi 如皋縣志 — Rugao county gazetteer, 1808

Shanxi tongzhi 山西通志 — Shanxi provincial gazetteer, 1564

Shexianzhi 歙縣志 — Shexian county gazetteer, 1771, 1937

Taizhouzhi 泰州志 — Taizhou department gazetteer, 1827

Tongzhou (zhilizhou) zhi 通州(直隸州)志 — Tongzhou department gazetteer, 1578, 1755

Xinghua xianzhi 興化縣志 — Xinghua county gazetteer, 1852

Yancheng xianzhi 鹽城縣志 — Yancheng county gazetteer, 1895

Yanfa tongzhi 鹽法通志 — Complete gazetteer of salt regulations, 1918

Yangzhou (Weiyang) fuzhi 揚州(維揚)府志 — Yangzhou prefectural gazetteer, 1542, 1601, 1733, 1810, 1874

Yizheng (Yizhen) xianzhi 儀徵(儀眞)縣志 — Yizhen[g] county gazetteer, 1566, 1808

Other Sources

Alexander, William, *Costumes et vues de la Chine*. 2 vols. Paris, 1815.

Andrews, Julia. "Zha Shibiao." In *Shadows of Mt. Huang*, ed. James Cahill, pp. 102–8. Berkeley, Calif.: University Art Museum, 1981.

Atwell, William S. "From Education to Politics." In *The Unfolding of Neo-Confucianism*, ed. Wm. Theodore De Bary, pp. 333–67. New York: Columbia University Press, 1975.

Aucourt, P. "Journal d'un bourgeois de Yang-tcheou (1645)." *Bulletin de l'Ecole française d'Extrême Orient* 7 (1907): 297–312.

Ayling, Alan, and Duncan Mackintosh. *A Collection of Chinese Lyrics*. London: Routledge and Kegan Paul, 1965.

Backhouse, E., and J. O. P. Bland. "The Sack of Yang Chou-fu." In Backhouse and Bland, *Annals and Memoirs of the Court of Peking*, pp. 105–208. Boston and New York: Houghton Mifflin, 1914.

Bai Jian 白堅 and Ding Zhian 丁志安. "Bian Shoumin san ti" (Three colophons by Bian Shoumin). In *Yangzhou baguai kao banji* 揚州八怪考辨集 (Collected studies of Yangzhou's eight eccentrics), ed. Xue Yongnian 薛永年, pp. 75–80. Nanjing: Jiangsu meishu chubanshe, 1992.

Bao Shichen 包世臣. *Anwu sizhong, Zhongqu yi shao, shang* 安吳四種: 中衢一勺, 上 (Four works from [Mr.] Anwu: suggesting the middle way, part A) [1872]. Reprinted—Taibei: Wenhai chubanshe, 1969.

Bao Zhao 鮑照. "Wucheng fu" 蕪城賦 (Ballad of the overgrown city). In *Baoshi ji* 鮑氏集 (Works of Mr. Bao). Shanghai: Shangwu yinshuguan, n.d.

Barkan, Lenore. "Nationalists, Communists and Rural Leaders: Political Dynamics in a Chinese County, 1927–1937." Ph.D. diss., University of Washington, 1983.

Barnhart, Richard M. *Peach Blossom Spring: Gardens and Flowers in Chinese Paintings*. New York: Metropolitan Museum of Art, 1985.

Barrow, John. *Travels in China*. London: T. Cadell and W. Davies, 1806.

Beattie, Hilary. "The Alternative to Resistance: The Case of T'ung-ch'eng, Anhui." In *From Ming to Ch'ing: Conquest, Region and Continuity in Seventeenth-Century China*, ed. Jonathan Spence and John Wills, pp. 241–76. New Haven: Yale Unversity Press, 1979.

Bell, Colonel Mark S. "The Great Central Asian Trade Route from Peking to Kashgaria." *Proceedings of the Royal Geographical Society and Monthly Record of Geography*, n.s., 12, no. 2 (1890): 57–93.

Bian Xiaoxuan 卞小萱. "Yangzhou baguai zhi yi de Gao Xiang" 揚州
八怪之一的高翔 (Gao Xiang: one of the Yangzhou eight eccentrics).
In *Yangzhou baguai kao banji* 揚州八怪考辨集 (Collected studies of
Yangzhou's eight eccentrics), ed. Xue Yongnian 薛永年, pp. 232–
42. Nanjing: Jiangsu meishu chubanshe, 1992.

Boorman, Howard L., ed. *Biographical Dictionary of Republican China*.
5 vols. New York: Columbia University Press, 1967.

Borota, Lucie. "*The Painted Barques of Yangzhou*: Excerpts, by Li Dou."
Renditions 46 (Autumn 1996): 58–65.

Braudel, Fernand. "In Bahia, Brazil." In idem, *On History*, trans. Sarah
Matthews, pp. 165–76. London: Weidenfeld and Nicholson, 1980.

Bray, Francesca. *Technology and Gender: Fabrics of Power in Late Imperial China*. Berkeley: University of California Press, Berkeley, 1997.

Brook, Timothy. *The Confusions of Pleasure: Commerce and Culture in
Ming China*. Berkeley: University of California Press, 1998.

Brouillion, Le R. P. *Mémoire sur l'état actuel de la mission du Kiang-nan,
1842–1855, suivi de lettres relatives à l'insurrection, 1851–1855*. Paris:
Julien, Lanier et cⁱᵉ, 1855.

Brunnert, H. S., and V. V. Hagelstrom. *Present Day Political Organization of China*. Trans. A. Beltchenko and E. E. Moran. Taibei: Ch'eng
Wen, 1978.

Bureau of Foreign Trade. *China Industrial Handbooks: Kiangsu*. Shanghai, 1933. Reprinted—Taibei, 1973.

Burke, Peter. "Res et verba: Conspicuous Consumption in the Early
Modern World." In *Consumption and the World of Goods*, ed. John
Brewer and Roy Porter, pp. 148–60. London: Routledge, 1993.

———. *Venice and Amsterdam: A Study of Seventeenth-Century Elites*.
Cambridge, Eng.: Polity Press, 1994.

Cahill, James. "Introduction." In *Shadows of Mt. Huang: Chinese Painting and Printing of the Anhui School*, ed. James Cahill, pp. 7–15.
Berkeley, Calif.: University Art Museum, 1981.

———. "The Older Anhui Masters." In *Shadows of Mt. Huang: Chinese
Painting and Printing of the Anhui School*, ed. James Cahill, pp. 67–
75. Berkeley, Calif.: University Art Museum, 1981.

———. *The Painter's Practice: How Artists Lived and Worked in Traditional China*. New York: Columbia University Press, 1994.

———. "Yuan Jiang and His School," pts I and II. *Ars Orientali* 5
(1963): 259–72; 6 (1967): 191–212.

Cahill, James, ed. *Shadows of Mt. Huang: Chinese Painting and Printing
of the Anhui School*. Berkeley, Calif.: University Art Museum, 1981.

Cai Guihua 蔡貴華. "Yangzhou Meihua shuyuan kao" 揚州梅花書
院考 (A study of Yangzhou's Meihua Academy). In *Yangzhou*

yanjiu: Jiangdu Chen Yiqun bailing mingdan jinian lunwenji 揚州研究: 江都陳軼群百齡冥誕紀念論文集 (Research on Yangzhou: collected essays to commemorate the hundredth anniversary of the birth of Chen Yiqun of Jiangdu), by Feng Erkang 馮爾康 et al., pp. 373–87. Taibei: Chen Jiexian, 1996.

Cao Xun 漕汛. "Ji Cheng yanjiu: wei jinian Ji Cheng dansheng sibai zhounian er zuo" 計成研究:爲紀念計成誕生四百周年而作 (A study of Ji Cheng: written in commemoration of the four hundredth anniversary of Ji Cheng's birth). *Jianzhu shi* 13 (1982): 1–16.

Carles, W. R. "The Grand Canal of China." *Journal of the Royal Asiatic Society* (China Branch) 31 (1896–97): 102–10.

Carlitz, Katherine. "The Social Uses of Female Virtue in Late Ming Editions of *Lienü Zhuan*." *Late Imperial China* 12, no. 2 (Dec. 1991): 117–52.

———. "Style and Suffering in Two Stories by 'Langxian.'" In *Culture and State in Chinese History: Conventions, Accommodation, and Critiques*, ed. Theodore Huters, R. Bin Wong, and Pauline Yu, pp. 207–35. Stanford: Stanford University Press, 1997.

Catalogue of the Exhibition of Individualists and Eccentrics: The Mr. and Mrs. R. W. Finlayson Collection of Chinese Paintings. Toronto: Royal Ontario Museum and University of Toronto; Seattle: Seattle Art Museum, 1963–64.

Chang, Sen-dou. "The Morphology of Walled Capitals." In *The City in Late Imperial China*, ed. G. W. Skinner, pp. 75–100. Stanford: Stanford University Press, 1977.

Chaves, Jonathan. "Moral Action in the Poetry of Wu Chia-chi (1618–1684)." *Harvard Journal of Asiatic Studies* 46, no. 1 (1986): 387–469.

Chen Congzhou 陳從周. *Shuo yuan* 説園 (Speaking of gardens). Shanghai: Tongji daxue chubanshe, 1984.

———. "Yangzhou yuanlin de yishu tese" 揚州園林的藝術特色 (The special characteristics of Yangzhou gardens). *Jianghai xuekan* 1962, no. 9 (Sept.): 45–49.

———. *Yuanlin tancong* 園林談叢 (Collected talks on gardens). Shanghai: Shanghai wenhua chubanshe, 1985.

Chen Lifang and Yu Sanglin. *The Garden Art of China*. Portland: Timber Press, 1986.

Chen Qubing 陳去病. *Wu shi zhi* 五石脂 (Five-mineral ointment). In Gu Gongxie 顧公燮, Anon., and Chen Qubing, *Danwu biji, Wucheng riji, Wu shi zhi* 丹午筆記, 吳城日記, 五石脂 (Occasional writings by Danwu; Suzhou diary; Five-mineral ointment). *Jiangsu difang wenxian congshu* 江蘇地方文獻叢書. Nanjing: Jiangsu guji chubanshe, 1999.

Chen Wannai 陳萬鼐. *Kong Shangren yanjiu* 孔尚壬研究 (Study of Kong Shangren). Taibei: Shangwu yinshuguan, 1971.

Chen Zhoushi 陳舟士. *Tianxia lucheng* 天下路程 (Routes of the empire). Jian'an: Bentang cangban, 1741.

Cheng Mengxing 程夢星. *Pingshantang xiaozhi* 平山堂小志 (Little gazetteer of Pingshan Hall). Wang Family Printers, 1752.

Cheung, Anthony, and Paul Gurofsky, eds. and trans. *Cheng Pan-ch'iao: Selected Poems, Calligraphy, Paintings and Seal Engravings.* Hong Kong: Joint Publishing, 1987.

Chou, Ju-hsi, and Claudia Brown. *The Elegant Brush: Chinese Painting Under the Qianlong Emperor.* Phoenix, Ariz.: Phoenix Art Museum, 1985.

Chu, Samuel C. *Reformer in Modern China: Chang Chien, 1853–1926.* New York: Columbia University Press, 1965.

Clunas, Craig. *Fruitful Sites: Garden Culture in Ming Dynasty China.* London: Reaktion Books, 1996.

Cohen, Monique, and Nathalie Monnet. *Impressions de Chine.* Paris: Bibliothèque Nationale, 1992.

Collis, Maurice. *The Great Within.* London: Faber and Faber, 1941.

Contag, Victoria. *Chinese Masters of the 17th Century.* Trans. Michael Bullock. London: Lund Humphries, 1969.

Cressey, George Babcock. *China's Geographic Foundations: A Survey of the Land and Its People.* New York: McGraw-Hill, 1934.

Crochet, Jules. "Histoire d'une Chrétienté." Archives [jésuites] de la Province de France, Vanves. FCH 370.

Crossley, Pamela Kyle. *The Manchus.* Cambridge, Mass.: Blackwell, 1997.

———. *A Translucent Mirror: History and Identity in Qing Imperial Ideology.* Berkeley: University of California Press, 1999.

da Cruz, Gaspar. "Treatise in Which the Things of China Are Related at Great Length." 1569. In *South China in the Sixteenth Century,* ed. C. R. Boxer, pp. 45–239. Nendeln, Liechtenstein: Kraus Reprint, 1967.

Dai Mingshi 戴名世. "Yangzhou chengshou jilüe" 揚州城守紀略 (Brief record of the defense of Yangzhou). In *Liangjie guazhong Shi Kefa* 諒解孤忠史可法 (The outstanding loyalist Shi Kefa), ed. Yangzhou wenwu yanjiu shi, Shi Kefa jinian guan 揚州文物研究室, 史可法紀念館, pp. 44–53. Yangzhou: Jiangsu wenyi chubanshe, 1993.

Dan Shumo 單樹模 and Wen Pengling 文朋陵. "'Kangxi liunian Jiangsu jian sheng' shuo queqie wuwu" "康熙六年江蘇建省"說確切無誤 (There is absolutely no mistake in the statement "Jiangsu province was established in the sixth year of the Kangxi reign"). *Jiangsu difang zhi* 6 (1990): 35–37.

Dan Shumo 單樹模, ed. *Zhonghua renmin gongheguo diming cidian: Jiangsu sheng* 中華人民共和國地名詞典：江蘇省 (Dictionary of place-names in the People's Republic of China: Jiangsu province). Beijing: Shangwu yinshuguan, 1987.

Dan Yizi 憺漪子. *Shishang yaolan: Tianxia shuilu lucheng* 士商要覽：天下水陸路程. Suzhou(?) n.d. (Ming dynasty).

Daqing lichao shilu 大清歷朝實錄 (Veritable records of the Qing dynasty). Taibei, 1965.

Davis, Richard L. *Wind Against the Mountain: The Crisis of Politics and Culture in Thirteenth-Century China.* Cambridge, Mass.: Harvard University, Council on East Asian Studies, 1996.

De Bary, Wm. Theodore D. "Individualism and Humanitarianism in Late Ming Thought." In *Self and Society in Ming Thought*, ed. Wm. Theodore De Bary and the Conference on Ming Thought, pp. 145–245. New York: Columbia University Press, 1970.

de Rada, Martin. "The Relation of Fr. Martin de Rada, O.E.S.A." 1575. In *South China in the Sixteenth Century*, ed. C. R. Boxer, pp. 241–310. Nendeln, Liechtenstein: Kraus Reprint, 1967.

Diény, J. P., trans. *Les "Lettres Familiales" de Tcheng Pan-k'iao.* Mélanges publiées par l'Institut des hautes études chinoises, vol. 2. Paris: Presses universitaires de France, 1960.

Dodgen, Randall A. "Hydraulic Evolution and Dynastic Decline: The Yellow River Conservancy, 1796–1855." *Late Imperial China* 12, no. 2 (Dec. 1991): 36–63.

Du Fuweng 杜負翁. "Shitao yu Yangzhou" 石濤與揚州 (Shitao and Yangzhou), pt. 1. *Jiangsu wenxian* 1975, no. 6 (Aug.): 4–6.

Du Halde, J. B. *The General History of China.* London: J. Watts, 1741.

Du Jinghua 杜景華. "Chai Shijin tousong citie an" 柴世進投送詞貼案 (The case of Chai Shijin presenting sheets of poetry). In *Qingdai wenzi yuan* 清代文字獄案 (Literary prosecutions of the Qing era), ed. Zhang Shucai 張書才 and Du Jinghua 杜景華, pp. 122–25. Beijing: Zijincheng chubanshe, 1991.

Eberhard, Wolfram. "What Is Beautiful in a Chinese Woman?" In *Moral and Social Values of the Chinese: Selected Essays*, by Wolfram Eberhard, pp. 271–304. Taibei: Chengwen, 1971.

Eisenstadt, S. N. *The Political System of Empires.* New York: Free Press, 1969.

The Elegant Brush: Chinese Painting Under the Qianlong Emperor. Hong Kong: The Council, 1986.

Elliott, Mark C. *The Manchu Way: The Eight Banners and Ethnic Identity in Late Imperial China*. Stanford: Stanford University Press, 2001.

Elman, Benjamin. *From Philosophy to Philology: Intellectual and Social Aspects of Change in Late Imperial China*. Cambridge, Mass.: Harvard University, Council on East Asian Studies, 1984.

Elvin, Mark. *The Pattern of the Chinese Past*. London: Eyre Methuen, 1973.

Ennin's Diary: The Record of Pilgrimage in Search of the Law. Trans. and ed. Edwin O. Reischauer. New York: Ronald Press, 1955.

Fan Changjiang 范長江. "You Yangzhou" 游揚州 (Traveling to Yangzhou). *Nanjing shifan xueyuan, Wenjiao ziliao jianbao* 1979, no. 5: 52.

Fang, Achilles, trans. *The Chronicle of the Three Kingdoms*. Cambridge, Mass.: Harvard Univerity Press, 1952.

Fang Chen 方晨. "Yangzhou Qing muke nianhua" 揚州清代木刻年畫 (Qing woodblock New Year's prints from Yangzhou). *Zhongguo minjian gongyi*, nos. 13–14 (June 1994): 85–88.

Fang Xiangying 方象瑛. *Xiuyuan ji* 休園記 (Memoir of the Garden of Retirement). In *Yangzhou fuzhi* 揚州府志 (1810), 31.55b–56b.

Fang Yujin 方裕僅. "Daoguang chunian Lianghuai siyan yanjiu" 道光初年兩淮私鹽研究 (Study of the private salt trade in Lianghuai in the early years of the Daoguang era). *Lishi dang'an* 1998, no. 4: 80–89.

Fang Yujin 方裕僅, ed. "Daoguang chunian Chu an yanchuan fenglun sanmai shiliao" 道光初年楚岸鹽船封輪散賣史料 (Historical materials on the "seal and rotate" and dispersed purchase [methods of salt sales from] the salt boats at the Hunan entrepôt), pt. 1. *Lishi dang'an* 1991, no. 1: 37–48.

Fang Zhuofen, Hu Tiewen, Jian Rui, and Fang Xing. "Hedong Lake Salt and Huainan Sea Salt." In *Chinese Capitalism, 1522–1840*, ed. Xu Dixin and Wu Chengming; ed. and abridged Peter Curwen, pp. 347–71. Basingstoke, Eng.: Macmillan, 1999.

Fei Xiaotong (Fei Hsiao Tung) 費孝通. "Small Towns in Northern Jiangsu." In *Small Towns in China: Functions and Propsects*, by Fei Hsiao Tung, pp. 88–132. Beijing: New World Press, 1986.

———. "Xiao chengzhen: Subei chutan" 小城鎮: 蘇北初談 (Small towns: a preliminary inquiry into Subei). In *Xiao chengzhen si ji* 小城鎮四記 (Four accounts of small towns), by Fei Xiaotong, pp. 75–114. Beijing: Xinhua chubanshe, 1985.

Feng Erkang 馮爾康 et al. *Yangzhou yanjiu: Jiangdu Chen Yiqun bailing mingdan jinian lunwenji* 揚州研究: 江都陳軼群百齡冥誕紀念論文集 (Research on Yangzhou: collected essays to commemorate the hundredth anniversary of the birth of Chen Yiqun of Jiangdu). Taibei: Chen Jiexian, 1996.

Feng Zikai 豐子愷. "Yangzhou meng" 揚州夢 (Dream of Yangzhou). In *Yuanyuantang suibiji* 緣緣堂隨筆集 (Collected random jottings from Yuanyuan Hall), by Feng Zikai, pp. 378–83. Hangzhou: Zhejiang wenyi chubanshe, 1983.

Feuchtwang, Stephen. "School-Temple and City God." In *The City in Late Imperial China*, ed. G. William Skinner, pp. 581–608. Stanford: Stanford University Press, 1977.

Finnane, Antonia. "Bureaucracy and Responsibility: A Reassessment of the River Administration Under the Qing." *Papers in Far Eastern History* (Australian National University, Department of Far Eastern History), no. 30 (Sept. 1984): 161–98.

———. "The Origins of Prejudice: The Malintegration of Subei in Late Imperial China." *Comparative Studies in Society and History* 35, no. 2 (Apr. 1993): 216–17.

———. "A Place in the Nation: Yangzhou and the Idle Talk Controversy of 1934." *Journal of Asian Studies* 53, no. 4 (Nov. 1994): 1150–74.

———. "Yangzhou: A Central Place in the Qing Empire." In *Cities of Jiangnan in Late Imperial China*, ed. Linda Cooke Johnson, pp. 117–49. Albany: State University of New York Press, 1993.

Fisher, Tom. "Loyalist Alternatives in the Early Ch'ing." *Harvard Journal of Asiatic Studies* 44, no. 1 (June 1984): 83–122.

Fu Zehong 傅澤洪, ed. *Xingshui jinjian* 行水金鑒 (Golden mirror of water management). 1725. Reprinted — Shanghai: Shangwu yinshuguan, 1936.

Fujii Hiroshi 藤井宏. "Mingdai yanshang de yi kaocha — bianshang, neishang, shuishang de yanjiu" 明代鹽商的一考察 — 邊商, 內商, 水商的研究 (An inquiry into salt merchants in the Ming dynasty: a study of frontier merchants, inland merchants, and waterway merchants). Trans. Liu Sen 劉森. In *Huizhou shehui jingji shi yanjiu yiwenji* 徽州社會經濟史研究譯文集 (Translated studies in the social and economic history of Huizhou), ed. Liu Sen, pp. 252–346. Hefei: Huangshan shushe, 1988.

Furth, Charlotte. *A Flourishing Yin: Gender in China's Medical History, 960–1665*. Berkeley: University of California Press, 1999.

Fushe xingshi 復社性氏 (Members of the Resoration Society). Anonymous and undated manuscript. Microfilm copy of an original held in the Beiping Rare Books Collection in the National Library, Taibei.

Gadoffre, F. "Le Pays des canaux: essai sur la province du Kiangsou." *Revue de Géographie* 50, no. 9 (Mar. 1902): 218–37.

Gandar, Dominic. *Le Canal impérial: étude historique et descriptive.* Variétés sinologiques, no. 4. Shanghai, 1894.

Gao Jin 高晉, ed. *Nanxun shengdian* 南巡盛典 (The imperial grace of the southern tours) [1771]. Reprinted—Yonghezhen: Wenhai chubanshe, 1971.

Giacalone, Vita, with Ginger Cheng-chi Hsü. *The Eccentric Painters of Yangzhou.* New York: China House Gallery, China Institute in America, 1990.

Gong Zizhen 龔自珍. *Gong Zizhen quanji* 龔自珍全集 (Complete works of Gong Zizhen). Shanghai: shanghai renmin chubanshe, 1975.

Goodman, Bryna. *Native Place, City, and Nation: Regional Networks and Identities in Shanghai, 1853–1937.* Berkeley: University of California Press, 1995.

Goodrich, L. Carrington, ed. *Dictionary of Ming Biography.* New York: Columbia University Press, 1976.

Gui Chaowan 桂超萬. *Huanyou jilüe* 宦遊紀略 (Brief account of official postings). Taibei: Guangwen shuju, 1972.

Guinness, Geraldine (Mrs. Howard Taylor). *In the Far East: Letters from Geraldine Guinness in China.* Melbourne: China Inland Mission, 1901.

Guy, R. Kent. *The Emperor's Four Treasuries: Scholars and the State in the Late Ch'ien-lung Era.* Cambridge, Mass.: Harvard University, Council on East Asian Studies, 1987.

Hahn, Emily. *China to Me.* Boston: Beacon Press, 1988.

Hanan, Patrick. "*Fengyue Meng* and the Courtesan Novel." *Harvard Journal of Asiatic Studies* 58, no. 2 (Dec. 1998): 345–72.

Hanshang mengren 邗上蒙人. *Fengyue meng* 風月夢 (Dreams of wind and moon). Preface dated 1848. Beijing: Beijing daxue chubanshe, 1988.

Hay, Jonathan S. *Shitao: Painting and Modernity in Early Qing China.* New York: Cambridge University Press, 2001.

———. "Shitao's Late Work (1679–1707): A Thematic Map." Ph.D. diss., Yale University, 1989.

He Bingdi, *see* Ho Ping-ti.

Hechter, Michael. *Internal Colonialism: The Celtic Fringe in British National Development, 1536–1966.* Berkeley: University of California Press, 1977.

Hershatter, Gail. *Dangerous Pleasures: Prostitution and Modernity in Twentieth-Century Shanghai.* Berkeley: University of California Press, 1997.

Hinton, Harold C. *The Grain Tribute System of China*. Cambridge, Mass.: Harvard University, Chinese Economic and Political Studies, 1956.

Ho, Clara Wing-chung. "Encouragement from the Opposite Gender: Male Scholars' Interests in Women's Publications in Ch'ing China—A Bibliographical Study." In *Chinese Women in the Imperial Past: New Perspectives*, ed. Harriet T. Zurndorfer, pp. 308-53. Leiden: Brill, 1999.

Ho Ping-ti (He Bingdi 何炳棣). *The Ladder of Success in Imperial China: Aspects of Social Mobility*. New York: Science Editions, 1964.

———. "The Salt Merchants of Yang-chou: A Study of Commercial Capitalism in Eighteenth-Century China." *Harvard Journal of Asiatic Studies* 17 (1954): 130-68.

———. *Studies in the Population of China*. Cambridge, Mass.: Harvard University Press, 1959.

———. *Zhongguo huiguan shilun* 中國會館史論 (On the history of Chinese native-place associations). Taibei: Xuesheng shuju, 1966.

Ho Wai-kam. "The Literary Gathering at a Yangzhou Garden." In *Eight Dynasties of Chinese Painting: The Collections of the Nelson Gallery–Atkins Museum, Kansas City, and the Cleveland Museum of Art*, ed. Sally W. Goodfellow, pp. 363-66. Cleveland, Ohio: Cleveland Museum of Art, 1980.

Hohenberg, Paul M., and Lynn Hollen Lees. *The Making of Urban Europe, 1000-1950*. Cambridge, Mass.: Harvard University Press, 1985.

Honig, Emily. *Creating Chinese Ethnicity: Subei People in Shanghai, 1850-1980*. New Haven: Yale University Press, 1992.

———. "The Politics of Prejudice: Subei People in Republican-Era Shanghai." *Modern China* 15, no. 3 (1989): 243-74.

———. "Pride and Prejudice: Subei People in Contemporary Shanghai." In *Unofficial China: Popular Culture and Thought in the People's Republic*, ed. Perry Link, Richard Madsen, and Paul G. Pickowicz, pp. 138-55. Boulder, Colo.: Westview Press, 1989.

Hou Hanshu (Later Han history). Beijing: Zhonghua shuju, 1971.

Hsiao, Ch'i-ch'ing. *The Military Establishment of the Yuan Dynasty*. Cambridge, Mass.: Harvard University, Council on East Asian Studies, 1978.

Hsieh, Bao-Hua. "The Acquisition of Concubines in China, 14–17th Centuries." *Jindai Zhongguo funü shi yanjiu* 1993, no. 1 (June): 125-200.

Hsieh, Winston. "Triads, Salt Smugglers and Local Uprisings." In *Popular Movements and Secret Societies in China, 1840–1950,* ed. Jean Chesneaux, pp. 144–64. Stanford: Stanford University Press, 1972.

Hsü, Ginger Cheng-chi. *A Bushel of Pearls: Painting for Sale in Eighteenth-Century Yangchow.* Stanford: Stanford University Press, 2001.

Hu, Ch'ang-tu. "The Yellow River Administration in the Ch'ing Dynasty." *Far Eastern Quarterly* 14, no. 4 (Aug. 1955): 505–13.

Hu Puan 胡朴安. *Zhonghua quanguo fengsu zhi* 中華全國風俗志 (All-China gazetteer of local customs). 1922. Zhengzhou: Zhongzhou guji chubanshe, 1990.

Hu Wenkai 胡文楷. *Lidai funü zhuzuo kao* 歷代婦女著作考 (Study of women writers by historical period). Shanghai: Shangwu yinshuguan, 1957.

Huang Junzai 黃鈞宰. "Jinhu langmo" 金壺浪墨 (Desultory writings from the golden inkpot). In idem, *Jinhu qimo* 金壺七墨 (Seven writings from the golden inkpot). Shanghai: Saoye shanfang, 1929.

Huang, Martin. *Literati and Self-Re/Presentation: Autobiographical Sensibility in the Eighteenth-Century Chinese Novel.* Stanford: Stanford University Press, 1995.

Huang, Ray. *Taxation and Governmental Finance in Sixteenth-Century Ming China.* London and New York: Cambridge University Press, 1974.

Huangchao jingshiwen tongbian 皇朝經世文統編 (Complete edition of statecraft writings of the Qing), ed. Runpu 潤浦. Shanghai: Baoshanzhai, 1901.

Huangchao Zhong wai yitong yutu 皇朝中外壹統輿圖 (Qing atlas: China and foreign places), comp. Zou Shiyi 鄒世詒, Yan Qizhen 宴起鎮, and Yan Shusen 嚴樹森. Wuchang(?): Hubei fushu Jinghuanlou, 1863.

Hummel, Arthur, ed. *Eminent Chinese of the Ch'ing Period.* Washington, D.C.: U.S. Government Printing Office, 1943.

Hung, Chang-tai. 1991. "Paper Bullets: Fan Changjiang and New Journalism in Wartime China." *Modern China* 17, no. 4 (Oct. 1991): 427–68.

Jang, Scarlett. "Cheng Sui." In *Shadows of Mt. Huang: Chinese Painting and Printing of the Anhui School,* ed. James Cahill, pp. 11–12. Berkeley: University Art Museum, 1981.

Ji Chao 嵇超. "Fangong ti de xingzhu ji qi zuoyong" 范公的興築堤及其作用 (The construction and use of the Lord Fan Dyke). *Fudan xuebao,* no. 8 (1980): 59–61.

Ji Cheng. *The Craft of Gardens*. Trans. Alison Hardie. New Haven: Yale University Press, 1988.

Ji Zhongqing 紀中慶. "Yangzhou guchengzhi bianqian chutan" 揚州古城址變遷初探 (Preliminary comments on changes in the foundations of the ancient wall of Yangzhou). *Wenwu* 1979, no. 9: 43–56.

Jiang Hua 蔣華. "Yangzhou gang yu Bosi wenhua de jiaoliu" 揚州港與波斯文化的交流 (Cultural exchanges between Persia and the port of Yangzhou). In *Yisilan jiao zai Yangzhou* 伊斯蘭教在揚州 (Islam in Yangzhou), ed. Wei Peichun 韋培春, pp. 73–79. Nanjing: Nanjing daxue, 1991.

Jiao Xun 焦循. *Yangzhou Beihu xiaozhi* 揚州北湖小志 (Little gazetteer of Yangzhou's North Lakes). Yangzhou, 1808.

———. *Yangzhou zuzheng lu* 揚州足徵錄 (Verified account of Yangzhou). Yangzhou, 1815. Reprinted in Beijing tushuguan guji zhenben congkan, 25. Beijing: Sumu wenxian chubanshe, ca. 1988.

Jiao Xun 焦循 and Jiang Fan 江藩. *Jiaqing Yangzhoufu tujing* 嘉慶揚州府圖經 (A survey of Yangzhou prefecture, [written in the] Jiaqing reign). Reprinted—Yangzhou: Jiangsu Guangling guji keyinshe, 1981.

Jiaodong Zhou Sheng 焦東周生. *Yangzhou meng* 揚州夢 (Dream of Yangzhou). Taibei: Shijie shuju, 1978.

Jin Fu 靳輔. *Jin Wenxiang gong Zhihe fanglüe* 靳文襄公治河方略 (Duke Jin Wenxiang's "Strategy for River Control"). 1867. Reprinted—Taibei: Wenhai chubanshe, n.d.

Jin Zhi 金埴. *Bu xia dai bian* 不下帶編 (Informal compositions). Beijing: Zhonghua shuju, 1982.

Jinshen quanshu 縉紳全書 (Complete book of official appointments). 1855 ed.

Jiu Tangshu 舊唐書 (Old history of the Tang). Beijing: Zhonghua shuju, 1973.

Johnson, Linda Cooke. *Shanghai: An Emerging Jiangnan Port*. Stanford: Stanford University Press, 1994.

Johnston, R. Stewart. *Scholar Gardens of China*. Cambridge, Eng.: Cambridge University Press, 1991.

Jones, Susan Mann, see Mann, Susan.

Kao, Mayching, ed. *Paintings by Yangzhou Artists of the Qing Dynasty from the Palace Museum*. Beijing: Palace Museum; Hong Kong: Art Gallery, Chinese University of Hong Kong, 1985.

Keenan, Barry C. *Imperial China's Last Classical Academies: Social Change in the Lower Yangzi, 1864–1911*. Berkeley: University of California, Institute of East Asian Studies, 1994.

Kennelly, M., S.J., trans. and rev. *L. Richard's Comprehensive Geography of China*. Shanghai: T'suwei Press, 1908.

Kim, Hongnam. "Zhou Liang-kung and His *Tu-hua-lu* Painters." In *Artists and Patrons: Some Social and Economic Aspects of Chinese Painting*, ed. Chu-Tsing Li, pp. 189–201. Lawrence: University of Kansas, Kress Foundation Department of Art History; Nelson-Atkins Museum of Art, with University of Washington Press, 1989.

Kinkley, Jeffrey. *The Odyssey of Shen Congwen*. Stanford: Stanford University Press, 1987.

Kinnane, Garry. *George Johnston: A Biography*. Melbourne: Nelson, 1986.

Ko, Dorothy. "Bondage in Time: Footbinding and Fashion Theory." *Fashion Theory* 1, no. 1 (1997): 3–27.

———. *Teachers of the Inner Chambers: Women and Culture in Seventeenth-Century China*. Stanford: Stanford University Press, 1994.

Kong Shangren (K'ung Shangren) 孔尚任. *Huhai ji* 湖海集 (Collection from the lakes and seas). Shanghai: Gudian wenxue chubanshe, 1957.

———. *Kong Shangren shiwenji* 孔尚任詩文集 (Collected writings of Kong Shangren), ed. Wang Weilin 汪蔚林. Beijing: Zhonghua shuju, 1962.

———. *The Peach Blossom Fan* (T'ao-hua-shan). Trans. Chen Shih-hsiang and Harold Acton. Berkeley: University of California Press, 1976.

———. *Taohua shan* 桃花扇 (Peach Blossom Fan). Ed. Wang Jisi 王季思 and Su Huanzhong 蘇寰中. Beijing: Renmin wenxue chubanshe, 1959.

Kuhn, Philip. *Soulstealers: The Chinese Sorcery Scare of 1768*. Cambridge, Mass.: Harvard University Press, 1990.

Lamouroux, Christian. "From the Yellow River to the Huai: New Representations of a River Network and the Hydraulic Crisis of 1128." In *Sediments of Time: Environment and Society in Chinese History*, ed. Mark Elvin and Liu Ts'ui-jung, pp. 545–84. New York: Cambridge University Press, 1998.

Legge, James. *The Chinese Classics*, vol. 3, *The Shoo King*. Reprinted — Hong Kong: Hong Kong University Press, 1960.

———. *The Chinese Classics*, vol. 5, *The Ch'un Tsew with the Tso Chuen*. Reprinted — Hong Kong: Hong Kong University Press, 1960.

———. *The Four Books*. Reprinted — Taibei: Culture Book, 1973.

Legouix, Susan. *Image of China: William Alexander*. London: Jupiter Books, 1980.

Leith, James A. *The Idea of Art as Propaganda in France, 1750–1799: A Study in the History of Ideas*. Toronto: University of Toronto Press, 1965.

Leonard, Jane Kate. *Wei Yuan and China's Rediscovery of the Maritime World*. Cambridge, Mass.: Harvard University, Council on East Asian Studies, 1984.

Leung Yuen-sang. *The Shanghai Taotai*. Singapore: Singapore University Press, 1987.

Lewis, Bernard. *The Arabs in History*. New York: Harper and Brothers, 1960.

Li Bozhang. "Changes in Climate, Land, and Human Efforts: The Production of Wet-Field Rice in Jiangnan During the Ming and Qing Dynasties." In *Sediments of Time: Environment and Society in Chinese History*, ed. Mark Elvin and Liu Ts'ui-jung, pp. 447–84. New York: Cambridge University Press, 1998.

Li Changfu 李長傅, ed. *Jiangsu sheng dizhi* 江蘇省地志 (Geographical gazetteer of Jiangsu province). Shanghai: Zhonghua shuju, 1936.

Li Cheng 李澄. *Huaizuo beiyao* 淮鹺備要 (Essentials of Huai salt). 1823.

Li Dou 李斗. *Yangzhou huafang lu* 揚州畫舫錄 (Chronicle of the painted barques of Yangzhou). Yangzhou: Jiangsu Guangling guji keyinshe, 1984.

Li Huan 李桓. *Guochao qixian leizheng* 國朝耆獻類徵 (Historical materials on scholars of the Qing). Taibei: Wenhai chubanshe, 1966.

Li Ju-chen (Li Ruzhen). *Flowers in the Mirror*. Trans. and ed. Lin Tai-yi. London: Peter Owen, 1965.

Li Ke 李珂. "Mingdai kaizhongzhi xia shang zao gouhsiao guanxi tuojie wenti zai tan" 明代開中制下商灶購銷關係脫節問題再談 (A further inquiry into the problem of the dislocation of the trading relationship between merchants and salters under the Ming dynasty's "salt for grain" system). *Lishi dang'an* 1992, no. 4: 33–39.

Li Linqi 李琳琦. "Ming Qing Huizhou liangshang shulun" 明清徽州糧商述論 (Discussion of the Huizhou grain merchants of the Ming-Qing period). *Jianghuai luntan* 1993, no. 4: 73–78.

Li Longqian 李龍潛. "Mingdai yan de kaizhongfa yu yanshang ziben de fazhan" 明代鹽的開中法與鹽商資本的發展 (The Ming dynasty's *kaizhong* system for salt and the growth of salt merchant capital). In *Ming Qing zibenzhuyi mengya yanjiu lunwen ji* 明清資本主義萌芽研究論文集 (Research on the sprouts of capitalism in the Ming and Qing dynasties), ed. Nanjing daxue, Lishixi, Ming Qing shi yanjiu shi 南京大學歷史系明清史研究室, pp. 498–537. Shanghai: Renmin chubanshe, 1981.

Li Shixu 李世序, ed. *Xuxingshui jinjian* 續行水金鑒 (Golden mirror of water management, continued). 1832. Reprinted—Shanghai: Shangwu yinshuguan, 1936.

Li Tingxian 李廷先. *Tangdai Yangzhou shikao* 唐代揚州史考 (Historical study of Tang Yangzhou). Yangzhou: Jiangsu guji chubanshe, 1992.

Liang Ch'i-ch'ao. *Intellectual Trends in the Ch'ing Period*. Trans. Immanuel C. Y. Hsü. Cambridge, Mass: Harvard University Press, 1959.

Liang Qizi 梁其姿. "Mingmo Qingchu minjian cishan huodong de xingqi: yi Jiang Zhe diqu wei li" 明末清初民間慈善活動的興起: 以江浙地區爲例 (The rise of popular philanthropy in the late Ming–early Qing: the example of Jiangsu and Zhejiang). *Shihuo yuekan*, supplement 15, no. 7–8 (Jan. 1986): 52–78.

———. *Shishan yu jiaohua: Ming Qing de cishan zuzhi* 施善與教化: 明清的慈善組織 (Charity and civilization: the organization of philanthropy in the Ming and Qing). Taibei: Lianjing, 1997.

Liang Shaoren 梁紹壬. *Liang ban qiuyüan suibi* 兩般秋雨盦隨筆 (Random notes from the Hut of Two Autumn Showers). Shanghai: Shanghai guji chubanshe, 1982.

Liang Zhangju 梁章鉅. *Guitian suoji* 歸田瑣記 (Collected fragments from my retirement). Taibei: Taiwan shangwu yinshuguan, 1976.

Lin Sumen 林蘇門. *Hanjiang sanbai yin* 邗江三百吟 (Three hundred sonnets from the Han River). Yangzhou, 1808.

Lin Xiuwei 林秀微, ed. *Yangzhou huapai* 揚州畫派 (The Yangzhou school of painters). Taibei: Yishu tushu gongsi, 1985.

Lin Yutang. *Translations from the Chinese (The Importance of Understanding)*. Cleveland, Ohio: Forum, 1960.

Liu Baonan 劉寶楠. *Baoying tujing* 寶應圖經 (Survey of Baoying). Yangzhou(?): 1848.

———. *Shengchao xun Yang lu* 勝朝殉揚錄 (Yangzhou martyrs of the vanquished dynasty). [Nanjing]: Huainan shuju, 1871.

Liu Binru 劉彬如 and Chen Dazuo 陳達祚. "Yangzhou 'Huihui tang' he Yuandai Alabowen de mubei" 揚州回回堂和元代阿拉伯文的墓碑 (The Yangzhou Mosque and Arabic tomb incriptions of the Yuan dynasty). *Jianghai xuekan* 1962, no. 2: 49–52.

Liu Jun 劉儁. "Daoguangchao Lianghuai feiyin gaipiao shimo" 道光朝兩淮廢引改票始末 (The abolition of the *yin* and change to the ticket [system] in Lianghuai during the Daoguang reign). *Zhongguo jindai jingjishi yanjiu jikan* 1, no. 2 (May 1933): 123–88.

Liu Rendao 劉人島, ed. *Zhongguo chuanshi renwu minghua quanji* 中國傳世人物名畫全集 (Collected famous figure paintings from China through the ages). 2 vols. Beijing: Zhongguo xiju chubanshe, 2001.

Liu Sen 劉森. "Huishang Bao Zhidao ji qi jiashi kaoshu" 徽商鮑志道及其家世考述 (Huizhou merchant Bao Zhidao and his family). *Jianghuai luntan* 1983, no. 3: 58–67.

Liu Sen 劉森, ed. *Huizhou shehui jingji shi yanjiu yiwenji* 徽州社會經濟史研究譯文集 (Translated studies in the social and economic history of Huizhou). Hefei: Huangshan shushe, 1988.

Liu Wenqi 劉文淇. *Yangzhou shuidao ji* 揚州水道記 (Record of Yangzhou's waterways). Huainan shuju, 1879.

Liu Yuxi 劉禹錫. "Wan bu Yangzi you nan wang sha wei" 晚步揚子游南望沙尾 (Evening stroll by the southward-flowingYangzi, gazing at where the sands die away). In *Quan Tang shi* 全唐詩, *juan* 355, vol. 6, p. 3992. Beijing: Zhonghua shuju.

Lo, Jung-pang. "The Controversy over Grain Conveyance During the Reign of Qubilai Qaqan, 1260–94." *Far Eastern Quarterly* 13, no. 3 (1954): 263–85.

Loewe, Michael. "Imperial Sovereignty: Dong Zhongshu's Contribution and His Predecessors." In *Foundations and Limits of State Power in China*, ed. Stuart Schram, pp. 33–57. London: School of Oriental and African Studies; Hong Kong: Chinese University Press, 1987.

Lopes, Robert S. "Nouveaux documents sur les marchands italiens en Chine a l'époque mongole." *Académie des insciptions et belles-lettres, comptes rendus des séances de l'année 1977*, Apr.–June 1977, pp. 454–67.

Lü Hao 盧浩, ed. *Yangzhou baguai huaji* 揚州八怪畫集 (Collected paintings of the eight eccentrics of Yangzhou). Nanjing: Jiangsu meishu chubanshe, 1990.

Lü Zuoxie 呂作燮. "Shilun Ming Qing shiqi huiguan de xingzhi he zuoyong" 試論明清時期會館的性質和作用 (Appraisal of the nature and functions of native-place associations in the Ming and Qing periods). In *Zhongguo zibenzhuyi mengya wenti lunwenji* 中國資本主義萌芽問題論文集 (Collected essays on the sprouts of capitalism in China), ed. Nanjing daxue, Lishixi, Ming Qing shi yanjiu shi 南京大學歷史系明清史研究室, pp. 172–211. Nanjing: Jiangsu renmin chubanshe, 1983.

Luo Weiwen 羅蔚文. "Shanguan ji jing yi zhu, linglong qishi you cun" 山館幾經易主, 玲瓏奇石猶存 (The Mountain Hall changes owners, the strange perforated rock still remains). *Yangzhou shiyuan xuebao (shehui kexue ban)* 1983, no. 3: 126–27.

Luo Zhenchang 羅振常. "Shi Kefa biezhuan" 史可法別傳 (Alternative biography of Shi Kefa). In *Liangjie guazhong Shi Kefa* 諒解孤忠 史可法 (The outstanding loyalist Shi Kefa), ed. Yangzhou wenwu yanjiu shi, Shi Kefa jinian guan 揚州文物研究室, 史可法紀念館, pp. 11–43. Yangzhou: Jiangsu wenyi chubanshe, 1993.

Lust, John. *Chinese Popular Prints.* Leiden: E. J. Brill, 1996.

Mackerras, Colin. *The Rise of the Peking Opera, 1770–1870: Social Aspects of the Theatre in Manchu China.* Oxford: Clarendon Press, 1972.

Maier, Charles S. "Consigning the Twentieth Century to History: Alternative Narratives for the Modern Era." *American Historical Review* 105, no. 3 (June 2000): 807–31.

Mann (Jones), Susan. "Historical Change in Female Biography from Song to Qing Times: The Case of Early Qing Jiangnan (Jiangsu and Anhui Provinces)." *Transactions of the International Conference of Orientalists in Japan* 30 (1985): 65–77.

———. *Local Merchants and the Chinese Bureaucracy, 1750–1950.* Stanford: Stanford University Press, 1987.

———. "The Ningpo *Pang* and Financial Power at Shanghai." In *The Chinese City Between Two Worlds*, ed. Mark Elvin and G. William Skinner, pp. 73–96. Stanford: Stanford University Press, 1974.

———. *Precious Records: Women in China's Long Eighteenth Century.* Stanford: Stanford University Press, 1997.

Mao, Lucien. "A Memoir of Ten Days' Massacre in Yangchow." *Tien-hsia Monthly* 4/5 (May 1937): 515–37.

Marmé, Michael. "Heaven on Earth: The Rise of Suzhou." In *Cities of Jiangnan in Late Imperial China*, ed. Linda Cooke Johnson, pp. 17–45. Albany: State University of New York Press, 1993.

Martini, Martino, S.J. *Novus atlas sinensis.* French ed. N.p.: Clausa Recludo(?), 1655.

Meng Shi. "Shi Tao's 'Brilliant Autumn in Huaiyang.'" *Chinese Literature*, Winter 1989, pp. 146–48.

Métailé, George. "Some Hints on 'Scholar Gardens' and Plants in Traditional China." *Studies in the History of Gardens and Designed Landscapes* 18, no. 3 (Autumn 1998): 248–56.

Metzger, Thomas. "The Organizational Capabilities of the Ch'ing State in the Field of Commerce: The Liang-huai Salt Monopoly, 1740–1840." In *Economic Organization in Chinese Society*, ed. W. E. Wilmott, pp. 9–45. Stanford: Stanford University Press, 1972.

———. "T'ao Chu's Reform of the Huai-pei Salt Monopoly (1831–1833)." *Papers on China* (Harvard University) 16 (1962): 1–39.

Meyer-Fong, Tobie. *Building Culture in Early Qing Yangzhou.* Stanford: Stanford University Press, 2003.

———. "Making a Place for Meaning in Early Qing Yangzhou." *Late Imperial China* 20, no. 1 (1999): 49–84.

Ming Qing like jinshi timing beilu 明清歷科進士提名碑錄 (Engraved record of *jinshi* scholars from Ming Qing examinations). Taibei: Huawen shuju, 1969.

Mingshi 明史 (History of the Ming). Beijing: Zhonghua shuju, 1974.

Mo Dongyin 莫東寅. *Manzushi luncong* 滿族史論叢 (Collected essays on Manchu history). 1958. Reprinted—Beijing: Sanlian shudian, 1979.

Mori Noriko 森紀子. "Yanchang de Taizhou xuepai" 鹽場的泰州學派 (The Taizhou school of the salt yards). Trans. Zheng Rong and Chen Bingshui. In *Taizhou xuepai xueshu taolunhui jinian lunwenji* 泰州學派學術討論會紀念 論文集 (Collected articles in commemoration of the conference on the scholarship of the Taizhou school), ed. Taizhou xuepai jinianguan choubeizu 泰州學派紀念館籌備組 and Taizhoushi zhengxie wenshi ziliao yanjiu weiyuanhui 泰州市政協文史委員會, pp. 46–77. Taizhou, 1987.

Moser, Leo J. *The Chinese Mosaic: The Peoples and Provinces of China.* Boulder, Colo.: Westview Press, 1985.

Mote, Frederick. "The Intellectual Climate in Eighteenth-Century China: Glimpses of Beijing, Suzhou and Yangzhou in the Qianlong Period." Special issue: *Chinese Painting Under the Qianlong Emperor*, vol. 1. *Phoebus* 6, no. 1 (1986): 17–55.

———. "The Transformation of Nanjing." In *The City in Late Imperial China*, ed. G. William Skinner, pp. 101–53. Stanford: Stanford University Press, 1977.

Mu Yiqin. "An Introduction to Painting in Yangzhou in the Qing Dynasty." In *Paintings by Yangzhou Artists of the Qing Dynasty from the Palace Museum*, ed. Mayching Kao, pp. 21–32. Beijing: Palace Museum; Hong Kong: Art Gallery, Chinese University of Hong Kong, 1985.

Nanjing bowuyuan 南京博物院. "Yangzhou gucheng 1978 nian diaocha fazhuo jianbao" 揚州古城 1978 年調查發掘簡報 (Brief report on the 1978 investigatory excavation of the ancient wall of Yangzhou). *Wenwu* 1979, no. 9: 33–42.

Naquin, Susan. *Peking: Temples and City Life, 1400–1900.* Berkeley: University of California Press, 2000.

Nelson, Susan E. "Rocks Beside a River: Ni Tsan and the Ching-kuan Style in the Eyes of Seventeenth Century Critics." *Archives of Asian Art* 32 (1980): 65–88.

North China Herald. Shanghai.

Oertling, Sewall, II. "Patronage in Anhui During the Wanli Period." In *Artists and Patrons: Some Social and Economic Aspects of Chinese Painting*, ed. Chu-tsing Li, pp. 165–75. Lawrence: University of Kansas, Kress Foundation Department of Art History, and Nelson-Atkins Museum of Art, with University of Washington Press, 1989.

Ono Kazuko 小野和子. "*Kyōka en* no sekai (Shinchō kōshō gakusha no yutopia zō)" 鏡花緣の世界（清朝考証学者のユウトピア像）(The world of *Destinies of Flowers in the Mirror*: utopian images from a Qing dynasty evidential scholar). *Shiso*, no. 721 (1982): 40–55.

Ōtani Toshio 大谷敏夫. "Yōshū Jōshū gakujutsu kō: sono shakaiteki kanren" 揚州常州学術考その社会的関連 (Study of the Yangzhou and Changzhou schools from point of view of their social relations). In *Min-Shin jidai no seiji to shakai* 明清時代の政治と社会, ed. Ono Kazuko 小野和子, pp. 313–45. Kyoto: Kyōto daigaku, Jinbun kagaku kenkyūjo, 1983.

Owen, Stephen. "Salvaging Poetry: The 'Poetic' in the Qing." In *Culture and State in Chinese History: Conventions, Accommodations, and Critiques*, ed. Theodore Huters, R. Bin Wong, and Pauline Yu, pp. 104–25. Stanford: Stanford University Press, 1997.

Pang, Mae Anna Quan. "Late Ming Painting Theory." In *The Restless Landscape: Chinese Painting of the Late Ming Period*, ed. James Cahill, pp. 22–28. Berkeley: University Art Museum, 1971.

Parsons, James B. *The Peasant Rebellions of the Late Ming Dynasty*. Tucson: University of Arizona Press, 1970.

Payne, Robert. *The White Pony: An Anthology of Chinese Poetry from Earliest Times to the Present Day*. London: Allen and Unwin, 1949.

Pemble, John. *Venice Rediscovered*. Oxford: Clarendon Press, 1995.

Perdue, Peter. *Exhausting the Earth: State and Peasant in Hunan, 1500–1850*. Cambridge, Mass.: Harvard University, Council on East Asian Studies, 1987.

———. "Water Control in the Dongting Lake Region During the Ming and Qing Periods." *Journal of Asian Studies* 41, no. 7 (Aug. 1982): 747–65.

Polachek, James. *The Inner Opium War*. Cambridge, Mass.: Harvard University, Council on East Asian Studies, 1992.

———. "Literati Groups and Literati Politics in Early Nineteenth Century China." Ph.D. diss., University of California, 1976.

Pollard, D. E., trans. "The Jades of Yangzhou." *Renditions* 40 (Autumn 1993): 160–62.

Pomeranz, Kenneth. *The Great Divergence: China, Europe, and the Making of the Modern World Economy*. Princeton: Princeton University Press, 2000.

————. *The Making of a Hinterland: State, Society and Economy in Inland North China, 1853–1937*. Berkeley: University of California Press, 1993.

Pruitt, Ida. *A Daughter of Han: The Autobiography of a Chinese Working Woman*. Stanford: Stanford University Press, 1967.

Public Record Office. China: F.O. Legation and embassy archives.

Qian Yiji 錢儀吉, comp. *Bei zhuan ji* 碑傳集 (Collected engraved biographies). Shanghai: Jiangsu shuju, 1893.

Qian Yong 錢永. *Lüyuan conghua* 履園叢話 (Collected chats from walking around gardens). Suzhou: Zhenxin shushe, 1870.

Qinding Daqing huidian shili 欽定大清會典事例 (Collected precedents of the Qing dynasty, by imperial command). 1899 ed. Taibei: Zhongwen shuju, n.d.

Qinding gongbu zeli 欽定工部則例 (Regulations of the Board of Works, by imperial command), comp. Weng Tonghe 翁同龢 [1875]. Reprinted—Taibei: Chengwen chubanshe, 1966.

Qinding Hubu caoyun quanshu 欽定戶部漕運全書 (The book of the grain transport, under the Board of Revenue, by imperial command). 1766 ed. Taibei: Chengwen chubanshe, 1969.

Qingchao tongdian 清朝通典 (Collected Qing statutes). Shitong 十通 ed. Shanghai: Zhonghua shuju, 1963.

Qingchao wenxian tongkao 清朝文獻通考 (Collected Qing documents). Shitong 十通 ed. Shanghai: Zhonghua shuju, 1963.

Qingchao yeshi daguan 清朝野史大觀 (Overview of unofficial histories of the Qing dynasty). Shanghai: Shanghai wenyi chubanshe, 1990.

Qingshi liezhuan 清史列傳 (Biographies from Qing history). Taibei: Zhonghua shuju, 1962.

Qiu Liangren 丘良任. "Yangzhou er Ma ji qi 'Xiao linglongshan guan tuji'" 揚州二馬及其"小玲瓏山館圖記" (The Ma brothers of Yangzhou and their "Note on the Picture of the Little Translucent Mountain Lodge"). *Yangzhou shiyuan xuebao (shehui kexue ban)* 1983, no. 3: 123–25.

Quan Hansheng 全漢昇. *Tang Song diguo yu yunhe* 唐宋帝國與運河 (The Tang-Song empire and the Grand Canal). Shanghai: Academia Sinica, 1946.

————. "Tang Song shidai Yangzhou jingji jingkuang de fanrong yu shuailuo" 唐宋時代揚州經濟景況的繁榮與衰落 (The flourishing and decline of Yangzhou's economic circumstances in the Tang and Song). *Guoli zhongyang yanjiuyuan, Lishi yuyan yanjiusuo jikan*, no. 11 (1947): 149–76.

Quan Zuwang 全祖望. "Meihualing ji" 梅花嶺記 (Record of Plum Blossom Hill). In *Liangjie guazhong Shi Kefa* 諒解孤忠史可法 (The outstanding loyalist Shi Kefa), ed. Yangzhou wenwu yanjiu shi, Shi Kefa jinian guan 揚州文物研究室, 史可法紀念館, pp. 211–12. Yangzhou: Jiangsu wenyi chubanshe, 1993.

Rawski, Evelyn Sakakida. *Education and Popular Literacy in Ch'ing China*. Ann Arbor: University of Michigan Press, 1979.

————. *The Last Emperors: A Social History of Qing Imperial Institutions*. Berkeley: University of California Press, 1998.

Relations de la mission de Nan-king confiée aux religieux de la Compagnie de Jésus. Shanghai: Imprimérie de la mission catholique, 1876.

Ren Zuyong 任祖鏞. "*Yuewei caotang biji* suo lu Ren Dachun yishi kao" "閱微草堂筆記" 所錄任大椿軼事考 (Study of anecdotes about Ren Dachun recorded in [Ji Yun's] *Yuewei Caotang Writings*). *Yangzhou shiyuan xuebao (shehui kexue ban)* 1983, no. 4: 118–22.

Roddy, Stephen J. *Literati Identity and Its Fictional Representations in Late Imperial China*. Stanford: Stanford University Press, 1998.

Rouleau, Francis A. "The Yangchow Latin Tombstone as a Landmark of Medieval Christianity in China." *Harvard Journal of Asiatic Studies* 17, nos. 3–4 (1954): 346–65.

Rowe, William T. *Hankow: Commerce and Society in a Chinese City, 1796–1889*. Stanford: Stanford University Press, 1984.

————. *Saving the World: Chen Hongmou and Elite Consciousness in Eighteenth-Century China*. Stanford: Stanford University Press, 2001.

Ruan Heng 阮亨. *Guangling mingshengtu* 廣陵名勝圖 (Illustrations of the famous sites of Yangzhou). Yangzhou(?), 1822.

Ruan Xian 阮先. *Yangzhou Beihu xuzhi* 揚州北湖續志 (Gazetteer of Yangzhou's North Lakes, continued). In *Yangzhou congke* 揚州叢刻, ed. Chen Henghe 陳恒和, fascicles 11–12. Jiangdu, 1936.

Ruan Yuan 阮元. *Dingxiangting bitan* 定香亭筆談 (Scribblings from Dingxiang Pavilion). 1800 ed. Reprinted—Taibei: Guangwen shu-ju, 1968.

————. *Guangling shishi* 廣陵詩事 (Matters concerning poetry in Guangling). Taibei: Wenyi yinshuguan, n.d.

————. *Huaihai yingling ji* 淮海英靈記 (Record of heroic spirits of Huai-Hai). 1798. Reprinted—Taibei: Taiwan shangwu yinshuguan, 1966.

————. "Ling mu Wang tairuren shou shi xu" 凌母王太孺人壽詩序 (Preface to birthday poems for Ling's mother, the Lady Wang). In idem, *Yanjingshi sanji* 研經室三集 (Third collection of writings from the Yanjing Studio), 5.2b. Yangzhou, 1842.

————. "*Yangzhou huafanglu* er ba" 揚州畫舫錄二跋 (Two afterwords to *Chronicle of the Painted Barques of Yangzhou*). In Li Dou 李斗,

Yangzhou huafang lu 揚州畫舫錄 (Chronicle of the painted barques of Yangzhou), pp. 7–8. Yangzhou: Jiangsu Guangling guji keyinshe, 1984.

Rudolph, Richard C. "A Second Fourteenth-Century Italian Tombstone in Yangchou." *Journal of Oriental Studies* (Hong Kong) 13 (1975): 133–37.

Saeki Tomi 佐伯富. *Shindai ensei no kenkyū* 清代塩政の研究 (Study of the Salt Administration under the Qing). Kyoto: Tōyōshi kenkyūkai, 1956.

———. "Yunshang de moluo he yanzheng de bihuai" 運商的沒落和鹽政的幣懷 (The decline of the transport merchants and corruption in the Salt Administration). Trans. Liu Sen 劉森. In *Huizhou shehui jingji shi yanjiu yiwenji* 徽州社會經濟史研究譯文集 (Translated studies of the social and economic history of Huizhou), ed. Liu Sen 劉森, pp. 368–416. Hefei: Huangshan shushe, 1988.

Schafer, E. H. *The Golden Peaches of Samarkand: A Study of T'ang Exotics*. Berkeley: University of California Press, 1963.

Scott, William Henry. "Yangzhou and Its Eight Eccentrics." In idem, *Hollow Ships on a Wine-Dark Sea and Other Essays*, pp. 53–68. Quezon City: New Day Publishers, 1976.

Shaanxi gudai daolu jiaotong shi 陝西古代道路交通史 (History of roads and communications in Shaanxi in premodern times). Beijing: Renmin chubanshe, 1989.

Shen Defu 沈德符. *Wanli yehuo bian* 萬歷野獲編 (Hunting in the wild: Wanli period). Beijing: Zhonghua shuju, 1959.

Shen Fu. *Chapters from a Floating Life: The Autobiography of a Chinese Artist*. Trans. Shirley Black. London: Oxford University Press, 1960.

Sheng Langxi 盛朗西. *Zhongguo shuyuan zhidu* 中國書院制度 (China's system of academies). Taibei: Huashi chubanshe, 1977.

Shi Dewei 史德威. *Weiyang xunjie jilüe* 維揚殉節紀略 (Overview of the fallen heroes of Yangzhou). 1812 ed. *Jieyue shanfang huichao* 借月山房彙鈔, comp. Zhang Haipeng 張海朋, no. 8. Taibei: Yishi shuju, 1968.

Shi Kefa pingjia wenti huibian: ji zi 1966 nian Shanghai Wenhuibao 史可法評價問題彙編：輯自 1966 年上海文匯報 (Compilation of criticisms of Shi Kefa: collected from the 1966 issues of the Wenhui newspaper, Shanghai). Hong Kong: Yangkai shubao gongyingshe, 1968.

Shiba, Yoshinobu. "Ningpo and Its Hinterland." In *The City in Late Imperial China*, ed. G. William Skinner, pp. 391–439. Stanford: Stanford University Press, 1977.

———. "Urbanization and the Development of Markets in the Lower Yangtze Valley." In *Crisis and Prosperity in Sung China*, ed. John Winthrop Haeger, pp. 13–48. Phoenix: University of Arizona Press, 1975.

Shimada Kenji. *Pioneer of the Chinese Revolution: Zhang Binglin and Confucianism*. Trans. Joshua A. Fogel. Stanford: Stanford University Press, 1990.

Skinner, G. William. "Cities and the Hierarchy of Local Systems." In *The City in Late Imperial China*, ed. G. William Skinner, pp. 275–351. Stanford: Stanford University Press, 1977.

———. "Introduction: Urban Social Structure in Ch'ing China." In *The City in Late Imperial China*, ed. G. William Skinner, pp. 521–53. Stanford: Stanford University Press, 1977.

———. "Regional Urbanization in Nineteenth-Century China." In *The City in Late Imperial China*, ed. G. William Skinner, pp. 211–49. Stanford: Stanford University Press, 1977.

Smith, Joanna E. Handlin. "Lü K'un's New Audence: The Influence of Women's Literature on Sixteenth-Century Thought." In *Women in Chinese Society*, ed. Margery Wolf and Roxane Witke, pp. 13–38. Stanford: Stanford University Press, 1975.

———. "Social Hierarchy and Merchant Philanthropy as Perceived in Several Late-Ming and Early-Qing Texts." *Journal of Economic and Social History of the Orient* 41, no. 3 (1998): 417–51.

So Kwan-wai. *Japanese Pirates in Ming China During the 16th Century*. East Lansing: Michigan State University Press, 1975.

Songshi 宋史 (History of the Song). Beijing: Zhonghua shuju, 1977.

Songshu 宋書 (History of the [Liu 劉] Song). Beijing: Zhonghua shuju, 1974.

Spence, Jonathan. "Opium." In idem, *Chinese Roundabout: Essays in History and Culture*, pp. 228–56. New York: Norton, 1992.

———. *Ts'ao Yin and the K'ang-hsi Emperor: Bondservant and Master*. New Haven: Yale University Press, 1966.

Staunton, Sir George. *An Authentic Account of an Embassy from the King of Great Britain to the Emperor of China*. 2 vols. London: W. Bulmer, 1797.

Strassberg, Richard E. *Inscribed Landscapes: Travel Writing from Imperial China*. Berkeley: University of California Press, 1994.

———. *The World of K'ung Shang-jen: A Man of Letters in Early Ch'ing China*. New York: Columbia University Press, 1983.

Struve, Lynn A. "Ambivalence and Action: Some Frustrated Scholars of the K'ang-hsi Period." In *From Ming to Ch'ing: Conquest, Region and Continuity in Seventeenth-Century China*, ed. Jonathan Spence and John E. Wills, pp. 323–65. New Haven: Yale University Press, 1979.

———. *The Southern Ming, 1644–1962*. New Haven: Yale University Press, 1984.

———. *Voices from the Ming-Qing Cataclysm: China in Tigers' Jaws*. New Haven: Yale University Press, 1993.

Suishu 隋書 (History of the Sui). Beijing: Zhonghua shuju, 1973.

Sun Xianjun 孫顯軍. "Gu jiaoyujia Hu Yuan" 古教育家胡瑗 (Hu Yuan, a teacher of antiquity). In *Yangzhou lidai renwu* 揚州歷代人物, ed. Wang Yu 王瑜, pp. 33–38. Yangzhou: Jiangsu guji chubanshe, 1992.

Sun Yatsen. *San Min Chu I: The Three Principles of the People*. Trans. F. W. Price. Chungking: Republic of China, Ministry of Information, 1943.

Suzuki Hiroyuki 鈴木博之. "Shindai Kishūfu no sōzoku to sonraku: Kiken no Kōson" 清代徽州府の宗族と村落: 徽県の江村 (Lineages and villages in Huizhou society during the Qing: Jiang village in Shexian). *Shigaku zasshi* 10, no. 4 (1992): 65–86.

Takino Shōjirō 瀧野正二郎. "Shindai Kenryū nenkan ni okeru kanryō to enshō (Ryōgai en'in an o chūshin to shite" 清代乾隆年間における官僚と鹽商 (兩淮鹽引を中心として) (The bureaucracy and the salt merchants in the Qing dynasty during the Qianlong reign: the Lianghuai salt case). *Kyūshū daigaku Tōyōshi ronji*, no. 15 (1986): 83–106.

———. "Shindai Waiankan no kōsei to kinō ni tsuite" 清代淮安関の構成と機能について (The structure and function of the Huai'an customs station in the Qing). *Kyūshū daigaku Tōyōshi ronji*, no. 14 (1985): 116–56.

Tanaka Masatoshi. "Rural Handicraft in Jiangnan." In *State and Society in China: Japanese Perspectives on Ming-Qing Social and Economic History*, ed. Linda Grove and Christian Daniels, pp. 79–100. Tokyo: University of Tokyo Press, 1984.

Tian Qiuye 田秋野 and Zhou Weiliang 周維亮. *Zhonghua yanye shi* 中華鹽業史 (A history of the salt industry in China). Taibei: Shangwu yinshuguan, 1979.

Tianran chisou 天然痴叟. "Jiangdushi xiaofu tu shen" 江都市孝婦屠身 (In the city of Jiangdu, a filial daughter-in-law has herself slaughtered). In idem, *Shi dian tou* 石點頭 (Rocks nodding heads), pp. 223–44. Jilin: Jinlin wenshi chubanshe, 1986.

Tong Jun 童寯. *Jiangnan yuanlin zhi* 江南園林志 (Gazetteer of Jiangnan gardens). Beijing: Zhongguo gongye chubanshe, 1963.

Tong, Te-kong, and Li Tsung-jen. *The Memoirs of Li Tsung-jen.* Boulder, Colo.: Westview Press, 1979.

Torbert, Preston M. *The Ch'ing Imperial Household Department: A Study of Its Organization and Principal Functions, 1662–1796.* Cambridge, Mass.: Harvard University, Council on East Asian Studies, 1977.

Tsang, Ka Bo. "Portraits of Hua Yan and the Problem of His Chronology." *Oriental Art* 28, no. 1 (1982): 64–79.

———. "The Relationships of Hua Yan and Some Leading Yangzhou Painters as Viewed from Literary and Pictorial Evidence." *Journal of Oriental Studies* (Hong Kong) 23 (1985): 1–28.

Twitchett, Denis. *Financial Administration Under the T'ang Dynasty.* Cambridge, Eng.: Cambridge University Press, 1970.

Vermeer, E. B. "P'an Chi-hsün's Solutions for the Yellow River Problems of the Later 16th Century." *T'oung Pao* 73, nos. 1–3 (1987): 33–67.

Wakeman, Frederic, Jr. *The Great Enterprise: The Manchu Reconstruction of Imperial Order in Seventeenth-Century China.* Berkeley: University of California Press, 1985.

Waley, Arthur, comp. and trans. *Chinese Poems.* London: George Allen and Unwin, 1976.

Waley-Cohen, Joanna. *The Sextants of Beijing: Global Currents in Chinese History.* New York: Norton, 1999.

Walker, Kathy Le Mons. *Chinese Modernity and the Peasant Path: Semicolonialism in the North Yangzi Delta.* Stanford: Stanford University Press, 1999.

Wang, David Der-wei. *Fin-de-Siècle Splendor: Repressed Modernities of Late Qing Fiction, 1849–1911.* Stanford: Stanford University Press, 1997.

Wang Fangzhong 王方中. "Qingdai qianqi de yanfa, yanshang, yu yanye shengchan" 清代前期的鹽法,鹽商,與鹽業生產 (The salt laws, salt merchants, and salt production in the first part of the Qing). *Qingshi luncong* 1982, no. 4: 1–48.

Wang Hong 王鴻. *Lao Yangzhou: yanhua mingyue* 老揚州: 煙花明月 (Old Yangzhou: mist and flowers [beneath] a bright moon). Nanjing: Jiangsu meishu chubanshe, 2001.

Wang Huzhen 王湖楨 and Wu Weizu 吳慰祖, comps. *Qingdai hechen zhuan* 清代河臣傳 (Biographies of Qing river officials). Taibei: Wenhai chubanshe, 1970.

Wang Muhan 王慕韓. *Jiangsu yankenqu tudi liyong wenti zhi yanjiu* 江蘇鹽墾區土地利用問題之研究 (Research on the problems of land utilization in the reclaimed salt lands of Jiangsu). *Minguo ershi niandai Zhongguo dalu tudi wenti ziliao* 民國二十年代中國大陸土地問題資料 (Materials on land problems in the Chinese mainland in the third decade of the Republican era), no. 45. Taibei: Chengwen chubanshe, 1977.

Wang Shiqing 汪世清. "Dong Qichang de jiaoyou" 董其昌的交游 (Dong Qichang's circle). In *The Century of Tung Ch'i-ch'ang, 1555–1636*, ed. Wai-Kam Ho, pp. 461–83. Kansas City: Nelson-Atkins Museum of Art, in association with the University of Washington Press, 1992.

Wang Shixing 王士性. *Guangzhi yi* 廣志繹 (General topography: a disquisition). Beijing: Zhonghua shuju, 1981.

Wang Shizhen 王士禛. *Yuyang jinghua lu jianzhu* 漁洋精華錄箋註 (The essential Yuyang, annotated). Ji'nan: Qilu shushe, 1992.

Wang Shunu 王書奴. *Zhongguo changji shi* 中國娼妓史 (History of Chinese sing-song girls). Shanghai: Sanlian shudian, 1988.

Wang Sizhi 王思治 and Jin Chengji 金成基. "Qingdai qianqi Lianghuai yanshang de shengshuai" 清代前期兩淮鹽商的盛衰 (The rise and fall of the Lianghuai salt merchants in the first part of the Qing). *Qingshi luncong* 1982, no. 4: 50–84.

Wang Xiang 王驤. *Jiangsu suishi fengsu tan* 江蘇歲時風俗談 (On festival customs in Jiangsu). Nantong: Jiangsu guji chubanshe, 1985.

Wang Xirong 王錫榮. "Zheng Banqiao jiaoyou, xingzong mankao" 鄭板橋交游, 行蹤漫考 (Full study of Zheng Banqiao's associates and movements). In *Yangzhou baguai kao banji* 揚州八怪考辨集 (Collected studies of Yangzhou's eight eccentrics), ed. Xue Yongnian 薛永年, pp. 300–324. Nanjing: Jiangsu meishu chubanshe, 1992.

Wang Xiuchu 王秀楚. "Yangzhou shiri ji" 揚州十日記 (Ten-day diary of Yangzhou). In *Zhongguo jindai neiluan waihuo lishi gushi congshu* 中國近代內亂外禍歷史故事叢書 (Collectanea of stories of internal disturbances and external calamities in modern China), 2: 229–43. Taibei: Guangwen shuju, 1964.

Wang Yinggeng 汪應庚. *Pingshantang lansheng zhi* 平山堂覽勝志 (The eminent sites of Pingshan). 1742. Reprinted—Yangzhou: Guangling guji keyinshe, 1988.

Wang Yun 王雲. *Yangzhou huayuan lu* 揚州畫苑錄 (Record of Yangzhou painting). In *Yangzhou congke* 揚州叢刻 (Collected works on Yangzhou), comp. Chen Henghe 陳恒和. Jiangdu, 1936.

Wang Zhangtao 王章濤. *Ruan Yuan zhuan* 阮元傳 (Biography of Ruan Yuan). Hefei: Huangshan shushe, 1994.

Wang Zhenzhong 王振忠. *Ming Qing Huishang yu Huai Yang shehui bianqian* 明清徽商與淮揚社會變遷 (Huizhou merchants and social changes in Huai-Yang during the Ming and Qing). Beijing: Sanlian shudian, 1996.

———. "Ming Qing Yangzhou yanshang shequ wenhua ji qi yingxiang" 明清揚州鹽商社區文化及其影響 (The local culture of Yangzhou salt merchants in the Ming-Qing period and its influence). *Zhongguoshi yanjiu* (*jing*) 1992, no. 2: 3–25.

Wang Zhong 汪中. "Guangling dui" 廣陵對 (Answers to questions about Guangling). In *Wang Zhong ji* 汪中集 (Collected works of Wang Zhong), ed. Jiang Qiuhua 蔣秋華 and Lin Qingzhang 林慶彰, with notes by Wang Qingxin 王清信 and Ye Chunfang 葉純芳, pp. 161–69. Taibei: Zhongyang yanjiuyuan, Zhongguo wenzhe yanjiusuo zhoubeichu, 2000.

———. *Guangling tongdian* 廣陵通典 (Complete records of Guangling). Yangzhou: Yangzhou shuju, 1860.

———. *Wang Zhong ji* 汪中集 (Collected works of Wang Zhong). Ed. Jiang Qiuhua 蔣秋華 and Lin Qingzhang 林慶彰, with notes by Wang Qingxin 王清信 and Ye Chunfang 葉純芳. Taibei: Zhongyang yanjiuyuan, Zhongguo wenzhe yanjiuso zhoubeichu, 2000.

———. "Xianmu Zou ruren lingbiao" 先母鄒孺人靈表 (Obituary for my deceased mother, the learned Zou). In *Wang Zhong ji* 汪中集 (Collected works of Wang Zhong), ed. Jiang Qiuhua 蔣秋華 and Lin Qingzhang 林慶彰, with notes by Wang Qingxin 王清信 and Ye Chunfang 葉純芳, pp. 240–42. Taibei: Zhongyang yanjiuyuan, Zhongguo wenzhe yanjiusuo zhoubeichu, 2000.

Wei Minghua 偉明鏵. "Gu 'Yangzhou pai': Yangzhou chuantong wenhua jingshen de de yu shi" 估揚州派—揚州傳統文化精神的得與失 (Evaluating "Yangzhou schools": the successes and failures of traditional Yangzhou culture). In idem, *Yangzhou wenhua tanpian* 揚州文化談片 (Fragments of talks on Yangzhou culture), pp. 123–49. Beijing: Sanlian shudian, 1994.

———. "Kao 'Yangzhou luantan'—guanyu Qingdai xiqushi de yige kaozheng" 考揚州亂彈—關於清代戲曲史的一個考証 (Researching the *luantan* of Yangzhou—a study in the history of Qing opera). In idem, *Yangzhou wenhua tanpian* 揚州文化談片 (Fragments of talks on Yangzhou culture), pp. 185–236. Beijing: Sanlian shudian, 1994.

———. "Shuo Yangzhou shiri—du Wang Xiuchu 'Yangzhou shiri ji'" 說揚州十日—讀王秀楚'揚州十日記' (On Yangzhou's ten days: reading Wang Xiuchu's "Ten-Day Diary of Yangzhou"). In idem, *Yangzhou wenhua tanpian* 揚州文化談片 (Fragments of talks on Yangzhou culture), pp. 168–84. Beijing: Sanlian shudian, 1994.

———. "Xi 'Yanghzou meng'—da shiren Du Mu he tade qiannian fengliu meng" 析"揚州夢"—大詩人杜牧和他的千年風流夢 (Analyzing "Dreams of Yangzhou"—the great poet Du Mu and his thousand-year erotic dream). In idem, *Yangzhou wenhua tanpian* 揚州文化談片 (Fragments of talks on Yangzhou culture), pp. 109–22. Beijing: Sanlian shudian, 1994.

————. *Yangzhou shouma* 揚州瘦馬 (The thin horses of Yangzhou). Fuzhou: Fujian renmin chubanshe, 1998.

Wei Minghua and Antonia Finnane. "The Thin Horses of Yangzhou." *East Asian History*, no. 10 (Dec. 1995): 47–66.

Wei Peh-t'i. "Juan Yuan: A Biographical Study with Special Reference to Mid-Ch'ing Security and Control in Southern China, 1799–1835." Ph.D. diss., University of Hong Kong, 1981.

Wei Xi 魏禧. "Shande jiwen lu" 善德紀聞錄 (Record of good and virtuous [deeds]). In *Yangzhou zuzheng lu* 揚州足徵錄, comp. Jiao Xun 焦循, 27.25b–34b. Yangzhou, 1815. Reprinted in Beijing tushuguan guji zhenben congkan, 25. Beijing: Sumu wenxian chubanshe, n.d.

Widmer, Ellen. "Tragedy or Travesty? Perspectives on Langxian's 'The Siege of Yangzhou.'" In *Paradoxes of Traditional Chinese Literature*, ed. Eva Hung, pp. 167–98. Hong Kong: Chinese University Press, 1994.

Wilhelm, Hellmut. 'The Po-hsüeh hong-ru Examination of 1679." *Journal of the American Oriental Society* 71 (1951): 60–66.

Wilkinson, Endymion. "Chinese Merchant Manuals and Route Books." *Ch'ing-shih wen-t'i* 2, no. 9 (Jan. 1973): 8–34.

Will, Pierre-Etienne. "On State Management of Water Conservancy in Late Imperial China." *Papers in Far Eastern History* (Australian National University, Department of Far Eastern History), no. 36 (Sept. 1987): 71–91.

————. "State Intervention in the Administration of a Hydraulic Infrastructure: The Example of Hubei Province in Premodern Times." In *The Scope of State Power in China*, ed. Stuart Schram, pp. 295–347. London: School of Oriental and African Studies; Hong Kong, Chinese University Press, 1985.

Wood, Frances. *Did Marco Polo Go to China?* London: Secker and Warburg, 1995.

Worthy, Edmund H. "Regional Control in the Southern Song Salt Administration." In *Crisis and Prosperity in Sung China*, ed. J. W. Haeger, pp. 101–41. Arizona: University of Arizona Press, 1975.

Wright, Arthur E. *The Sui Dynasty*. Knopf, New York, 1978.

Wu Chengming. "Introduction: On Embryonic Capitalism." In *Chinese Capitalism, 1522–1840*, ed. Xu Dixin and Wu Chengming, pp. 1–20. Houndmills, Eng.: Macmillan, 2000.

Wu Jiaji 吳嘉紀. *Louxuan shiji* 陋軒詩集 (Collected poems of Lou-xuan). Taibei: Wenhai chubanshe, 1966.

———. "Song Wang Zuoyan gui Xin'an" 送汪左嚴歸新安 (Seeing off Wang Zuoyan on his return to Huizhou). In *Wu Jiaji shi jianjiao* 吳嘉紀詩箋校 (The poems of Wu Jiaji: an annotated collation), ed. Yang Jiqing 楊積慶, p. 80. Shanghai: Shanghai guji chubanshe, 1980.

Wu Jianyong 吳建雍. "Qing qianqi queguan ji qi guanli zhidu" 清前期榷關及其管理制度 (Customs management in the early Qing). *Zhongguoshi yanjiu* 1984, no. 1: 85–96.

Wu Jingzi (Wu Ching-tzu) 吳敬梓. *Rulin waishi* 儒林外史 (The scholars). Hong Kong: Taiping shuju, 1969.

———. *The Scholars*. Trans. Yang Hsien-yi and Gladys Yang. New York: Grosset and Dunlap, 1972.

Wu, Silas. *Passage to Power: K'ang-hsi and His Heir Apparent, 1661–1722*. Cambridge, Mass.: Harvard University Press, 1979.

Wu Tongju 武同舉, comp. *Huaixi nianbiao* 淮系年表 (Chronology of the Huai River system). 1928. Reprinted—Taibei: Wenhai chuban-she, n.d.

Wu, William Ding Yee. "Kung Hsien (ca. 1619–1689)." Ph.D. diss., Princeton University, 1979.

Wu, Yenna. "Her Hide for Barter." *Tamkang Review* 27, no. 2 (Winter 1996): 129–82.

Wu Zhaozhao 吳肇釗. "Ji Cheng yu Yingyuan xingzao" 計成與影園興造 (Ji Cheng and the creation of the Garden of Reflections). *Jianzhushi* 1985, no. 23: 167–77.

Wu Zhefu 吳哲夫. *Siku quanshu zuanxiu zhi yanjiu* 四庫全書纂修之研究 (Studies on the compilation of the Four Treasuries collection). Taibei: National Palace Museum, 1990.

Xiao Guoliang 蕭國亮. "Lun Qingdai gangyan zhidu" 論清代綱鹽制度 (On the salt shipment system of the Qing dynasty). *Lishi yanjiu* 1988, no. 1: 64–73.

Xie Guozhen 謝國楨. *Ming Qing zhi ji dangshe yundong kao* 明清之際黨社運動考 (Study of party movements in the Ming-Qing period). Beijing: Zhonghua shuju: 1982.

Xie Zhaozhe 謝肇淛. *Wu za zu* 五雜組 (Five miscellanies). Taibei: Weiwen, 1977.

Xu Fengyi 許風儀 and Zhu Fugui 朱福娃. *Yangzhou fengwu zhi* 揚州風物志 (Local culture in Yangzhou). Yangzhou: Jiangsu renmin chubanshe, 1980.

Xu Hong 徐泓. *Qingdai Lianghuai yanshang de yanjiu* 清代兩淮鹽商的研究 (Study of Lianghuai salt merchants under the Qing). Taibei: Jiaxin shuini gongsi, 1972.

Xu Ke 徐珂. *Qingbai leichao* 清稗類鈔 (Gleanings from the Qing). Taibei: Shangwu yinshuguan, 1966.

Xu, Yinong. *The Chinese City in Space and Time: The Development of Urban Form in Suzhou.* Honolulu: University of Hawai'i Press, 2000.

Xu Yuanchong 許淵仲, Lu Peixuan 陸佩弦, and Wu Diaotao 吳鈞陶, eds. *Tangshi sanbaishou xinyi* 唐詩三百首新譯 (Three hundred Tang poems: a new translation). Hong Kong: Zhongguo duiwai fanyi chuban gongsi, 1991.

Xue Yongnian 薛永年 and Xue Feng 薛鋒. *Yangzhou baguai yu Yangzhou shangye* 揚州八怪與揚州商業 (The eight eccentrics and commerce in Yangzhou). Beijing: Renmin meishu chubanshe, 1991.

Xue Zongzheng 薛宗正. "Mingdai yanshang de lishi yanbian" 明代鹽商的歷史演變 (Progressive changes in the history of the salt merchants of the Ming). *Zhongguoshi yanjiu* 1980, no. 2: 27–37.

Yang Dequan 樣德泉. "Qingdai qianqi Lianghuai yanshang ziliao chuji" 清代前期兩淮鹽商資料初集 (Preminary collection of materials on the Lianghuai salt merchants in the early Qing). *Jianghuai xuekan*, no. 45 (Nov. 1962): 45–49.

Yang Hongxun 楊鴻勛. *Jiangnan yuanlin lun* 江南園林論 (Treatise on Jiangnan gardens). Shanghai: Renmin chubanshe, 1994.

Yang Jiqing 楊積慶, ed. *Wu Jiaji jianjiao* 吳嘉紀箋校 (Annotated works of Wu Jiaji). Shanghai: Shanghai guji chubanshe, 1980.

Yang Lien-sheng. "Government Control of Urban Merchants in Traditional China." *Tsing Hua Journal of Chinese Studies*, n.s. 8, nos. 1–2 (1970): 186–206.

"Yangzhou bianlüe" 揚州變略 (Outline of changes in Yangzhou). In *Zhongguo jindai neiluan waihuo lishi gushi congshu* 中國近代內亂外禍歷史故事叢書 (Collectanea of stories of internal disturbances and external calamities in modern China), 3: 127–30. Taibei: Guangwen shuju, 1964.

"Yangzhou jingji zhuangkuang" 揚州經濟狀況 (Yangzhou's economic circumstances). *Zhongwai jingji zhoukan*, no. 221 (July 23, 1927): 15–21.

Yangzhou wenwu yanjiushi, Shi Kefa jinian guan 揚州文物研究室, 史可法紀念館, ed. *Liangjie guazhong Shi Kefa* 諒解孤忠史可法 (The outstanding loyalist Shi Kefa). Yangzhou: Jiangsu wenyi chubanshe, 1993.

Yangzhou yingzhi 揚州營志 (Gazetteer of the Yangzhou garrison), ed. Chen Shuzu 陳述祖 and Li Beishan 李北山. Yangzhou, 1831.

Reprinted in Beijing tushuguan guji zhenben congkan, no. 48. Beijing: Sumu wenxian chubanshe, n.d.

Yao Wentian 姚文田. *Guangling shilüe* 廣陵事略 (Outline of events in Guangling). Kaifeng, preface dated 1812.

Ye Sen 葉森 and Zhu Feng 朱峰. "Yangzhou jiexiang zatan" 揚州街巷雜談 (Various comments on the streets and lanes of Yangzhou). *Zhongguo minjian gongyi*, no. 13–14 (June 1994): 143–46.

Yi Junzuo 易君左. *Xianhua Yangzhou* 閒話揚州 (Idle talk on Yangzhou). Shanghai: Zhonghua shuju, 1934.

Yoshida, Tora. *Salt Production Techniques in Ancient China: The Aobo Tu*. Trans. and rev. Hans Ulrich Vogel. Leiden: E. J. Brill, 1993.

Yu Dafu 郁達夫. "Yangzhou jiu meng ji Yutang" 揚州舊夢寄語堂 (The old dream of Yangzhou: to [Lin] Yutang). *Renjianshi*, no. 28 (1935): 3–6.

Yu Yuchu 俞玉儲. "Xu Shukui 'Yi zhu lou shi' an" 徐述夔"一柱樓詩"案 (The case of Xu Shukui's "Poems from the Single Pillar Mansion"). In *Qingdai wenzi yu'an* 清代文字獄案 (Literary prosecutions of the Qing era), ed. Zhang Shucai 張書才 and Du Jinghua 杜景華, pp. 178–86. Beijing: Zijincheng chubanshe, 1991.

Yuanshi 元史 (History of the Yuan). Beijing: Zhonghua shuju, 1976.

Yue Shi 樂史, comp. *Taiping huanyu ji* 太平寰宇記 (Record of the world in an age of great peace). Taibei: Wenhai chubanshe, 1963.

Yule, Henry. *The Book of Ser Marco Polo*. 1871. Rev. Henri Cordier. London: John Murray, 1903.

———. *Cathay and the Way Thither: Being a Collection of Medieval Notices of China*. 1866. Rev. Henri Cordier, 1913. Reprinted—Nendeln, Liechtenstein: Kraus Reprint, 1967.

Zhang Dai 張岱. *Tao'an mengyi* 陶庵夢憶 (The dream memories of Tao'an). Shanghai: Xin wenhua shushe, 1934.

Zhang Geng 張庚. *Guochao huazheng lu* 國朝畫徵錄 (Painting under the [Qing] dynasty). In *Huashi congshu* 畫史叢書 (Collected works on the history of painting), vol. 3, comp. Yu Anlan 于安瀾. Shanghai: Shanghi meishu chubanshe, 1963.

Zhang Jian 張鑑. *Leitang'an zhu dizi ji* 雷塘庵諸弟子記 (Record by disciples of the Master of Leitang Temple). Published under the title *Ruan Yuan nianpu* 阮元年譜 (Yearly chronology of Ruan Yuan). Beijing: Zhonghua shuju, 1995.

Zhang Liansheng 張連生. "Qingdai Yangzhou yanshang de xingshuai yu yapian yuru" 清代揚州鹽商的興衰與鴉片輸入 (The rise and fall of the Yangzhou salt merchants and the introduction of opium under the Qing). *Yangzhou shiyuan xuebao* 1982, no. 2: 21–26.

Zhang Shucai 張書才 and Du Jinghua 杜景華, eds. *Qingdai wenzi yu'an* 清代文字獄案 (Literary prosecutions of the Qing era). Beijing: Zijincheng chubanshe, 1991.

Zhang Shunhui 張舜徽. *Qingdai Yangzhouxue ji* 清代揚州學記 (Notes on Yangzhou scholarship of the Qing period). Shanghai: Renmin chubanshe, 1962.

———. *Qing ruxue ji* 清儒學記 (Notes on Confucian scholarship of the Qing era). Ji'nan: Jilu shushe, 1991.

Zhang Zhengming 張正明. *Jinshang xingshuai shi* 晉商興衰史 (The rise and fall of the Shanxi merchants). Taiyuan: Shanxi guji chubanshe, 1995.

Zhao Erxun 趙爾巽. *Qing shi gao* 清史稿 (Draft history of the Qing). Beijing: Zhonghua shuju, 1977.

Zhao Hang 趙航. *Yangzhou xuepai xinlun* 揚州學派新論 (A new treatise on the Yangzhou school). Jintan: Jiangsu wenyi chubanshe, 1991.

Zhao Hongen 趙宏恩. *Yuhua ji* 玉華集 (Yuhua's collected writings). Preface dated 1734.

Zhao Xueli 趙學禮 and Cao Yongquan 曹永全. "Tan Yangzhou hunsu" 談揚州婚俗 (On wedding customs in Yangzhou). In *Yangzhou fengqing* 揚州風情 (added title in English: *The Flavour of Yangzhou*), ed. Cao Yongsen 曹永森, pp. 63–74. Yangzhou: Jiangsu wenyi chubanshe, 1991.

Zhao Yi 趙翼. *Gaiyu congkao* 陔餘叢考 (Collected researches of Gaiyu). Kyoto: Chūbun shuppansha, 1979.

Zhao Zhibi 趙之壁. *Pingshantang tuzhi* 平山堂圖志 (Illustrated gazetteer of Pingshan Hall). Yangzhou, preface dated 1765.

Zheng Changgan 鄭昌淦. *Ming Qing nongcun shangpin jingji* 明清農村商品經濟 (The rural commodity economy of the Ming Ching period). Beijing: Zhongguo renmin daxue chubanshe, 1989.

Zheng Qingyou 鄭慶祐. *Yangzhou Xiuyuan zhi* 揚州休園志 (Gazetteer of the Garden of Retirement). Yangzhou(?): Caishitang, 1773.

Zheng Xie 鄭燮. "Si jia" 思家 (Thinking of home). In *Chêng Pan-Ch'iao: Selected Poems, Caligraphy, Paintings and Seal Engravings* (parallel text), pp. 74–75. Ed. and trans. Anthony Gurofsky and Paul Gurofsky. Hong Kong, Joint Publishing Company, n.d.

———. *Zheng Banqiao quanji* 鄭板橋全集 (Complete works of Zheng Banqiao). Beijing: Zhongguo shudian, 1994.

Zheng Yuan 鄭遠, ed. *Qianlong huangdi quanzhuan* 乾隆皇帝全傳 (Complete biography of the Qianlong emperor). Beijing: Xueyuan chubanshe, 1994.

Zheng Yuanxun 鄭元勳. "Yingyuan zizhi" 影園自志 (My own memoir on the Garden of Reflection). In *Yangzhou fuzhi* 揚州府志 (1810), 31.51b–52a.

Zheng Zhaojing 鄭肇經. *Zhongguo shuili shi* 中國水利史 (History of water control in China). Changsha: Shangwu yinshuguan, 1939.

Zhongguo meishu quanji 中國美術全集 (Collected fine arts of China), vol. 10, *Qingdai huihua* 清代繪畫 (Qing painting), ed. Yang Han 楊涵, Gong Jixian 龔繼先, and Hu Haichao 胡海超. Shanghai: Shanghai meishu renmin chubanshe, 1988.

Zhou Cun 周邨. *Taipingjun zai Yangzhou* 太平軍在揚州 (The Taiping army in Yangzhou). Shanghai: Shanghai renmin chubanshe, 1957.

Zhou Lianggong 周亮工. *Du hua lu* 讀畫錄 (Studying paintings). Shanghai: Shanghai renmin chubanshe, 1963.

Zhou Weiquan 周維權. *Zhongguo gudian yuanlin shi* 中國古典園林史 (History of Chinese traditional gardens). Beijing: Qinghua daxue chubanshe, 1990.

Zhou Zhenhe 周振鶴 and You Rujie 游汝杰. *Fangyan yu Zhongguo wenhua* 方言與中國文化 (Dialects and Chinese culture). Shanghai: Renmin chubanshe.

Zhu Fugui 朱福烓. "Yisilanjiao yu Yangzhou" 伊斯蘭教與揚州 (Islam and Yangzhou). In *Yisilanjiao zai Yangzhou* 伊斯蘭教在揚州 (Islam in Yangzhou), ed. Wei Peichun 韋培春, pp. 40–45. Nanjing: Nanjing daxue, 1991.

Zhu Fugui 朱福烓 and Xu Fengyi 許風儀. *Yangzhou shihua* 揚州史話 (On the history of Yangzhou). Yangzhou: Jiangsu guji chubanshe, 1985.

Zhu Jiang 朱江. "Jewish Traces in Yangzhou." In *Jews in Old China: Studies by Chinese Scholars*, ed. Sidney Shapiro, pp. 143–58. New York: Hippocrene Books, 1984.

——. *Yangzhou yuanlin pinshang lu* 揚州園林品賞錄 (Various aspects of Yangzhou gardens). Shanghai: Shanghai wenwu chubanshe, 1984.

Zhu Kebao 諸可寶. *Jiangsu quansheng yutu* 江蘇全省輿圖 (Maps of Jiangsu province). Jiangsu: Jiangsu shuju, 1895.

Zhu Xie 朱偰. *Zhongguo yunhe shiliao xuanji* 中國運河史料選集 (Selected documents on the Grand Canal of China). Beijing: Zhonghua shuju, 1962.

Zhu Ziqing (Chu Tzu-ch'ing) 朱自清. "My Wife and Children." Trans. Ernst Wolff. In *Chinese Civilization and Society: A Source Book*, ed. Patricia Ebrey Buckley, pp. 294–95. New York: Free Press, 1981.

——. "Shuo Yangzhou" 説揚州 (Speaking of Yangzhou). *Renjianshi*, no. 16 (1934): 35–36.

———. "Wo shi Yangzhouren" 我是揚州人 (I am a man of Yang-zhou). In *Zhu Ziqing* 朱自清, ed. Jiangsu sheng zhengxie wenshi ziliao weiyuanhui 江蘇省政協文史資料委員會 and Yangzhou shi zhengxie wenshi ziliao weiyuanhui 揚州市政協文史資料委員會, pp. 222–25. Nanjing: Jiangsu wenshi ziliao bianjibu, 1992.

———. "Yangzhou de xiari" 揚州的夏日 (Summer days in Yang-zhou). In idem, *Ni, wo* 你,我 (You and I), pp. 38–47. Beijing: San-lian shudian, 1984.

———. "Ze'ou ji" 擇偶記 (Making a match). In *Zhu Ziqing* 朱自清, ed. Jiangsu sheng zhengxie wenshi ziliao weiyuanhui 江蘇省政協文史資料委員會 and Yangzhou shi zhengxie wenshi ziliao weiyuanhui 揚州市政協文史資料委員會, pp. 245–47. Nanjing: Jiangsu wenshi ziliao bianjibu, 1992.

Zong Yuanding 宗元鼎. "Maihua laoren zhuan" 賣花老人傳 (Tale of the old flowerseller). In *Yuchu xinzhi* 虞初新志 (New annals of Yu Chu), ed. Zhang Chao 張潮, pp. 60–61. Shanghai: Shanghai shu-dian, 1986.

Zurndorfer, Harriet T. *Change and Continuity in Chinese Local History.* Leiden: E. J. Brill, 1989.

———. "From Local History to Cultural History: Reflections on Some Recent Publications." *T'oung Pao* 83 (1997): 386–424.

———. "The *Hsin-an ta-tsu chih* and the Development of Chinese Gentry Society, 800–1600." *T'oung Pao* 67, no. 3–5 (1981): 154–214.

———. "Local Lineages and Local Development: A Case Study of the Fan Lineage, Hsiu-ning hsien, Huizhou, 800–1500." *T'oung Pao* 70 (1984): 18–59.

C3

Character List

The entries are ordered alphabetically ignoring syllable and word breaks, with the exception of personal names, which are ordered first by the surname and then by the given name.

An Qi (b. 1683) 安岐
Anding Academy 安定書院
Anqing 安慶

ba 壩
Bai Zhongshan (d. 1761) 白鍾山
Baita Canal 白塔河
Bao Shichen (1775–1855) 包世臣
Bao Shufang 鮑漱芳
Bao Youheng 鮑有恆
Bao Zhao (414?–66) 鮑照
Bao Zhidao (1743–1801) 鮑志道
baoxiao 報效
Baoying 寶應
Baozhang Lake 保障湖
beifang 北房
Beihu 北湖
Beiliu xiang 北柳巷
bense 本色
Bi Shiduo (d. 888) 畢師鐸
Bian 卞
bian chang mo ji 鞭長末及

bianshang 邊商
bingmin 兵民
Bo Sea 渤海

caicun huoyou zhouche 裁存火
优舟車
caishen 財神
Cao Zhenyong (1755–1835)
曹振鏞
cao ting 草亭
Chai Binchen 柴賓臣
Chai Shijin 柴士進
Chai Yiqin 柴宜琴
chang 場
Chang'an 長安
Changde 常德
Changlu 長蘆
changmao 長毛
changpao 長袍
changshang 場商
changyun geshang juankuan
場運各商捐款

Changzhou 常州
Chanzhi Temple 禪智寺
chaofan 炒飯
Chaozhou 潮州
chayuan 查員
Chen Hongmou (1696–1771)
　陳宏謀
Chen Jiru (1558–1639)
　陳繼儒
Chen Qubing (1874–1933)
　陳去病
Chen Shouyi 陳授衣
Chen Weisong (1626–82)
　陳維崧
Chen Zhuan (fl. 1670–1740)
　陳撰
Cheng (surname) 程
Cheng Jiasui 程嘉遂
Cheng Liangru 程量入
Cheng Mengxing (*js* 1712)
　程夢星
Cheng Sui (1605–91) 程邃
Cheng Yourong 程有容
Cheng Zhiquan 程之銓
Cheng Zhiying 程之韺
chi 尺
Chu 楚
Chuanchanghe 串場河
Cixi 慈溪
Cui Hua 崔崋
Cuihua jie 翠花街

da cheng 大城
Dade 大德
Dadu 大都
da gu 大賈
Dai Benxiao (1621–93) 戴本孝
Dai Zhen (1724–77) 戴震
daishang 代商
Danyang 丹陽
daocheng 倒城
Daoguang 道光

Datian 大田
Datong 大同
daxing 大姓
dian 典
difang dayuan 地方大員
Ding Gao (Hezhou) 丁皋 (鶴洲)
Ding Yicheng (Yimen) 丁以誠
　(義門)
dingchou 丁丑
Dong Qichang (1555–1636)
　董其昌
Dong Zhongshu (ca. 179–ca. 104
　BCE) 董仲舒
Donglin 東林
Dongri deng Hailing Dexiang
　ge 冬日登海陵德香閣
Dongtai 東泰
Dongting Lake 洞廷湖
Dongyuan 東園
doufugan 豆腐乾
Du Fu (712–70) 杜甫
Du Jun (1611–87) 杜濬
Du Mu (803–52) 杜牧
Duan Yucai (1735–1815)
　段玉裁
Duanzi jie 緞子街
Du hua lu 讀畫錄
Duoduo (1614–49) 多鐸
Duozijie 多子街

er jing 二勁
erduobian 耳朵邊
erhuang 二簧
Erlüeding 爾略丁

Fahaiqiao 法海橋
Fahai Temple 法海寺
Fan Changjiang (1909–70)
　范長江
Fan Quan 范荃
Fan Zhongyan (989–1052)
　范仲淹

Fang (surname) 方
Fang Dongshu (1772–1851)
　方東樹
Fang Junyi (1815–89) 方濬頤
Fang Ruting 方如斑
Fang Shijie 方士庶
Fang Shishu (1692–1751)
　方世庶
Fangong Dike 范公堤
fanli 藩籬
Fanli Monastery 蕃釐觀
Fei Chun (ca. 1739–1811) 費淳
Fei Mi (1625–1701) 費密
Fei Xiaotong (b.1910) 費孝通
Feng Zikai (1898–1975) 豐子愷
fenglun 封輪
fensi 分司
fu (prefecture) 府
fu (supplementary) 附
Fu Chai (r. 495–476 BCE) 夫差
Funing 阜寧

gaihu 丐戶
gang 綱
gang'an 綱岸
gangce 綱策
gangfa 綱法
Ganquan 甘泉
gansi 干絲
Gansu 甘肅
Ganting 甘亭
Gao (surname) 高
Gao Bin (1683–1755) 高斌
Gao Fenghan (1683–ca. 1748)
　高鳳翰
Gao Heng (d. 1768) 高恆
Gao Jie (d. 1645) 高傑
Gao Jin (1707–79) 高晉
Gao Pian (d. 887) 高駢
Gao Xiang (1688–1753) 高翔
gaohuo 膏火
Gaojiayan 高家堰

Gaomin Temple 高旻寺
Gaoyan 高堰
Gaoyou 高郵
Ge Garden 个園
Gengzi jie 埂子街
genwo 根窩
Gong Jilan (js 1637) 宮繼蘭
Gong Weiliu (js 1643) 宮偉鏐
Gong Xian (ca. 1619–89) 龔賢
Gong Zizhen (1792–1841)
　龔自珍
gongchai kongque huo weizhi
　公差空缺或位置
Gongdaoqiao 公道橋
Gongdeshan 功德山
Gu Guangqi (1776–1835)
　顧廣圻
Gu Yanwu (1613–1682) 顧炎武
Guangchu Gate 廣儲門
Guangling 廣陵
Guangling dui 廣陵對
Guangling shuyuan 廣陵
　書院
guanji 官妓
guanshang 官商
Guanyin 觀音
Guazhou 瓜洲
Guiren 歸仁
Guizhou 貴州
Gujin tushu jicheng 古今圖書
　集成
gunba 滾壩
Guo (surname) 郭
Guo Hanzhang guan 郭漢章館
Guo Moruo (1892–1978)
　郭沫若
Guocui xuebao 國粹學報

Haian 海安
haifang hewu tongzhi 海防河務
　同知
Hailing 海陵

Haimen 海門
Haizhou 海州
Han 漢
Han Gou 邗溝
Hancheng 邗城
Hang Shijun (1696–1773)
　杭世駿
Hangzhou 杭州
Hanshang mengren 邗上
　蒙人
Hanyuan 韓園
haoyou 豪右
He 河
He Cheng 何城
He Junzhao 賀君召
Hejin 河津
Henan 河南
hewu 河務
Hexinyuan 合欣園
Hong Chongshi 洪充實
Hong Liangji (1746–1809)
　洪亮吉
Hong Zhengzhi 洪徵治
Hong Zhenke 洪振珂
Hong Zhenyuan 洪箴遠
Hongdong 洪洞
Hongqiao 紅/虹橋
Hongqiao shuyuan 紅橋書院
Hongren 泓仁
Hongshuiwang 紅水汪
Hongze Lake 洪澤湖
Hongzhi 弘治
Hu Qiheng 胡期恒
Hu Qizhong 胡蘄忠
Hu Yuan (Anding) (993–1059)
　胡瑗 (安定)
hua 畫
Hua Yan (1682–1756) 華嵒
huabu 花部
Huai 淮
Huai'an 淮安
Huaibei 淮北

Huai, Hai wei Yangzhou 淮海
　惟(維)揚州
Huaihai yingling ji 淮海英靈集
Huainan 淮南
Huainanzi 淮南子
Huaiqing 懷慶
Huaiyang 淮揚
Huan Jing 桓景
Huang Chao (d. 884) 黃巢
Huang Chengji (1771–1824)
　黃承吉
Huang Degong (d. 1645)
　黃得功
Huang Lühao 黃履昊
Huang Lümao 黃履昴
Huang Lüxian 黃履暹
Huang Shen (1687–1770) 黃慎
Huang Yingtai 黃瀯泰
Huang Yuande 黃源德
Huang Zhiyun (Yingtai) 黃至筠
　(瀯泰)
Huangjin Embankment 黃金壩
Huangjue Bridge 黃玨橋
Huangshan 黃山
Huankui 闤闠
Huguang 湖廣
Hui Dong (1697–1758) 惠棟
huiguan 會館
Huizhou 徽州
Hunan 湖南
Huo (surname) 火
huo nong huo gu 或農或賈
Huozhou 霍州
Huzhou 湖州

Ji Cheng (b. 1582) 計成
Ji Huang (1711–94) 稽璜
Ji Yun (1724–1805) 紀昀
Jiading 嘉定
jianba 減壩
Jiang Chun 江春
Jiang Fan (1761–1831) 江藩

Jiang Guangda 江廣達
Jiang Kui 姜夔
Jiang Yong (1681–1762)
　江永
Jiang Zhuzhou 江助周
Jiang'an men 江安門
Jiangbei 江北
Jiangdu ji 江都籍
Jiangdu ren 江都人
Jiangfang tongzhi 江防同知
Jianghuai 江淮
Jiangnan 江南
Jiangnan hedao zongdu 江南
　河道總督
Jiangning 江寧
Jiangsu 江蘇
Jiangxi 江西
Jiankang 建康
Jianli huiguan beiji 建立會館
　碑記
Jiao Xun (1763–1820) 焦循
Jiaoli xueshe 角里學舍
Jiaqing 嘉慶
Jiashan 嘉善
Jiaxing 嘉興
Jienan shuwu 街南書屋
jin 斤
Jin (dynasty) 金
Jin Fu (1633–92) 靳輔
Jin Nong (1687–1764) 金農
Jin Zhen 金鎮
Jin Zhi (1663–1740) 金埴
Jing hua yuan 鏡花緣
Jingjiang 靖江
Jingu Garden 金谷園
Jingxiangyuan 淨香園
Jingyang 涇陽
Jinling 金陵
Jinshan 金山
jinshi 進士
Jiqi 績溪
Jiufengyuan 九峰園

Jiuri xing'an wenyan 九日行庵
　文讌
Jiyang 濟陽
junren 郡人
junshen 君紳
juren 舉人
Jurong 句容

kai bian 開邊
kaizhong fa 開中法
kai zhong na mi 開中納米
Kang Garden 亢園
Kangshan caotang 康山草堂
keshang 客商
Kong Shangren (1648–1718)
　孔尚任
Kunming 昆明
Kunqu 昆曲
Kunshan 昆山
Kunshanqiang 昆山腔

Lanzhou 蘭州
laoshao 老少
lao Yangzhou 老揚州
li (distance measure) 里
li (subordinate) 隸
Li Bai (701–62) 李白
Li Changfu 李長傅
Li Cheng 李澄
Li Daonan (js 1759) 李道南
Li E (1692–1752) 厲鶚
Li Fayuan 李發元
Li Gan 李淦
Li Quan (d. 1231) 李全
Li Ruzhen (1763–ca. 1830)
　李汝珍
Li Shan (1686–1760) 李鱓
Li Tingzhi (1217?–76) 李庭芝
Li Yi 李沂
Li Yu (1625–84) 李漁
Li Zhi (js 1577) 李植
Li Zicheng (1605?–45) 李自成

Li Zongkong (*js* 1645) 李宗孔
Liang Qichao (1873–1929)
　梁啓超
Liang Qizi 梁其姿
Liang Yusi (d. 1645) 梁于涘
Liang Zhangju (1775–1849)
　梁章鉅
Lianghuai yanfa zhi 兩淮鹽法志
Lianghuai yanyicang 兩淮鹽
　義倉
Liangzhe 兩浙
Lianxing Temple 蓮性寺
Liaodong 遼東
Lin Daoyuan 林道源
Lin Ruhai 林如海
Lin Sumen (1749–after 1809)
　林蘇門
Lin Yutang (1895–1976) 林語堂
Linfen 臨汾
Ling Shu (1775–1829) 淩曙
Ling Tingkan (1757–1809)
　淩廷堪
Lingnan 嶺南
Lin Shen Shitian bi 臨沈石田筆
Linshui hongxia 臨水紅霞
Lintong 臨潼
Liu (surname) 劉
Liu Baonan (1791–1855) 劉寶楠
Liu Bi 劉濞
Liu Chongxuan (*js* 1737)
　劉重選
Liu Daguan 劉大觀
Liu Shipei (1884–1919) 劉師培
Liu Shouzeng (1838–82) 劉壽曾
Liu Taigong (1751–1805)
　劉台拱
Liu Wenqi (1789–1856) 劉文淇
Liu Xizai 劉熙載
Liu Yan 劉晏
Liu Yong (1720–1805) 劉墉
Liu Yusong (1818–67) 劉毓崧
Liu Zhaoji 劉肇基

Liubu 留步
Liushang 流觴
liuyu 留遇
Lizhentang 立貞堂
Longmen Academy 龍門書院
Louxuan 陋軒
Lu (surname) 魯
Lu Jianzeng (Yayu) (1690–1768)
　盧見曾(雅雨)
Lu Shu 陸書
Lu Tinglun 陸廷掄
Lu Xun 盧詢
Lu Zhonghui 陸鍾輝
luantan 亂彈
Luo Ping (1733–99) 羅聘
Luo Youlong 羅优龍
Luoluo 羅羅
Lupu 蘆浦

Ma Wulu (d. 1645) 馬鳴騄
Ma Yueguan (1688–1755)
　馬曰琯
Ma Yuelu (1697–after 1766)
　馬曰璐
Ma Zhenbo 馬振伯
Maluo 馬邏
Man 滿
Mangdao Canal 芒稻河
Mao Xiang (1611–93) 冒襄
Meihua Academy 梅花書院
meiren 美人
Meiyouge 媚幽閣
Meng Haoran (689–740) 孟浩然
mianguan 麵館
Miao 苗
min 民
Min Ding 閔鼎
Min Hua 閔華
Min Shizhang (b. 1607) 閔世章
Min Tingzuo 閔廷佐
Ming (dynasty) 明
mingji 名妓

Nanchang 南昌
nanfang 南房
nan geng, nü zhi 男耕女織
Nanhe 南河
Nanhexia Street 南河下街
Nanjing 南京
Nantong 南通
Nan Zhili 南值隸
neishang 内商
Ni Zan (1301–74) 倪瓚
Nian 捻
nianwo 年窩
Ningbo 寧波
nügong 女工

Ouyang Xiu (1007–72) 歐陽修

Pan Jixun (1521–95) 潘季馴
Pan Shien (1770–1854) 潘世恩
Pan Zengshou 潘曾綬
Pianshi shanfang 片石山房
Pingliang 平涼
Pingshantang 平山堂
Pucheng 浦城
Pucheng (Map 3) 蒲城
Pugangchun 撲缸春
Puhading 普哈丁
Pujitang 普濟堂
Puzhou 蒲州

Qi (state) 齊
Qian Yong (1759–1844) 錢泳
Qianlong 乾隆
Qiao (surname) 喬
Qiao Chengwang 喬承望
Qiao Yuan zhi sanhao 喬元之
　三好
Qimen 祁門
qin 琴
Qin (dynasty) 秦
Qin Enfu (1760–1843) 秦恩復
Qinggou 清溝

Qinghua 清化
Qingjiang 清江
Qingjiangpu 清江鋪
Qinglianzhai 青蓮齋
Qingming 清明
Qingshan 青山
Qingshuitan 清水潭
Qinhuai Canal 秦淮河
qinying 親迎
Qin you qu er shou 秦郵曲二首
Qiu Wenbo (fl. 10th c.) 丘文播
Qiu Ying (1494/95–1522) 仇英
Qu Fu 屈復
Quan Zuwang (1705–55)
　全祖望
Quekou Gate 缺口門

Rangpu 讓圃
Ren Dachun (1738–89) 任大春
Ren Minyu (d. 1645) 任民育
Ruan Bingqian 阮秉謙
Ruan Changsheng (1788?–1833)
　阮常生
Ruan Heng 阮亨
Ruan Xian 阮先
Ruan Yuan (1764–1849) 阮元
Ruan Yutang (1695–1759)
　阮玉堂
Ruan Zuo 阮祚
Rugao 如皋

Sanbao 三保
Sancha (he) Canal 三汊河
san cheng 三城
Sanfan 三藩
Sanjiangying 三江營
sanxian dalian 三鮮大連
Sanyuan 三原
shang 商
shangji 商籍
shangtun 商屯
Shangyuan 上元

Shan-Shaan 山陝
Shanyang 山陽
shaobing 燒餅
Shaobo 邵伯
Shao Garden 勺園
Shaoxing 紹興
Shen (surname) 沈
Shen Defu (1578–1642) 沈德符
Shen Fu (1763–?) 沈復
Shen Kuo (1031–95) 沈括
Shen Zhou (Shitian) (1427–1509)
　沈周 (石田)
Shengzu 聖祖
shenshi 紳士
Shexian 歙縣
shi (market) 市
shi (yes) 是
Shi (surname) 史
Shi Chong 石崇
Shi Dewei 史德威
Shi Kefa (1604–45) 史可法
Shi Maohua 石茂華
Shi Panzi 施胖子
Shi Yuan 施原
shi'an 食岸
shidafu 士大夫
shijing wulai 市井無賴
shimin 士民
Shishang yaolan 士商要覽
shisi 食肆
Shitao (1644–1707) 石濤
Shizhuang (d. 1732) 石莊
Shouxihu 瘦西湖
shouzong 首總
Shuangtong shushi 雙桐書室
Shuangying 雙英
Shugang 蜀崗
Shugang chaoxu 蜀岡朝旭
shuili 水利
shuili tongzhi 水利同知
shuishang 水商
shuixiang 水鄉

Sichuan 四川
sigua 絲瓜
Siku quanshu 四庫全書
siqiao yinyu 四橋烟雨
Si River 泗河
si yuan bao 四元寶
Song (dynasty) 宋
Songjiang 松江
Su Gaosan 蘇高三
Su Shi (1036–1101) 蘇軾
Suchangjie 蘇唱街
sui 歲
Suide 綏德
Sui Yangdi 隋煬帝
Sun Dacheng 孫大成
Sun Mo (1613–78) 孫默
Sun Ru (d. 892) 孫儒
Sun Zhiwei (1620–87) 孫枝蔚
Sushi xiaoyin 蘇式小飲
Suzhou 蘇州

Taihang 太行
Taiping 太平
Taixing 泰興
Taiyuan 太原
Taizhou 泰州
Tang (dynasty) 唐
Tang Jianzhong 唐建中
Tang Laihe 湯來賀
tao 桃
Tao Yi (d. 1778) 陶易
Tao Yuanming (365–427)
　陶淵明
Tao Zhu (1779–1839) 陶澍
Tian Han (1898–1968) 田漢
Tianbao Wall 天保城
Tianjin 天津
Tianning Temple 天寧寺
Tianshouan 天壽庵
tianxia 天下
Tiebao (1752–1824) 鐵保
Tongbai Mountains 桐柏山

Tongcheng 桐城
Tongrentang 同仁堂
Tongzhi 同治
tu 圖
Tu Yuelong (d. 1798) 涂躍籠
tuchang 土昌
tuzhu 土著

Waichengjiao 外城腳
waifan michuan 外販米船
waiji 外妓
wan 萬
Wan'an 萬安
Wan'an (dialect) 皖安
Wang (surname) 汪
Wang Chang (1725–1806) 王昶
Wang Changxin 汪長馨
Wang Daokun (1525–93)
　汪道昆
Wang Fangqi 王方岐
Wang Fangwei 王方魏
Wang Fangzhong 王方中
Wang Gen (1483–1541) 王艮
Wang Hongxu (1645–1723)
　王鴻緒
Wang Ji (1636–99) 汪楫
Wang Jiaoru 汪交如
Wang Lütai 王履泰
Wang Maohong (1664–1741)
　王懋竑
Wang Maolin (1640?–88)
　汪懋麟
Wang Nalian (js 1607) 王納諫
Wang Niansun (1744–1832)
　王念孫
Wang Shiduo (1802–89) 汪士鐸
Wang Shimin (1592–1680)
　王時敏
Wang Shishen (1685–1759)
　汪士慎
Wang Shixing (1547–98) 王士性
Wang Shiyu 汪士裕

Wang Shizhen (Yuanting)
　(1634–1711) 王士禎 (阮亭)
Wang Shizhen fang xian 王世禎
　放鷴
Wang Tianfu 王天福
Wang Tingzhang 汪廷璋
Wang Wei (710–61) 王維
Wang Wentong (js 1733)
　王文充
Wang Xihou 王錫侯
Wang Xisun (1786–1847)
　汪喜荀
Wang Xiuchu 王秀楚
Wang Xiwen 王希文
Wang Xuejiang 汪雪礓
Wang Yinggeng 汪應庚
Wang Yinzhi (1766–1834)
　王引之
Wang Yongji 王永吉
Wang Yun (1652–after 1735)
　王雲
Wang Yun (1816–after 1883)
　汪鋆
Wang Yushu 汪玉樞
Wang Yuzao (js 1643) 王玉藻
Wang Zao 王藻
Wang Zheng (1571–1644) 王徵
Wang Zhenzhong 王振忠
Wang Zhong (1745–94) 汪中
Wang Zuoyan 汪左嚴
Wan School 皖派
Wanshiyuan 萬石園
Wanshou 萬壽
Wei (state) 魏
Wei River 渭河
Wei Xi (1624–81) 魏禧
Wei Yuan (1794–1856) 魏源
Wei Zhongxian (1568–1627)
　魏忠賢
Weixian 濰縣
Weiyang 維揚
Wenchang Tower 文昌閣

Wenfengta 文峰塔
Wenhuige 文匯閣
Wenhui tu 文會圖
wenren 文人
Wenxuan Mansion 文選樓
Wenzheng Academy 文正
　書院
wo 窩
wodan 窩單
wu 無
Wu (state, surname) 吳
Wu Bicheng 吳必長
Wu Jiaji (1618–84) 吳嘉紀
Wu Jingzi (1701–54) 吳敬梓
Wu Lunyuan 鄔掄元
Wu Qi (1619–94) 吳綺
Wu Sangui (1612–78) 吳三桂
Wu Sheng (1589–after 1644)
　吳甡
Wu Shihuang 吳世璜
Wu Woyao (1867–1910) 吳沃堯
Wu Ziliang 吳自亮
Wuchang 武昌
Wucheng 蕪城
Wuhu 蕪湖
Wuliesi 五烈寺
Wuling 武陵
wu mai ri huo 毋買日貨
Wu pai 吳派
Wu School 吳派
wu shi 烏師
Wuxi 無錫
wu xiang 吾鄉
Wuyuan 婺源

Xia (dynasty) 夏
Xia Zhifang (js 1723) 夏之芳
Xiahe 下河
Xia maimai jie 下買賣街
Xi'an 西安
xiancheng 縣丞
Xianfeng 咸豐

xiang fa Guangdong cai 想發廣
　東財
Xiangling 襄陵
xiangmin 鄉民
Xianliang jie 賢良街
Xiannümiao 仙女廟
xian shou qing feng 賢守清風
xiao jiaochang 小校場
Xiao linglong shanguan 小玲瓏
　山館
xiaomin 小民
xiao nan bi xiang, da nan bi
　cheng 小難避鄉, 大難避城
Xiaoyuan 篠園
Xie Qikun (1737–1802) 謝啓昆
Xie Rongsheng (js 1745) 謝溶生
Xie Shisong 謝士松
Xie Yong (Jinpu) (1719–95) 謝墉
　(金圃)
Xie Zhaozhe (1567–1624)
　謝肇淛
Xie qin shinü 攜琴仕女
Xin'an 新安
Xing'an 行庵
xinggong 行宮
xing yance 行鹽筴
Xingyuan 杏園
Xinsheng jie 新勝街
xinshishi 新時式
xinwei 辛未
xin yang 新樣
Xiong Wencan (d. 1640) 熊文燦
xishang 西商
Xiuning 休寧
Xiuyuan 休園
Xixi 西溪
xiyanghua 西洋話
xiyangjing 西洋鏡
Xu Chengzong 許承宗
Xu Duan (d. 1812) 徐端
Xu Hong 徐泓
Xu Huaizu (d. 1777) 徐懷祖

Xu Ke 徐珂
Xu Rong (1686–1751) 許容
Xu Shangzhi 徐尚志
Xu Shiqi 徐石麒
Xu Shukui (*jr* 1738) 徐述夔
Xue Fubao (1840–81) 薛福保
Xue Shou 薛壽
xuegong 學宮
Xuli gonghui 恤嫠公會
xun 汛
xunshang 巡商
xunyan yushi 巡鹽御史
xunyi 巡役

yajia 芽茄
Yan Jin 閻金
Yan Ruoju (1636–1704) 閻若璩
Yancheng 鹽城
Yang Cheng 楊成
Yang Guang 楊廣
Yang Xianming 楊顯名
Yang Xingmi (d. 905) 楊行密
yangbu 洋布
Yangdi 煬帝
yan'ge 沿革
Yangjiyuan 養濟院
yanglianyin 養廉銀
Yangren 揚人
yang shouma 養瘦馬
Yangzhai 羊寨
Yangzhou 揚州
Yangzhou bangzi 揚州梆子
Yangzhoufu ji 揚州府籍
Yangzhou huafang lu 揚州畫舫錄
Yangzhou junyi 揚州郡邑
Yangzhou lishi wenhua fengsu
    neibu ziliao 揚州歷史文化風
    俗內部資料
Yangzhou xuepai 揚州學派
Yangziqiao 揚子橋
yanjin 鹽斤
yankesi dashi 鹽課司大使

yanwudao 鹽務道
yanyunshi 鹽運使
yanyunsi yunpan 鹽運司運判
Yanzhou 延州
Yao Nai (1732–1815) 姚鼐
Yao Wentian (1758–1827)
    姚文田
Ye Fanglin 葉芳林
Ye Qi 葉淇
Yechun jueju shier shou 冶春絕
    句十二首
yemingzhu 夜明珠
yeshang 業商
yeshi 野食
Yi (surname) 喬
Yi Junzuo (1898–1976) 易君左
Yiling 宜陵
yimin 遺民
yin 引
Yin Huiyi (1691–1748) 尹會一
Yin Qi 殷起
Yin Xiaoyuan 飲篠園
ying 營
Yingyuan 影園
Yinjishan (1696–1771) 尹繼善
yinyi 隱逸
yiren 邑人
yishen 邑紳
Yishuang Tavern 把爽酒肆
Yixian 黟縣
Yiyang 弋陽
yi yance zhanji 以鹽筴占籍
yi yan gai ji 以鹽改籍
Yizheng 儀徵
Yongle 永樂
Yu (Prince; 1614–49) 豫
Yu (sage-king) 禹
Yu Chenglong (1638–1700)
    于成龍
Yu Dafu (1896–1945) 郁達夫
Yu Guande 余觀德
Yu Jimei 尉濟美

Yu Yuanjia 余元甲
Yu Zhiding (1647–after 1709)
　禹之鼎
Yu'an 余岸
yuan 遠
Yuan (dynasty) 元
Yuan Guotang 員果堂
Yuan Hongxiu 員洪麻
Yuan Jiang (ca. 1671–ca. 1746)
　袁江
Yuan Jixian (1598–1646)
　袁繼咸
Yuan Mei (1716–97) 袁枚
Yuan Yao (fl. 1739–78) 袁耀
Yuanye 園冶
yuehu 樂戶
Yukou 喻口
Yulin 榆林
yunji 運籍
yunku 運庫
yunshang 運商
Yuyang Mountain 漁洋山
yuyin 余引

zao 灶
zaoding 灶丁
zaohu 灶戶
zaozhi gongshi zhi fa 造製宮室
　之發
Zha Shibiao (1615–98) 查士標
zhafu 閘夫
zhang 丈
Zhang (surname) 張
Zhang Dai (Tao'an; 1597–1689)
　張岱 (陶庵)
Zhang Geng (1685–1760) 張庚
Zhang Guohua 張國華
Zhang Liansheng 張連生
Zhang Pengge (1649–1725)
　張鵬翮
Zhang Shicheng (1321–67)
　張士誠

Zhang Shike 張士科
Zhang Shimeng 張師孟
Zhang Shunhui 張舜徽
Zhang Sike 張四可
Zhang Sixiang 張嗣祥
Zhang Xianzhong (1605–47)
　張獻忠
Zhang Zeduan 張擇端
Zhao Zhibi 趙之壁
Zhejiang 浙江
zhen 鎮
zheng 正
Zheng (Xie) Banqiao (1694–1765)
　鄭 (燮) 板橋
Zheng Chenggong (1624–62)
　鄭成功
Zheng Jinglian 鄭景濂
Zheng Weiguang (js 1659)
　鄭爲光
Zheng Weihong (d. 1645)
　鄭爲虹
Zheng Xiaru (1610–73) 鄭俠如
Zheng Yuanbi 鄭元弼
Zheng Yuanhua (d. after 1655)
　鄭元化
Zheng Yuansi 鄭元嗣
Zheng Yuanxi (js 1631)
　鄭元禧
Zheng Yuanxun (1598–1644)
　鄭元勳
Zheng Zhilun 鄭之綸
Zheng Zhiyan 鄭之彥
Zhengde 正德
Zhengxin Academy 正心書院
Zhengyi Academy 正誼書院
zhengyin 正引
Zhenzhou 眞州
Zhili 值隸
zhilizhou 直隸州
Zhong, Beihexia 中, 北河下
zhong yan 中鹽
Zhou (dynasty) 周

Zhou Lianggong (1612–72)
周亮工
Zhou Sheng 周生
zhoutong 州同
Zhu Jiang 朱江
Zhu Shi (1665–1736) 朱軾
Zhu Xi (1130–1200) 朱熹
Zhu Xiaochun (jr 1762) 朱孝純
Zhu Yihai (1618–62) 朱以海
Zhu Yuanzhang (1328–98)
朱元璋
Zhu Ziqing (1898–1948) 朱自清
Zhuangzi 莊子
Zhutang 竹堂
Zhuxi xushe 竹西續社
zhu zhou xian huiguan 諸州縣
會館
Zong Bu 宗部

Zong Guan (jr 1702) 宗觀
Zong Hao (js 1643) 宗灝
Zong Jie 宗節
Zong Mingshi (js 1589) 宗名世
Zong Wanguo 宗萬國
Zong Wanhua (jr 1609)
宗萬化
Zong Yuanding (1620–98)
宗元鼎
Zong Yuanyu 宗元豫
zongshang 總商
Zou Weizhen (d. 1787)
鄒維貞
zui you fusheng 最有富盛
zun 遵
Zuoweijie 佐衛街
zuoza 佐雜
zushang 租商

Index

academies, 125, 268; Meihua and Anding, 247–48; outside Yangzhou, 250–51; and student stipends, 379*n*53. *See also individual academies by name*
Alexander, William, 149, 335–36
Anding Academy, 247–48, 279–80
Anfeng, 32
Anhui: divided from Jiangsu, 29, 120
An Lushan Rebellion, 22, 44
artists, 5, 202, 208, 334, 370*n*96; residences of, in Yangzhou, 179, 366*n*30, 369*n*90; periodization of, 208, 371*n*126. *See also* painting

Bai Zhongshan, 163
Baita, *see* White Stupa
Bamboo Garden, 192, 201, 209
Bao Shichen, 276–77, 304–5
Bao Shufang, 124
Bao Zhao, 20
Bao Zhidao, 124, 249

baoxiao, 121, 302
Baoying, 32, 140
Baoying Lake, 149
Baozhang Lake, 102, 115
Barrow, John, 18
bathhouses, 207; locations of, 371*n*124
Beihu, 80, 90–91, 93–94, 96, 279; prominent scholar-officials in, 349*n*62
Beijing, 8
Bourke, Peter, 17
Braudel, Fernand, 315
Brook, Timothy, 60
brothels, 205
Brown, Claudia, 259

Cahill, James, 68
Canton, 8, 292
Cao Zhenyong, 304
central place theory, 37
Chai Shijin case, 271
Chang'an, 52
Changlu salt monopoly, 45, 48
Chaves, Jonathan, 95
Chen Congzhou, 188

Chen Hongmou, 246, 250, 270
Chen Jiru, 68
Chen Qubing, 281
Chen Weisong, 192
Chen Zhuan, 369n90
Cheng Liangru, 123
Cheng Mengxing, 11, 123, 191–
    92, 193, 201, 257
Cheng Sui, 92, 95
Cheng Zhiying, 123
Chou, Ju-hsi, 259
*Chronicle of the Painted Barques of
    Yangzhou*, 4, 12–13, 284, 295.
    See also Li Dou
Chuanchang he, *see* Salt Yards
    Canal
Clavelin, Stanislaus, 308
cotton, 225–27
courtesans, 178, 219–21. *See also*
    prostitutes
Crossley, Pamela, 273
Cui Hua, 118–19
customs revenue, 35
customs stations, 176, 340n79;
    scene at, in Yangzhou, 299

Dai Mingshi, 73, 78
Dai Zhen, 268, 274–75
Datong, 47, 51, 52
De Bary, Wm. Theodore, 277
Dong Qichang, 68
Dong Zhongshu, 117–18; Tem-
    ple of, 97, 117
Donglin Academy, 70
Dongtai, 32–33, 130–31, 147
dream of Yangzhou, 4–5
Dreams of Wind and Moon, 13,
    177, 186, 293, 298
Du Fu, 44
Du Jun, 92
Du Mu, 4
Du Halde, J. P., 175, 365n9

Duoduo, 70, 77

Eastern Garden (Dongyuan),
    192, 209
eight eccentrics, 371n125. *See
    also* artists; painting
Elman, Benjamin, 256, 276, 278,
    282
examinations, 52, 120; Yang-
    zhou successes in, 99, 251–53

famine, *see* natural disasters
famine-relief granaries, 244–45
Fan Changjiang, 5
Fan Quan, 94
Fan Zhongyan, 229
Fang Dongshu, 265
Fang Shishu, 256
Fangong Dike, 37, 127; as social
    divider, 130
fashion, 222–24, 298
Fei Mi, 92, 93
Fei Xiaotong, 152
Fengyue meng, see *Dreams of
    Wind and Moon*
floods, Chap. 7 *passim*, 307–8.
    See also natural disasters
food culture, 290–91
footbinding, 55
Fountain of Letters Pavilion:
    construction of, 269–70; sack-
    ing of, 309
Four Treasuries, 106–74, 269
Fu Chai, 19
Funing, 33

Gandar, Dominic, 148–49, 309
Ganquan, 31–32, 175
Gao Heng, 126, 132
Gao Jie, 73–75, 81
Gao Xiang, 259
Gaoyou, 32, 140; floods in, 150

Garden of Nine Peaks, 196–97, 297

Garden of Reflections, 64–68

Garden of Retirement, 65–66, 193–94; gazetteer of, 194, 368n68

Garden of Ten Thousand Stones, 193, 209; and Shitao, 368n64

gardens, 64–68, 188–203; ownership of, 189; literary, 193–94, 199; and Southern Tours, 195–97, 210, Appendix F; palace style of, 198–99, access to, 205–6; decline of, 209–10, 297; and periodization of, in Yangzhou's history, 210; women in, 259

Ge Garden, 124, 209

gentry: and water control, 156; and merchants, 261–63

genwo, 122

Gong Weiliu, 80–81, 93

Gong Xian, 9, 92, 97

Gong Zizhen, 299

grain trade, 126

Grand Canal, 25–26, 32, 34–36, 111; and water control in Xiahe, 148–50, 153–54, 157–58, 176; in nineteenth century, 307

Guangling, 20

Guangzhou, see Canton

Guazhou, 15, 32, 36, 53, 74, 104; hospice in, 247, 250

Guo Moruo, 79

Hahn, Emily, 313

Hai'an, 36

Hailing, 18

Haimen, 33

hairstyles, 223

Han Gou, 19

Hanan, Patrick, 293–94

Hancheng, 19, 336n7

Hankou, 8, 145, 305, 309

Hanshang mengren, 13. See also Dreams of Wind and Moon

Hay, Jonathan, 9, 96, 335

He Cheng, 53–54

He Junzhao, 192, 201

Ho Ping-ti, 6–7, 63; on native-place associations, 241–43; on examination successes in Yangzhou prefecture, 251; on social mobility, 253, 255; on head merchants, 356n27

Ho Wai-kam, 253, 256

Hong Zhenke, 259

Hongze Lake, 153, 156, 307–8. See also Huai River

Honig, Emily, 5

Hsü, Ginger Cheng-chi, 201, 256

Hu Qizhong, 99

Hu Yuan, 229

Hua Yan, 186, 259, 367n39

huafang, see under pleasure craft

Huai River, 19, 33; confluence with Yellow River and the Grand Canal, 153, 362

Huai'an, 26, 30, 147

Huaibei, 33, 144

Huainan, 29, 144

Huang brothers, 189, 198, 249

Huang Chao rebellion, 22, 337

Huang Chengji, 384n58

Huang Lümao, 249

Huang Shen, 219

Huang Sheng, 191

Huang Yingtai, see under Huang Zhiyun

Huang Zhiyun (Yingtai), 124, 209, 304

Hui Dong, 268

Huizhou salt merchants: rise of, in Yangzhou, 56–68 *passim*; distinctive surnames of, 238, 380*n*87; residential patterns of, 239: native-place connections of, 239–40; and philanthropic activities, 237, 250–51; as garden owners, 260; and Yangzhou scholars, 283

Huizhou, 8, 57–59; contrasted with Yangzhou, 228; and early nineteenth-century changes, 301

Ilioni, Catherine and Antonio, 44

Ji Cheng, 65, 204
Ji Huang, 166, 362*n*17
Ji Yun, 273
Jiang Chun, 124, 201, 209, 219, 278, 287
Jiang Fan, 273, 384*n*58; scholarly interests of, 283, 284
Jiang Kui, 95
Jiang Yong, 275
Jiangbei: definition of, 29–30; urbanization in, 34–37
Jiangdu: meaning of, 21
Jianghuai (province), 311
Jiangnan: definition of, 29–30, 339*n*55
Jiangnan Director-General of River Administration, 29
Jiangnan hedao zongdu, *see* Jiangnan Director-General of River Administration
Jiangsu: creation of province, 29

Jiao Xun, 84, 266, 273–87 *passim*, 309; social origins of, 279–80; scholarly interests of, 283, 284

jiaochang, *see* Parade Ground
Jiaodong Zhou Sheng, *see* Zhou Sheng
Jin Fu, 111, 150, 159
Jin Nong, 208, 266
Jin Zhen, 112
Jingjiang, 33–34
Jingyang, 50
Jiqi, 58–59; pronunciation of, 344*n*62
Jiufengyuan, *see* Garden of Nine Peaks
Jiuri xing'an wenyan, *see* Ninth Day Literary Gathering in the Temporary Retreat
Jixi, *see* Jiqi
Johnson, Linda Cooke, 7

Kangxi emperor, 109, 111. *See also* Southern Tours
Keenan, Barry, 312
Kinkley, Jeffrey, 28
Kong Shangren, 108, 113, 119, 263

Li Cheng, 124, 136, 143, 303
Li Daonan, 278
Li Dou, 4, 11; cited, 38, 117, 176–77, 207, 199–200, 259, 288, 300; on Zheng Yuanxun, 84; on courtesans, 219–21; on fashions in Yangzhou, 223. *See also Chronicle of the Painted Barques of Yangzhou*
Li E, 257–58, 289
Li Fayuan, 97
Li Gan, 222, 374*n*41
Li Ruzhen, 293
Li Tingzhi, 23, 95
Li Yi, 106
Li Yu (936–78), 22
Li Yu (1611–80?), 55

Li Zicheng, 71, 97
Li Zongkong, 110–11, 237, 262
Liang Qichao, 267, 275
Liang Qizi, 237, 243, 245–46; on the blurring of social boundaries, 254–55, 380n77
Liang Yusi, 64, 80–81, 84, 90
Liang Zhangju, 298
Lianghuai salt monopoly, 8, 25–26, 32, 44–47; marketing area of, 37, 46; Huainan and Huaibei sectors of, 45; and *kaizhong fa*, 47–48; evolution in the Ming, 47–49; and shipment system (*gangfa*), 62, 122; resumption of, in the Qing, 97; and philanthropy, 111; gazetteers of, 118–19; officials of, 131–35; regional significance in Jiangbei, 136–44; reforms in Yongzheng period, 138–39; and river conservancy, 164–69; problems of, in the early nineteenth century, 301–8; ticket system (*piaofa*), 306–7
Liangzhe salt monopoly, 45, 59
Lianxing Temple, 192, 194
Lin Daoyuan, 219
Lin Sumen, 13, 284–92, 300
Lin Yutang, 313
Linfen, 52
Ling Shu, 276–77
Ling Shun, 282
Ling Tingkan, 266, 273, 278, 281, 283; social origins of, 280; scholarly interests of, 293
Literary Gathering at a Yangzhou Garden, *see under* Ninth Day Literary Gathering in the Temporary Retreat

literary inquisition, 101, 270–73
Little Qinhuai Canal, 179, 189, 205
Little Translucent Mountain Lodge, 199–201, 209, 369–70n90; Zhang Geng's painting of, 209, 372n134
Liu Baonan, 275, 283, 284
Liu Bi, 20
Liu Chongxuan, 247
Liu Sen, 249
Liu Shipei, 266
Liu Shouzeng, 266
Liu Taigong, 283
Liu Wenqi, 275, 276, 282
Liu Xizai, 312
Liu Zhaoji, 77
Liuqiu, 100
"long eighteenth century," 110, 120
Lotus Flower Bridge, 191, 194
loyalism, 90–96 *passim*, 105, 108–10
Lu Jianzeng, 126, 132, 209, 268, 273
Luo Ping, 201
Luo Youlong, 88

Ma brothers, 186, 193, 200–201, 249, 253, 289, 369n90; 1743 literary gathering of, 256–60. *See also* Little Translucent Mountain Lodge; Ma Yueguan; Ma Yuelu; Ma Zhenbo
Ma Wulu, 77, 82–83
Ma Yueguan, 257; and Meihua Academy, 247–48. *See also* Ma brothers
Ma Yuelu, 200. *See also* Ma brothers
Ma Zhenbo, 248
Macartney embassy, 18

Mackerras, Colin, 286
macroregions, 33
Manchus, 69–70, 100; ethnic identity of, 273; and Banner garrisons, 365n9
Mann, Susan, 244, 255, 262
Mao Xiang, 93
Marco Polo, 3, 23
merchants, 6; from Shanxi and Shaanxi, 8; from Ningbo, 8–9; Arab, Persian, and Jewish, 43. See also salt merchants
Metzger, Thomas, 299, 302
Meyer-Fong, Tobie, 11
Min Ding, 98
Min Shizhang, 98, 237, 243, 262
Min Tingzuo, 237
Ming dynasty: fall of, 41–42, 70–72, 88–89
Ming History, 78, 101
Mongols: as rulers of China, 44, 119
Mote, Frederick, 270
music, see opera
Muslims, 43–44

Nan Zhili, 29
Nanjing, 47
Nantong, see Tongzhou
Naquin, Susan, 8
native place, 6, Chap. 11 passim; consciousness of, 274, 285
native-place associations, 178, 240–43
natural disasters 30, 69, 71, 111
network system theory, 37
New City, 54, 56, 72, 172, 173–88
Ni Zan, 95
nianwo, 122
Ningbo, 8, 35

Ninth Day Literary Gathering in the Temporary Retreat, 253–54, 256–60
North Lakes, see Beihu
Nurhaci, 69–70

Oderic of Pordenone, 45
Old City, 173, 175
Ono Kazuko, 293
opera, 55, 286–88, 305
opium, 303
orphanage, 237, 243; conflicting accounts of, 376–77n4
Otani Toshio, 276, 278
Ouyang Xiu, 112

painted barques, 4, 115–16, 187, 204–5
painting, 57, 68, 98; topics of, 208–9. See also artists; eight eccentrics
Pan Jixun, 153–54
Pan Shi'en, 308
Pan Zengshou, 308
Parade Ground, 176, 185–86
Peach Blossom Fan, 109–10, 216, 286
philanthropic institutions, 236–37, 243–44, 245–46, 249
Pingshan Hall, 112, 189, 190
pirates, 53–54, 343n44
pleasure craft, see painted barques
Pomeranz, Kenneth, 222, 311
prostitutes, 214–16, 218, 219, hierarchy of, 221; and local origins of, in Yangzhou, 232–34
Puhading, 44

Qian Yong, 297

Qianlong emperor, 12; and water control, 162. *See also* Southern Tours
Qiao lineage, 72
Qimen, 58, 256
Quan Zuwang, 77

Rainbow Bridge, 268. *See also* Red Bridge
Red Bridge, 105, 107, 109, 189
Ren Dachun, 266, 272–73, 275
Ren Minyu, 77, 88
restaurants, 205, 207
Restoration Society, 64
River Administration, 156–63; and relations with the Lianghuai salt monopoly, 164–68
river conservancy, 158, 159–60, 163. *See also* River Administration; Xiahe
Roddy, Stephen, 293
Rowe, William, 144, 255, 305; on Chen Hongmou and philanthropy in Yangzhou, 378–79n47
Ruan Heng, 284
Ruan Xian, 284
Ruan Yuan, 13, 90, 92–94 *passim*, 266, 273; social origins of, 278–79; and local scholarship, 282–83; and family temple, 294; and decline of Yangzhou, 297–98
Rugao, 33

Saeki Tomi, 302
salt controller: yamen of, 118, 133; duties of, 133, 167–68
salters, 85–86, 98, 128–31; native place origins of, 122

salt merchants, 47; native-place origins of, in the Ming, 47, 49; in the Qing, 122; from Shanxi and Shaanxi, 49–56, 120; categories of, 49, 121–23; participation of, in the examination system, 52, 120; defense of the city by, 54; from Huizhou, 57–68; change in household registration of, 61; social mobility of, 63; in the early Qing, 98, 120; relations with the dynasty, 119–21; and examinations, 120; different categories of, 123; wealth of, 124–26; other business activities of, 125–26; relations with salters, 128, 131–35; place in the salt administration, 145; contributions to water control, 164–66, 302–3; residences of, 178; as garden owners, 189; West and Huizhou distinguished, 238; and social mobility, 261–64, and Four Treasuries project, 270; downward mobility of, 291; decline in numbers of, 300, 301, 307. *See also* Huizhou salt merchants
salt peddlers, 136–37
salt production, 129–30, 136
salt receivers, 134–35
salt smuggling, 36, 130, 135–44, 301
salt trade, as equivalent to farming, 118–19
Salt Transport Canal, 164
salt yards, 127–31, 133–35, 358n55, 359nn81–82
Salt Yards Canal, 36, 164

Sanfan Rebellion, 112
Sanyuan, 50, 97
Scott, William Henry, 14, 335
Shaanxi, 48, 50
Shanghai, 312
Shanxi, 47, 50
Shanyang, 33
Shaobo, 207; hospice in, 247, 250
Shen Defu, 86, 217
Shen Fu, 232
Shen Kuo, 23
shenshang, see gentry
Shexian, 56, 60, 62
Shi Dewei, 77
Shi Kefa, 5, 41, 71, 74–75, 109;
    death of, 77–79
Shiba Yoshinobu, 8
Shimada Kenji, 280
Shitao, 9, 13, 15, 21, 96–97, 191
shopping, 178
shouma, see under thin horses
Shrine to the Five Constant
    Ones, 228–30
Skinner, G. W., 76. See also
    macroregions
Smith, Joanna Handlin, 243, 261
social boundaries: blurring of, 6,
    254; and native place, 259–60,
    261–62, 276–83
social mobility, 253. See also so-
    cial boundaries
sojourners, 12, 144
Southern Ming, 110
Southern Tours, of Kangxi, 111,
    120–21, 149; of Qianlong, 12,
    121, 124, 172, 195
Su Shi, 112
Subei, 5. See also Jiangbei
Sui Yangdi, 6
Sun Mo, 92, 107
Sun Zhiwei, 92, 95, 97, 107

Suzhou, 47, 53, 68; contrasted
    with Yangzhou, 170, 172–73,
    231–32; and Yangzhou opera,
    287

Taiping Rebellion, 308–10
Taixing, 33
Taiyuan, 47, 52
Taizhou, 18, 32
Tang Laihe, 82
Tao Zhu, 124, 305, 306
taverns, 205
teahouses, 205, 206–7
Temporary Retreat (Xing'an),
    193, 199, 202–3, 253. See also
    Ninth Day Literary Gather-
    ing in the Temporary Retreat
"Ten-Day Diary of Yangzhou,"
    6, 70, 72–78, 88–89
thin horses 214, 216, 218–19, 235
Tian Han, 79
Tiebao, 302
Tongzhou, 33
tourism, 204–9
Tu Yuelong, 273

urbanization, 36–37

Venice, 3–4

Wang Chang, 273
Wang Fangqi, 91, 93
Wang Fangwei, 91, 94, 96, 102
Wang Gen, 85, 277
Wang Ji, 100, 106–7
Wang Jiaoru, 179, 366n29
Wang Maohong, 266
Wang Maolin, 97–98, 107
Wang Nalian, 91
Wang Niansun, 266, 268, 275,
    283, 284

Wang Shiduo, 59
Wang Shimin, 98
Wang Shixing, 216, 217
Wang Shizhen, 79, 102–8, 113, 119
Wang Tingzhang, 209, 366*n*29
Wang Xiuchu, 6, 72–77 *passim*
Wang Yinggeng, 11, 123, 189, 229, 249
Wang Yongji, 83
Wang Yun, 208
Wang Yuzao, 80–81, 84, 91, 96
Wang Zheng, 70
Wang Zhenzhong, 8, 122
Wang Zhong, 11–12, 266, 268, 273–83 *passim*; social origins of, 280–82
Wan school, 275
Wanshiyuan, *see* Garden of Ten Thousand Stones
water control, Chap. 7 *passim*; and local gentry, 156, distinguished from river conservancy, 158
weddings, 289–90, 300
Wei Minghua, 223
Wei Xi, 112–13, 243
Wei Yuan, 303
Wei Zhongxian, 70
Weiyang, 24
Wenhui ge, *see* Fountain of Letters Pavilion
West merchants: as philanthropists, 237
White Stupa, 195, 197
Will, Pierre-Etienne, 168, 169, 244
women, 38, 55; as Ming martyrs, 76–77; in iconography of Yangzhou, 213–15; literacy of, 217; and household economic activity, 224–27; and virtue/

piety, 228–30; from Huizhou and Yangzhou compared, 231; and communal boundaries, 234; and urban institutions for, 236; and the education of sons, 281–82. *See also* courtesans; prostitutes
Wu Jiaji, 79, 93, 95, 119; and Wang Shizhen, 106–7; works by, 351*n*11
Wu Jingzi, 219, 230
Wu Qi, 99, 218
Wu Sheng, 80
Wu Woyao, 205
Wucheng, 15, 21
Wulie si, *see* Shrine to the Five Constant Ones
Wu school, 275
Wuyuan, 58

Xia Zhifang, 168
Xiahe, 152, 154, 157; river conservancy officials in, 158–62; water control in, 302, 307
Xiangling, 52
Xiao linglong shanguan, *see* Little Translucent Mountain Lodge
Xiaoyuan, *see* Bamboo Garden
Xie Qikun 271, 273
Xie Rongsheng, 270
Xie Shisong, 270
Xie Zhaozhe, 29, 52, 216
Xin'an, *see* Huizhou
Xing'an, *see* Ninth Day Literary Gathering in the Temporary Retreat; Temporary Retreat
Xinghua, 32, 54
Xiong Wencan, 86–87
Xiuning, 58
Xiuyuan, *see* Garden of Retirement

Xixi, 36
Xu Duan, 302
Xu Ke, 219
Xu Rong, 162
Xu Shiqi, 94
Xu Shukui case, 271–73
Xu Yinong, 172, 176
Xue Fubao, 311

Yan Ruoju, 274
Yancheng, 33; as source of prostitutes, 233–34
Yang Cheng, 82–83
Yang Lien-sheng, 119
Yangzhou: in the Sui-Tang, 3, 21–22, 43–44; dreams of, 4–5; painters in eighteenth century, 9; rural hinterland of, 9–10, 27–28, 30–31, 148–52; visual depictions of, 13–15; famous sites in, 15, 79, 107, 108; meaning of the name, 18–19, 27; early history of, 18–24; walls of, 20, 22–24, 54, 173–74, strategic significance of, 24–25; urban population of, 30, 357n51; composition of prefecture, 31–33; markets in, 35, 187–88; as customs port, 35; in the 1930s, 38; New City, 54, 56; opera in, 55; compared to Yizheng, 60–61, 63–64; affected by late Ming disturbances, 71; native-place divisions in, 72–73; massacre of residents, 75–76; Ming loyalists in, 95–96; restoration in the early Qing, 99–101; threatened by Zheng Chenggong, 104; customs in, 112; impact of salt merchant wealth on, 126–27; consumption of regional products, 127; surveillance of salt sales in, 137–41; garrisons in, 175; maps of, 176–77; spatial differentiation in, 176–88; gardens in, 188–203; industry in, 224–27; local customs, 228, 289–91; famine relief granaries in, 244–45; scholarship in, 266; lineages in, 285; dialects in, 288–89, 296; decline of, Chap. 12 passim; in Taiping Rebellion, 308–10; in Republican era, 312–15
Yangzhou baguai, see eight eccentrics
Yangzhou huafang lu, see Chronicle of the Painted Barques of Yangzhou
Yangzhou meng, see dream of Yangzhou
Yangzhou school, 265–83, Appendix G; and merchant connections, 278–81
"Yangzhou shiri ji," see "Ten-Day Diary of Yangzhou"
yankesi dashi, see salt receivers
Yao Nai, 248, 268
Yao Wentian, 24
Ye Fanglin, 256
Yellow River, 50–51, 310
Yi Junzuo, 79, 314–15
yimin, see loyalism
Yin Huiyi, 250
Yin Qi, 83
Yingyuan, see Garden of Reflections
yinyi, see loyalism
Yixian, 58–59
Yizheng, 26, 32, 36, 71, 104, 136, 147; confused with Yangzhou, 60–61, 345–46n90

Yongzheng emperor, 159; and academies, 247; and philanthropic institutions, 243–44
Yu Chenglong, 248
Yu Dafu, 4, 312
Yu Yuanjia, 193, 368n64
Yuan Jixian, 80
Yuan Mei, 79, 94
Yuan Yao, 193, 201
Yulin, 53

zaohu, see salters
Zha Shibiao, 92, 97
Zhang Banqiao, 266
Zhang Dai (Tao'an), 176, 190, 214, 217–18
Zhang Geng, 199, 201
Zhang Liansheng, 303
Zhang Pengge, 154–55; hydraulic strategy of, 362n16
Zhang Shicheng, 33, 128
Zhang Shike, 259; distinguished from Zhang Sike, 381n92
Zhang Shimeng, 237
Zhang Shunhui, 275, 283, 293
Zhang Xianzhong, 71, 73
Zhao Zhibi, 11–12, 194, 271
Zheng Banqiao, 4, 101, 125, 201, 219
Zheng Chenggong, 92, 104
Zheng family, 62–63, 179, 345–46n90. See also Zheng Yuanxun
Zheng Jinglian, 62–63
Zheng Weihong, 64, 73, 80–81, 84, 99

Zheng Xiaru, 63, 65, 80, 100
Zheng Xie, see Zheng Banqiao
Zheng Yuanhua, 63; as benefactor of orphanage, 243, 376n4
Zheng Yuansi, 63
Zheng Yuanxun, 63–68, 70–71, 99, 257; and the fall of Yangzhou, 80–84; household registration of, 345n89; and relatives in Yizheng, 346
Zheng Zhiyan, 63
Zhenjiang, 38
Zhenzhou, 26, 338n45. See also Yizheng
Zhou Lianggong, 76, 101–2, 106, 119
Zhou Sheng, 4, 38, 95, 221, 232, 227–28, 232, 265, 289, 303; identity of, 341n89
Zhu Jiang, 65
Zhu Shi, 50
Zhu Xiaochun, 248
Zhu Yuanzhang, 23, 128
Zhu Ziqing, 1–2, 79, 213
Zhuangzi, 4
Zhuxi xushe, 64, 34n93
Zong Bu, 85
Zong Guan, 87
Zong Hao, 80–81, 85–87
Zong Jie, 85
Zong Mingshi, 85–86
Zong Wanhua, 86–87
Zong Yuanding, 87, 93, 191
Zong Yuanyu, 93
Zou Weizhen, 281
Zurndorfer, Harriet, 59, 61

Harvard East Asian Monographs
(* out-of-print)

*1. Liang Fang-chung, *The Single-Whip Method of Taxation in China*

*2. Harold C. Hinton, *The Grain Tribute System of China, 1845–1911*

3. Ellsworth C. Carlson, *The Kaiping Mines, 1877–1912*

*4. Chao Kuo-chün, *Agrarian Policies of Mainland China: A Documentary Study, 1949–1956*

*5. Edgar Snow, *Random Notes on Red China, 1936–1945*

*6. Edwin George Beal, Jr., *The Origin of Likin, 1835–1864*

7. Chao Kuo-chün, *Economic Planning and Organization in Mainland China: A Documentary Study, 1949–1957*

*8. John K. Fairbank, *Ching Documents: An Introductory Syllabus*

*9. Helen Yin and Yi-chang Yin, *Economic Statistics of Mainland China, 1949–1957*

*10. Wolfgang Franke, *The Reform and Abolition of the Traditional Chinese Examination System*

11. Albert Feuerwerker and S. Cheng, *Chinese Communist Studies of Modern Chinese History*

12. C. John Stanley, *Late Ching Finance: Hu Kuang-yung as an Innovator*

13. S. M. Meng, *The Tsungli Yamen: Its Organization and Functions*

*14. Ssu-yü Teng, *Historiography of the Taiping Rebellion*

15. Chun-Jo Liu, *Controversies in Modern Chinese Intellectual History: An Analytic Bibliography of Periodical Articles, Mainly of the May Fourth and Post-May Fourth Era*

*16. Edward J. M. Rhoads, *The Chinese Red Army, 1927–1963: An Annotated Bibliography*

17. Andrew J. Nathan, *A History of the China International Famine Relief Commission*

*18. Frank H. H. King (ed.) and Prescott Clarke, *A Research Guide to China-Coast Newspapers, 1822–1911*

19. Ellis Joffe, *Party and Army: Professionalism and Political Control in the Chinese Officer Corps, 1949–1964*

*20. Toshio G. Tsukahira, *Feudal Control in Tokugawa Japan: The Sankin Kōtai System*

21. Kwang-Ching Liu, ed., *American Missionaries in China: Papers from Harvard Seminars*

22. George Moseley, *A Sino-Soviet Cultural Frontier: The Ili Kazakh Autonomous Chou*

23. Carl F. Nathan, *Plague Prevention and Politics in Manchuria, 1910–1931*

*24. Adrian Arthur Bennett, *John Fryer: The Introduction of Western Science and Technology into Nineteenth-Century China*

25. Donald J. Friedman, *The Road from Isolation: The Campaign of the American Committee for Non-Participation in Japanese Aggression, 1938–1941*

*26. Edward LeFevour, *Western Enterprise in Late Ching China: A Selective Survey of Jardine, Matheson and Company's Operations, 1842–1895*

27. Charles Neuhauser, *Third World Politics: China and the Afro-Asian People's Solidarity Organization, 1957–1967*

28. Kungtu C. Sun, assisted by Ralph W. Huenemann, *The Economic Development of Manchuria in the First Half of the Twentieth Century*

*29. Shahid Javed Burki, *A Study of Chinese Communes, 1965*

30. John Carter Vincent, *The Extraterritorial System in China: Final Phase*

31. Madeleine Chi, *China Diplomacy, 1914–1918*

*32. Clifton Jackson Phillips, *Protestant America and the Pagan World: The First Half Century of the American Board of Commissioners for Foreign Missions, 1810–1860*

33. James Pusey, *Wu Han: Attacking the Present Through the Past*

34. Ying-wan Cheng, *Postal Communication in China and Its Modernization, 1860–1896*

35. Tuvia Blumenthal, *Saving in Postwar Japan*

36. Peter Frost, *The Bakumatsu Currency Crisis*

37. Stephen C. Lockwood, *Augustine Heard and Company, 1858–1862*

38. Robert R. Campbell, *James Duncan Campbell: A Memoir by His Son*

39. Jerome Alan Cohen, ed., *The Dynamics of China's Foreign Relations*

40. V. V. Vishnyakova-Akimova, *Two Years in Revolutionary China, 1925–1927*, trans. Steven L. Levine

*41. Meron Medzini, *French Policy in Japan During the Closing Years of the Tokugawa Regime*

42. Ezra Vogel, Margie Sargent, Vivienne B. Shue, Thomas Jay Mathews, and Deborah S. Davis, *The Cultural Revolution in the Provinces*

*43. Sidney A. Forsythe, *An American Missionary Community in China, 1895–1905*

Harvard East Asian Monographs

*44. Benjamin I. Schwartz, ed., *Reflections on the May Fourth Movement.: A Symposium*

*45. Ching Young Choe, *The Rule of the Taewŏngun, 1864–1873: Restoration in Yi Korea*

46. W. P. J. Hall, *A Bibliographical Guide to Japanese Research on the Chinese Economy, 1958–1970*

47. Jack J. Gerson, *Horatio Nelson Lay and Sino-British Relations, 1854–1864*

48. Paul Richard Bohr, *Famine and the Missionary: Timothy Richard as Relief Administrator and Advocate of National Reform*

49. Endymion Wilkinson, *The History of Imperial China: A Research Guide*

50. Britten Dean, *China and Great Britain: The Diplomacy of Commercial Relations, 1860–1864*

51. Ellsworth C. Carlson, *The Foochow Missionaries, 1847–1880*

52. Yeh-chien Wang, *An Estimate of the Land-Tax Collection in China, 1753 and 1908*

53. Richard M. Pfeffer, *Understanding Business Contracts in China, 1949–1963*

54. Han-sheng Chuan and Richard Kraus, *Mid-Ching Rice Markets and Trade: An Essay in Price History*

55. Ranbir Vohra, *Lao She and the Chinese Revolution*

56. Liang-lin Hsiao, *China's Foreign Trade Statistics, 1864–1949*

*57. Lee-hsia Hsu Ting, *Government Control of the Press in Modern China, 1900–1949*

58. Edward W. Wagner, *The Literati Purges: Political Conflict in Early Yi Korea*

*59. Joungwon A. Kim, *Divided Korea: The Politics of Development, 1945–1972*

*60. Noriko Kamachi, John K. Fairbank, and Chūzō Ichiko, *Japanese Studies of Modern China Since 1953: A Bibliographical Guide to Historical and Social-Science Research on the Nineteenth and Twentieth Centuries, Supplementary Volume for 1953–1969*

61. Donald A. Gibbs and Yun-chen Li, *A Bibliography of Studies and Translations of Modern Chinese Literature, 1918–1942*

62. Robert H. Silin, *Leadership and Values: The Organization of Large-Scale Taiwanese Enterprises*

63. David Pong, *A Critical Guide to the Kwangtung Provincial Archives Deposited at the Public Record Office of London*

*64. Fred W. Drake, *China Charts the World: Hsu Chi-yü and His Geography of 1848*

*65. William A. Brown and Urgrunge Onon, trans. and annots., *History of the Mongolian People's Republic*

66. Edward L. Farmer, *Early Ming Government: The Evolution of Dual Capitals*

*67. Ralph C. Croizier, *Koxinga and Chinese Nationalism: History, Myth, and the Hero*

*68. William J. Tyler, trans., *The Psychological World of Natsume Sōseki*, by Doi Takeo

69. Eric Widmer, *The Russian Ecclesiastical Mission in Peking During the Eighteenth Century*

*70. Charlton M. Lewis, *Prologue to the Chinese Revolution: The Transformation of Ideas and Institutions in Hunan Province, 1891–1907*

71. Preston Torbert, *The Ching Imperial Household Department: A Study of Its Organization and Principal Functions, 1662–1796*

72. Paul A. Cohen and John E. Schrecker, eds., *Reform in Nineteenth-Century China*

73. Jon Sigurdson, *Rural Industrialism in China*

74. Kang Chao, *The Development of Cotton Textile Production in China*

75. Valentin Rabe, *The Home Base of American China Missions, 1880–1920*

*76. Sarasin Viraphol, *Tribute and Profit: Sino-Siamese Trade, 1652–1853*

77. Ch'i-ch'ing Hsiao, *The Military Establishment of the Yuan Dynasty*

78. Meishi Tsai, *Contemporary Chinese Novels and Short Stories, 1949–1974: An Annotated Bibliography*

*79. Wellington K. K. Chan, *Merchants, Mandarins and Modern Enterprise in Late Ching China*

80. Endymion Wilkinson, *Landlord and Labor in Late Imperial China: Case Studies from Shandong by Jing Su and Luo Lun*

*81. Barry Keenan, *The Dewey Experiment in China: Educational Reform and Political Power in the Early Republic*

*82. George A. Hayden, *Crime and Punishment in Medieval Chinese Drama: Three Judge Pao Plays*

*83. Sang-Chul Suh, *Growth and Structural Changes in the Korean Economy, 1910–1940*

84. J. W. Dower, *Empire and Aftermath: Yoshida Shigeru and the Japanese Experience, 1878–1954*

85. Martin Collcutt, *Five Mountains: The Rinzai Zen Monastic Institution in Medieval Japan*

86. Kwang Suk Kim and Michael Roemer, *Growth and Structural Transformation*

87. Anne O. Krueger, *The Developmental Role of the Foreign Sector and Aid*

*88. Edwin S. Mills and Byung-Nak Song, *Urbanization and Urban Problems*

89. Sung Hwan Ban, Pal Yong Moon, and Dwight H. Perkins, *Rural Development*

*90. Noel F. McGinn, Donald R. Snodgrass, Yung Bong Kim, Shin-Bok Kim, and Quee-Young Kim, *Education and Development in Korea*

Harvard East Asian Monographs

91. Leroy P. Jones and Il SaKong, *Government, Business, and Entrepreneurship in Economic Development: The Korean Case*

92. Edward S. Mason, Dwight H. Perkins, Kwang Suk Kim, David C. Cole, Mahn Je Kim et al., *The Economic and Social Modernization of the Republic of Korea*

93. Robert Repetto, Tai Hwan Kwon, Son-Ung Kim, Dae Young Kim, John E. Sloboda, and Peter J. Donaldson, *Economic Development, Population Policy, and Demographic Transition in the Republic of Korea*

94. Parks M. Coble, Jr., *The Shanghai Capitalists and the Nationalist Government, 1927–1937*

95. Noriko Kamachi, *Reform in China: Huang Tsun-hsien and the Japanese Model*

96. Richard Wich, *Sino-Soviet Crisis Politics: A Study of Political Change and Communication*

97. Lillian M. Li, *China's Silk Trade: Traditional Industry in the Modern World, 1842–1937*

98. R. David Arkush, *Fei Xiaotong and Sociology in Revolutionary China*

*99. Kenneth Alan Grossberg, *Japan's Renaissance: The Politics of the Muromachi Bakufu*

100. James Reeve Pusey, *China and Charles Darwin*

101. Hoyt Cleveland Tillman, *Utilitarian Confucianism: Chen Liang's Challenge to Chu Hsi*

102. Thomas A. Stanley, *Ōsugi Sakae, Anarchist in Taishō Japan: The Creativity of the Ego*

103. Jonathan K. Ocko, *Bureaucratic Reform in Provincial China: Ting Jih-ch'ang in Restoration Kiangsu, 1867–1870*

104. James Reed, *The Missionary Mind and American East Asia Policy, 1911–1915*

105. Neil L. Waters, *Japan's Local Pragmatists: The Transition from Bakumatsu to Meiji in the Kawasaki Region*

106. David C. Cole and Yung Chul Park, *Financial Development in Korea, 1945–1978*

107. Roy Bahl, Chuk Kyo Kim, and Chong Kee Park, *Public Finances During the Korean Modernization Process*

108. William D. Wray, *Mitsubishi and the N.Y.K, 1870–1914: Business Strategy in the Japanese Shipping Industry*

109. Ralph William Huenemann, *The Dragon and the Iron Horse: The Economics of Railroads in China, 1876–1937*

110. Benjamin A. Elman, *From Philosophy to Philology: Intellectual and Social Aspects of Change in Late Imperial China*

111. Jane Kate Leonard, *Wei Yüan and China's Rediscovery of the Maritime World*

Harvard East Asian Monographs

112. Luke S. K. Kwong, *A Mosaic of the Hundred Days:. Personalities, Politics, and Ideas of 1898*

113. John E. Wills, Jr., *Embassies and Illusions: Dutch and Portuguese Envoys to K'ang-hsi, 1666–1687*

114. Joshua A. Fogel, *Politics and Sinology: The Case of Naitō Konan (1866–1934)*

*115. Jeffrey C. Kinkley, ed., *After Mao: Chinese Literature and Society, 1978–1981*

116. C. Andrew Gerstle, *Circles of Fantasy: Convention in the Plays of Chikamatsu*

117. Andrew Gordon, *The Evolution of Labor Relations in Japan: Heavy Industry, 1853–1955*

*118. Daniel K. Gardner, *Chu Hsi and the "Ta Hsueh": Neo-Confucian Reflection on the Confucian Canon*

119. Christine Guth Kanda, *Shinzō: Hachiman Imagery and Its Development*

*120. Robert Borgen, *Sugawara no Michizane and the Early Heian Court*

121. Chang-tai Hung, *Going to the People: Chinese Intellectual and Folk Literature, 1918–1937*

*122. Michael A. Cusumano, *The Japanese Automobile Industry: Technology and Management at Nissan and Toyota*

123. Richard von Glahn, *The Country of Streams and Grottoes: Expansion, Settlement, and the Civilizing of the Sichuan Frontier in Song Times*

124. Steven D. Carter, *The Road to Komatsubara: A Classical Reading of the Renga Hyakuin*

125. Katherine F. Bruner, John K. Fairbank, and Richard T. Smith, *Entering China's Service: Robert Hart's Journals, 1854–1863*

126. Bob Tadashi Wakabayashi, *Anti-Foreignism and Western Learning in Early-Modern Japan: The "New Theses" of 1825*

127. Atsuko Hirai, *Individualism and Socialism: The Life and Thought of Kawai Eijirō (1891–1944)*

128. Ellen Widmer, *The Margins of Utopia: "Shui-hu hou-chuan" and the Literature of Ming Loyalism*

129. R. Kent Guy, *The Emperor's Four Treasuries: Scholars and the State in the Late Chien-lung Era*

130. Peter C. Perdue, *Exhausting the Earth: State and Peasant in Hunan, 1500–1850*

131. Susan Chan Egan, *A Latterday Confucian: Reminiscences of William Hung (1893–1980)*

132. James T. C. Liu, *China Turning Inward: Intellectual-Political Changes in the Early Twelfth Century*

133. Paul A. Cohen, *Between Tradition and Modernity: Wang T'ao and Reform in Late Ching China*

Harvard East Asian Monographs

134. Kate Wildman Nakai, *Shogunal Politics: Arai Hakuseki and the Premises of Tokugawa Rule*

135. Parks M. Coble, *Facing Japan: Chinese Politics and Japanese Imperialism, 1931-1937*

136. Jon L. Saari, *Legacies of Childhood: Growing Up Chinese in a Time of Crisis, 1890-1920*

137. Susan Downing Videen, *Tales of Heichū*

138. Heinz Morioka and Miyoko Sasaki, *Rakugo: The Popular Narrative Art of Japan*

139. Joshua A. Fogel, *Nakae Ushikichi in China: The Mourning of Spirit*

140. Alexander Barton Woodside, *Vietnam and the Chinese Model.: A Comparative Study of Vietnamese and Chinese Government in the First Half of the Nineteenth Century*

141. George Elision, *Deus Destroyed: The Image of Christianity in Early Modern Japan*

142. William D. Wray, ed., *Managing Industrial Enterprise: Cases from Japan's Prewar Experience*

143. T'ung-tsu Ch'ü, *Local Government in China Under the Ching*

144. Marie Anchordoguy, *Computers, Inc.: Japan's Challenge to IBM*

145. Barbara Molony, *Technology and Investment: The Prewar Japanese Chemical Industry*

146. Mary Elizabeth Berry, *Hideyoshi*

147. Laura E. Hein, *Fueling Growth: The Energy Revolution and Economic Policy in Postwar Japan*

148. Wen-hsin Yeh, *The Alienated Academy: Culture and Politics in Republican China, 1919-1937*

149. Dru C. Gladney, *Muslim Chinese: Ethnic Nationalism in the People's Republic*

150. Merle Goldman and Paul A. Cohen, eds., *Ideas Across Cultures: Essays on Chinese Thought in Honor of Benjamin L Schwartz*

151. James M. Polachek, *The Inner Opium War*

152. Gail Lee Bernstein, *Japanese Marxist: A Portrait of Kawakami Hajime, 1879-1946*

153. Lloyd E. Eastman, *The Abortive Revolution: China Under Nationalist Rule, 1927-1937*

154. Mark Mason, *American Multinationals and Japan: The Political Economy of Japanese Capital Controls, 1899-1980*

155. Richard J. Smith, John K. Fairbank, and Katherine F. Bruner, *Robert Hart and China's Early Modernization: His Journals, 1863-1866*

156. George J. Tanabe, Jr., *Myōe the Dreamkeeper: Fantasy and Knowledge in Kamakura Buddhism*

157. William Wayne Farris, *Heavenly Warriors: The Evolution of Japan's Military, 500–1300*
158. Yu-ming Shaw, *An American Missionary in China: John Leighton Stuart and Chinese-American Relations*
159. James B. Palais, *Politics and Policy in Traditional Korea*
160. Douglas Reynolds, *China, 1898–1912: The Xinzheng Revolution and Japan*
161. Roger R. Thompson, *China's Local Councils in the Age of Constitutional Reform, 1898–1911*
162. William Johnston, *The Modern Epidemic: History of Tuberculosis in Japan*
163. Constantine Nomikos Vaporis, *Breaking Barriers: Travel and the State in Early Modern Japan*
164. Irmela Hijiya-Kirschnereit, *Rituals of Self-Revelation: Shishōsetsu as Literary Genre and Socio-Cultural Phenomenon*
165. James C. Baxter, *The Meiji Unification Through the Lens of Ishikawa Prefecture*
166. Thomas R. H. Havens, *Architects of Affluence: The Tsutsumi Family and the Seibu-Saison Enterprises in Twentieth-Century Japan*
167. Anthony Hood Chambers, *The Secret Window: Ideal Worlds in Tanizaki's Fiction*
168. Steven J. Ericson, *The Sound of the Whistle: Railroads and the State in Meiji Japan*
169. Andrew Edmund Goble, *Kenmu: Go-Daigo's Revolution*
170. Denise Potrzeba Lett, *In Pursuit of Status: The Making of South Korea's "New" Urban Middle Class*
171. Mimi Hall Yiengpruksawan, *Hiraizumi: Buddhist Art and Regional Politics in Twelfth-Century Japan*
172. Charles Shirō Inouye, *The Similitude of Blossoms: A Critical Biography of Izumi Kyōka (1873–1939), Japanese Novelist and Playwright*
173. Aviad E. Raz, *Riding the Black Ship: Japan and Tokyo Disneyland*
174. Deborah J. Milly, *Poverty, Equality, and Growth: The Politics of Economic Need in Postwar Japan*
175. See Heng Teow, *Japan's Cultural Policy Toward China, 1918–1931: A Comparative Perspective*
176. Michael A. Fuller, *An Introduction to Literary Chinese*
177. Frederick R. Dickinson, *War and National Reinvention: Japan in the Great War, 1914–1919*
178. John Solt, *Shredding the Tapestry of Meaning: The Poetry and Poetics of Kitasono Katue (1902–1978)*
179. Edward Pratt, *Japan's Protoindustrial Elite: The Economic Foundations of the Gōnō*
180. Atsuko Sakaki, *Recontextualizing Texts: Narrative Performance in Modern Japanese Fiction*

Harvard East Asian Monographs

181. Soon-Won Park, *Colonial Industrialization and Labor in Korea: The Onoda Cement Factory*

182. JaHyun Kim Haboush and Martina Deuchler, *Culture and the State in Late Chosŏn Korea*

183. John W. Chaffee, *Branches of Heaven: A History of the Imperial Clan of Sung China*

184. Gi-Wook Shin and Michael Robinson, eds., *Colonial Modernity in Korea*

185. Nam-lin Hur, *Prayer and Play in Late Tokugawa Japan: Asakusa Sensōji and Edo Society*

186. Kristin Stapleton, *Civilizing Chengdu: Chinese Urban Reform, 1895–1937*

187. Hyung Il Pai, *Constructing "Korean" Origins: A Critical Review of Archaeology, Historiography, and Racial Myth in Korean State-Formation Theories*

188. Brian D. Ruppert, *Jewel in the Ashes: Buddha Relics and Power in Early Medieval Japan*

189. Susan Daruvala, *Zhou Zuoren and an Alternative Chinese Response to Modernity*

190. James Z. Lee, *The Political Economy of a Frontier: Southwest China, 1250–1850*

191. Kerry Smith, *A Time of Crisis: Japan, the Great Depression, and Rural Revitalization*

192. Michael Lewis, *Becoming Apart: National Power and Local Politics in Toyama, 1868–1945*

193. William C. Kirby, Man-houng Lin, James Chin Shih, and David A. Pietz, eds., *State and Economy in Republican China: A Handbook for Scholars*

194. Timothy S. George, *Minamata: Pollution and the Struggle for Democracy in Postwar Japan*

195. Billy K. L. So, *Prosperity, Region, and Institutions in Maritime China: The South Fukien Pattern, 946–1368*

196. Yoshihisa Tak Matsusaka, *The Making of Japanese Manchuria, 1904–1932*

197. Maram Epstein, *Competing Discourses: Orthodoxy, Authenticity, and Engendered Meanings in Late Imperial Chinese Fiction*

198. Curtis J. Milhaupt, J. Mark Ramseyer, and Michael K. Young, eds. and comps., *Japanese Law in Context: Readings in Society, the Economy, and Politics*

199. Haruo Iguchi, *Unfinished Business: Ayukawa Yoshisuke and U.S.-Japan Relations, 1937–1952*

200. Scott Pearce, Audrey Spiro, and Patricia Ebrey, *Culture and Power in the Reconstitution of the Chinese Realm, 200–600*

201. Terry Kawashima, *Writing Margins: The Textual Construction of Gender in Heian and Kamakura Japan*

202. Martin W. Huang, *Desire and Fictional Narrative in Late Imperial China*

203. Robert S. Ross and Jiang Changbin, eds., *Re-examining the Cold War: U.S.-China Diplomacy, 1954–1973*

204. Guanhua Wang, *In Search of Justice: The 1905–1906 Chinese Anti-American Boycott*

205. David Schaberg, *A Patterned Past: Form and Thought in Early Chinese Historiography*

206. Christine Yano, *Tears of Longing: Nostalgia and the Nation in Japanese Popular Song*

207. Milena Doleželová-Velingerová and Oldřich Král, with Graham Sanders, eds., *The Appropriation of Cultural Capital: China's May Fourth Project*

208. Robert N. Huey, *The Making of 'Shinkokinshū'*

209. Lee Butler, *Emperor and Aristocracy in Japan, 1467–1680: Resilience and Renewal*

210. Suzanne Ogden, *Inklings of Democracy in China*

211. Kenneth J. Ruoff, *The People's Emperor: Democracy and the Japanese Monarchy, 1945–1995*

212. Haun Saussy, *Great Walls of Discourse and Other Adventures in Cultural China*

213. Aviad E. Raz, *Emotions at Work: Normative Control, Organizations, and Culture in Japan and America*

214. Rebecca E. Karl and Peter Zarrow, eds., *Rethinking the 1898 Reform Period: Political and Cultural Change in Late Qing China*

215. Kevin O'Rourke, *The Book of Korean Shijo*

216. Ezra F. Vogel, ed., *The Golden Age of the U.S.-China-Japan Triangle, 1972–1989*

217. Thomas A Wilson, ed., *On Sacred Grounds: Culture, Society, Politics, and the Formation of the Cult of Confucius*

218. Donald S. Sutton, *Steps of Perfection: Exorcistic Performers and Chinese Religion in Twentieth-Century Taiwan*

219. Daqing Yang, *Technology of Empire: Telecommunications and Japanese Imperialism, 1930–1945*

220. Qianshen Bai, *Fu Shan's World: The Transformation of Chinese Calligraphy in the Seventeenth Century*

221. Paul Jakov Smith and Richard von Glahn, eds., *The Song-Yuan-Ming Transition in Chinese History*

222. Rania Huntington, *Alien Kind: Foxes and Late Imperial Chinese Narrative*

223. Jordan Sand, *House and Home in Modern Japan: Architecture, Domestic Space, and Bourgeois Culture, 1880–1930*

224. Karl Gerth, *China Made: Consumer Culture and the Creation of the Nation*

225. Xiaoshan Yang, *Metamorphosis of the Private Sphere: Gardens and Objects in Tang-Song Poetry*

Harvard East Asian Monographs

226. Barbara Mittler, *A Newspaper for China? Power, Identity, and Change in Shanghai's News Media, 1872–1912*

227. Joyce A. Madancy, *The Troublesome Legacy of Commissioner Lin: The Opium Trade and Opium Suppression in Fujian Province, 1820s to 1920s*

228. John Makeham, *Transmitters and Creators: Chinese Commentators and Commentaries on the Analects*

229. Elisabeth Köll, *From Cotton Mill to Business Empire: The Emergence of Regional Enterprises in Modern China*

230. Emma Teng, *Taiwan's Imagined Geography: Chinese Colonial Travel Writing and Pictures, 1683–1895*

231. Wilt Idema and Beata Grant, *The Red Brush: Writing Women of Imperial China*

232. Eric C. Rath, *The Ethos of Noh: Actors and Their Art*

233. Elizabeth J. Remick, *Building Local States: China During the Republican and Post-Mao Eras*

234. Lynn Struve, ed., *The Qing Formation in World-Historical Time*

235. D. Max Moerman, *Localizing Paradise: Kumano Pilgrimage and the Religious Landscape of Premodern Japan*

236. Antonia Finnane, *Speaking of Yangzhou: A Chinese City, 1550–1850*